D1068244

INTERSCIENCE MONOGRAPHS ON CHEMISTRY

Interscience Monographs on Chemistry

INORGANIC CHEMISTRY SECTION

Edited by F. Albert Cotton and G. Wilkinson

GOODENOUGH	Magnetism and the Chemical Bond
COLTON	The Chemistry of Rhenium and Technetium
GRIFFITH	The Chemistry of the Rarer Platinum Metals
MALATESTA AND BONATI	Isocyanine Complexes of Metals
HAWKINS	Absolute Configuration of Metal Complexes

Other volumes to follow

Absolute Configuration of Metal Complexes

CLIFFORD J. HAWKINS
University of Queensland

WILEY-INTERSCIENCE
a Division of John Wiley & Sons, Inc.
New York · London · Sydney · Toronto

10 9 8 7 6 5 4 3 2 1

Library of Congress Catalog Card Number: 78-129975

ISBN 0-471-36280-8

Printed in the United States of America

Preface

Since the pioneering work of Pásteur, considerable chemical research has been directed toward understanding molecular dissymmetry. The studies of Alfred Werner on stereochemical aspects of coordination compounds early in the present century opened up new horizons in this field. Many different kinds of compound were found to be resolvable into optical isomers, presenting the problem of how to assign their absolute configuration. It is only in recent years, with the advent of sophisticated physical techniques and readily available instrumentation, that this problem has been satisfactorily overcome.

The present work has been written to describe how the various methods have been applied to the determination of the absolute configuration of metal complexes. At the time of writing no work that dealt exclusively with this subject was available, although some excellent ones covering the general applications of physical methods in inorganic chemistry have recently appeared. Here, the basic aspects of the theory of the different techniques have been covered only briefly, but it is hoped that sufficient background has been provided to make it unnecessary for the reader to go elsewhere for the essential facts. Nevertheless, the reader should be familiar with the principles of symmetry and group theory as covered in F. A. Cotton's *Chemical Applications of Group Theory*, and it would be beneficial to have a more advanced knowledge of the theory and general aspects of the various techniques. Further, it would appear to be almost essential for the reader to have on hand a set of molecular models, preferably the Dreiding type.

The determination of the absolute configuration of certain molecules requires a distinction to be made among their various geometrical isomers. Because of this, methods that have been used to study geometrical isomerism have been included. However, the discussion has mostly been limited to kinds of compounds of direct relevance to the main theme of the work.

Throughout the history of the subject numerous methods have been proposed for denoting a molecule's absolute configuration. The great variety of symbols and the different and often contradictory way in which they have been used have led to considerable confusion. The various methods are discussed in some detail in Chap. 2, and it is hoped that this will aid the reader when referring to original papers in the field.

Because the determination of conformational energies of metal complexes by classical approaches is becoming more common, it was thought worthwhile to review in some detail the various energy terms used in the calculations (see Chap. 3). It is hoped that this will not only emphasize the limitations of the calculations but may also help future workers in the field to make a more educated choice of equations and parameters.

The published work has been reviewed as critically as possible, and, when necessary, the results are reinterpreted in line with current thinking. The literature up to the beginning of 1969 has been covered and in some areas into the first half of 1969, but it must be emphasized that the main objective was to show how the various methods could be applied to absolute-configuration determination rather than to list every paper that has concerned itself with absolute configuration. The number of inorganic research groups actively working in the field of molecular dissymmetry, and, consequently, the rate of publications in the field, have increased dramatically in the last few years. It is hoped that this book will stimulate further interest in the field and will act as a basis for future research.

I should like to record my gratitude to my students and colleagues in the department of chemistry of the University of Queensland and elsewhere in Australia who have contributed to the final manuscript by their constructive criticism and suggestions. In particular, I want to thank Dr. A. McL. Mathieson of the Division of Chemical Physics, C.S.I.R.O., for his much needed help in drafting Chap. 4, and Dr. J. R. Gollogly, who, as a research student in this laboratory, carried out a literature survey for Chap. 3, and performed most of the calculations discussed there. I am also indebted to Mrs. M. Plooy for carrying out the onerous task of typing the manuscript.

A number of the figures and tables have been taken from other texts and journals, and I should like to thank the editors and publishers of the following for permission to reproduce material: Acta Crystallographica; Angewandte Chemie; Australian Journal of Chemistry; Bulletin of the Chemical Society of Japan; Chemical Society of London; Inorganic Chemistry; Journal of the American Chemical Society; Journal of Chemical Physics; Journal of Physical Chemistry; Molecular Physics; Pergamon Press Ltd.; Theoretica Chimica Acta.

C. J. HAWKINS

St. Lucia
May 1970

Contents

Ligand Abbreviations

en	ethylenediamine
pn	propylenediamine
tn	1,3-diaminopropane (trimethylenediamine)
bn	2,3-diaminobutane
Meen	*N*-methylethylenediamine
trien	*N*,*N*′-bis(2′-aminoethyl)-1,2-diaminoethane (triethylenetetramine)
dimetrien	dimethyltriethylenetetramine
dien	*N*-(2′-aminoethyl)-1,2-diaminoethane (diethylenetriamine)
cyclen	1,4,7,10-tetraazacyclododecane
penten	*N*,*N*,*N*′,*N*′-tetrakis(2′-aminoethyl)-1,2-diaminoethane
mepenten	*N*,*N*,*N*′,*N*′-tetrakis(2′-aminoethyl)-1,2-diaminopropane
chxn	*trans*-1,2-cyclohexanediamine
py	pyridine
bipy	2,2′-bipyridine
phen	1,10-phenanthroline
bgH	biguanide
EDTA	ethylenediaminetetraacetate
PDTA	propylenediaminetetraacetate
EDDA	ethylenediamine-*N*,*N*′-diacetate
EDDP	ethylenediamine-*N*,*N*′-di-α-propionate
IDA	iminodiacetate
MEDTA	*N*-methylethylenediaminetriacetate
YOH	*N*-2-hydroxyethylethylenediaminetriacetate
am	α-aminocarboxylate
gly	glycinate
ala	alaninate
glu	glutamate
sar	sarcosinate
pro	prolinate
asp	aspartate
asn	asparaginate
his	histidinate

leu	leucinate
val	valinate
$(cys)_2$	cystine
lys	lysinate
norleu	norleucinate
ileu	isoleucinate
met	methioninate
hypro	hydroxyprolinate
ser	serinate
phg	phenylglycinate
phe	phenylalaninate
trp	tryptophanate
tyr	tyrosinate
thr	threoninate
pen	1-aminocyclopentanecarboxylate
NH_2but	aminobutyrate
N-Me-L-leu	*N*-methyl-L-leucinate
ox	oxalate
mal	malonate
acac	acetylacetonate
sal	salicylate
(X-sal)$_2$bmp	2,2′-bis(salicylideneamino)-6,6′-dimethylbiphenyl

INTERSCIENCE MONOGRAPHS ON CHEMISTRY

1

Introduction

STEREOISOMERISM AND ABSOLUTE CONFIGURATION

Of the many kinds of isomerism exhibited by coordination compounds, that which is called *stereoisomerism* has attracted the most attention. As the name implies, stereoisomerism is concerned with the spatial distribution of atoms in a molecule. There are three forms: *geometrical isomerism*, *optical isomerism*, and *diastereoisomerism*. The determination of the exact three-dimensional structure of a particular isomer, so that the position of each atom in the complex is known relative to all other atoms, is a great challenge that has only recently been effectively met with the advent of the necessary instrumentation and techniques.

This structure is referred to as a compound's *absolute configuration*. Although the term is commonly restricted to cases of optical isomerism and diastereoisomerism, methods of studying all three kinds of stereoisomerism are dealt with in the present work. The term is often further restricted to cases where a breakage of bonds is necessary for the conversion of one isomer to another. No such restriction is used here. The term is applied to the structure of a complex in its entirety even if, by some mechanism, such as the Rây and Dutt [15], the Bailar [1,9,16], or the Springer and Sievers [18] twist mechanism, it is possible to go from one structure to another without the breakage of bonds. Without this limitation there is no distinction between absolute configuration and conformation, as it is normally defined. In the present work, however, conformation is not applied to a complex as a whole but rather to the distribution of atoms in individual chelate rings and to the various *rotamers* of the ligands, that is, different *conformers* obtained by rotation about a single bond.

GEOMETRICAL ISOMERISM

The so-called geometrical isomers have structures that are nonsuperimposable, because some groups in the molecule or ion occupy different positions relative to one another. The phenomenon is sometimes called *cis-trans isomerism,* because the groups in question can be either adjacent or opposed to one another about the central metal ion.

It is best illustrated by considering some general kinds of compounds.

Octahedral Complexes

The compounds Ma_4b_2 and Ma_3b_3 both exist in two isomeric forms, **I, II** and **III, IV**:

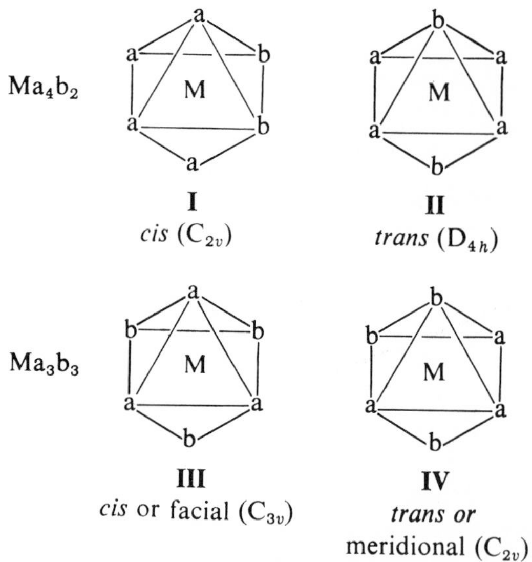

I *cis* (C_{2v}) **II** *trans* (D_{4h})

III *cis* or facial (C_{3v}) **IV** *trans or* meridional (C_{2v})

The number of isomers usually, but not always, increases with an increase in the number of different ligands coordinated. For example, $Ma_2b_2c_2$ has 5 geometrical isomers (**V** to **IX**), Mabcdef has 15, but Ma_4bc has only 2:

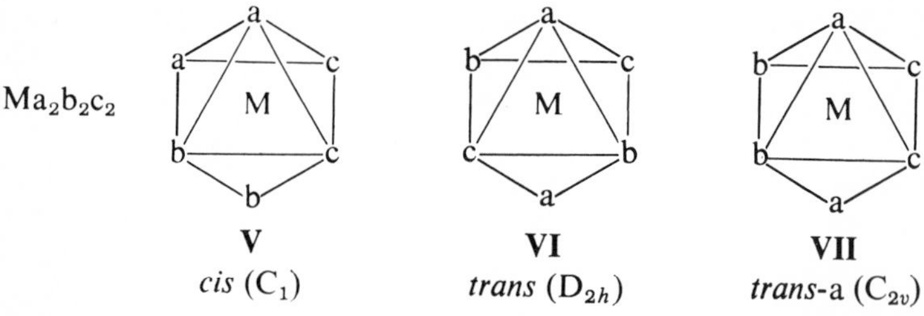

V *cis* (C_1) **VI** *trans* (D_{2h}) **VII** *trans*-a (C_{2v})

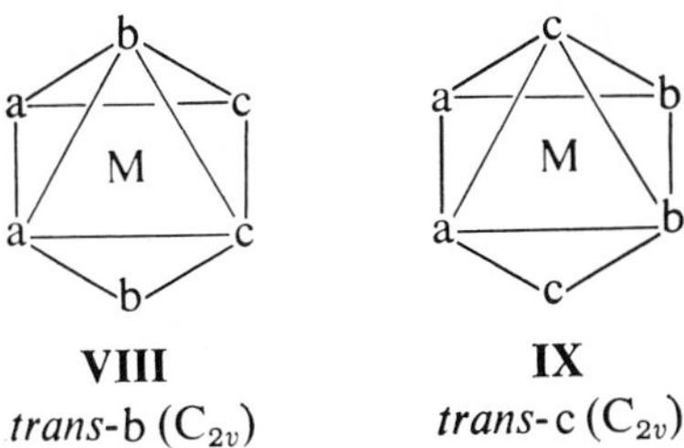

VIII *trans*-b (C_{2v}) **IX** *trans*-c (C_{2v})

If a symmetrical bidentate ligand aa replaces two a ligands from $Ma_2b_2c_2$, the number of isomers is reduced to three, because a normal bidentate, such as ethylenediamine, is unable to span the trans positions:

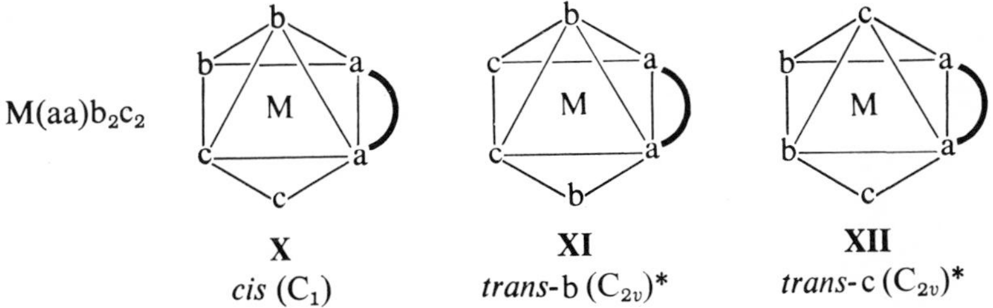

X *cis* (C_1) **XI** *trans*-b (C_{2v})* **XII** *trans*-c (C_{2v})*

An important example of this is $[Coen(NH_3)_2Cl_2]^+$, the three isomers of which were first prepared by Bailar and Peppard [2].

If an unsymmetrical chelate ab, such as 1,2-diaminopropane, is present instead of aa, there are four isomers:

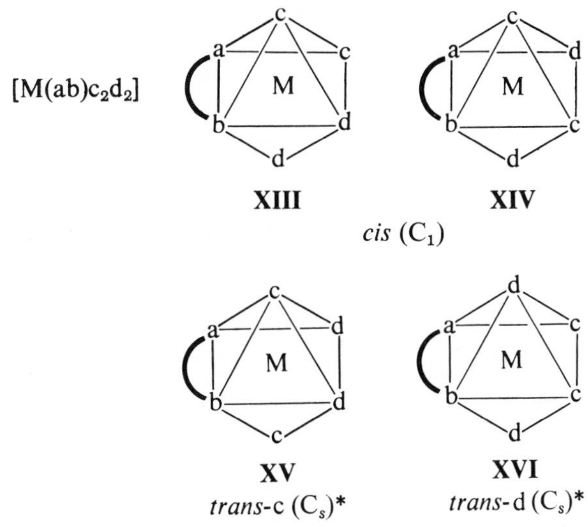

XIII **XIV** *cis* (C_1)

XV *trans*-c (C_s)* **XVI** *trans*-d (C_s)*

* This symmetry is correct only if the ligand is planar.

The bis-chelated complexes with symmetrical ligands have the normal cis and trans isomers (**XVII** and **XVIII**), but when the ligand is unsymmetrical, there are five geometrical isomers (**XIX** to **XXIII**):

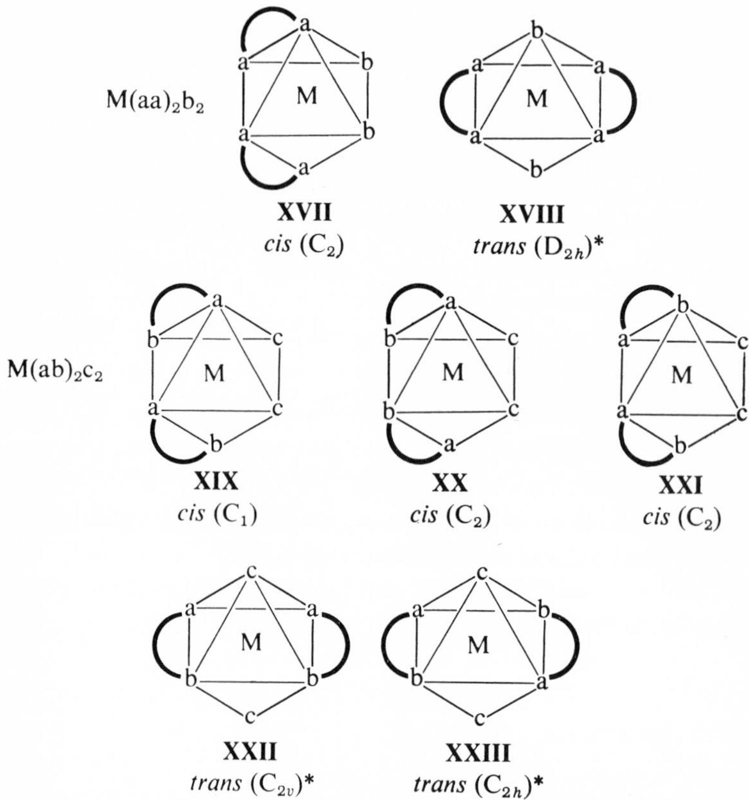

The tris-chelated complexes with symmetrical ligands do not have geometrical isomers, but those with an unsymmetrical ligand have isomers similar to those of Ma_3b_3 (see **III** and **IV**):

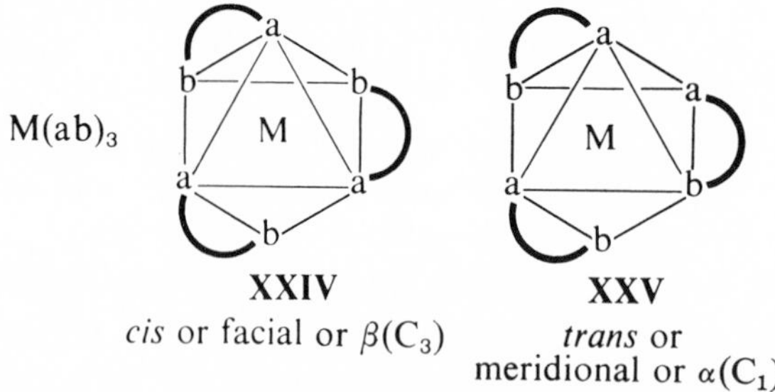

Multidentate ligands also lead to possible geometrical isomerism; for example, *N*,*N*′-bis(2′-aminoethyl)-1,2-diaminoethane (trien,*NNNN*) forms three geometrical isomers with octahedral metal ions (**XXVI** to **XXVIII**):

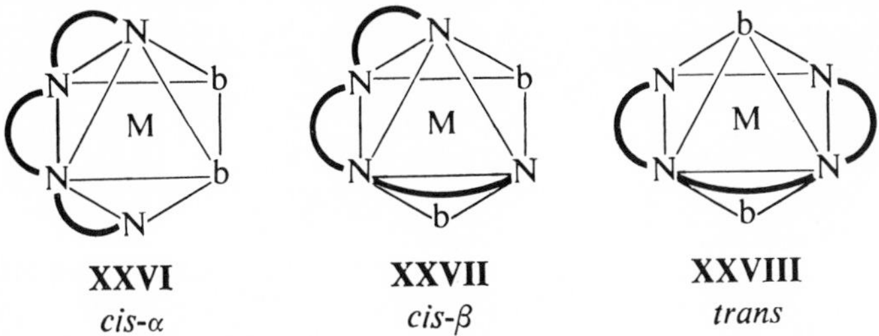

XXVI *cis*-α **XXVII** *cis*-β **XXVIII** *trans*

Square Planar Complexes

Some square planar complexes have geometrical isomers. The complexes Ma_2b_2 and $M(ab)_2$ are examples of this:

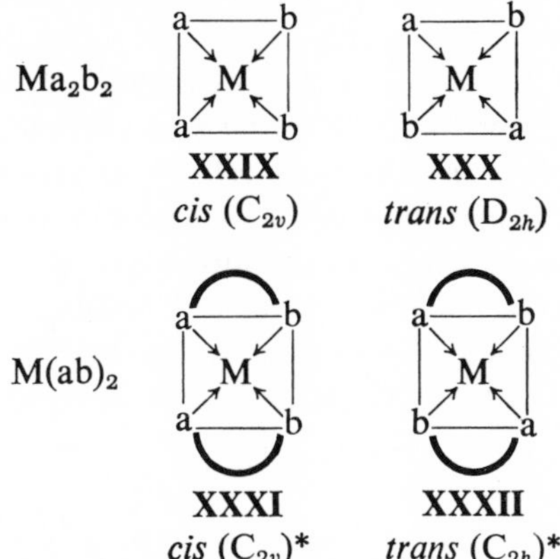

XXIX *cis* (C_{2v}) **XXX** *trans* (D_{2h})

XXXI *cis* (C_{2v})* **XXXII** *trans* (C_{2h})*

Tetrahedral Complexes

Tetrahedral complexes normally do not have geometrical isomers, because the position of each donor atom is cis to all other positions. Unsymmetrical chelates, however, such as R-1,2-diaminopropane can give rise to these kinds of isomers: the complex in Fig. 1-1 has two geometrical isomers with the substituent in position 1 or 2.

Often the structural differences between the geometrical isomers are so great that the isomers have significantly different physical and chemical properties. This has facilitated the successful application of a number of techniques to the identification of the isomers—electronic absorption, infrared, and magnetic resonance spectroscopy, for example. These are discussed in Chaps. 6 and 8.

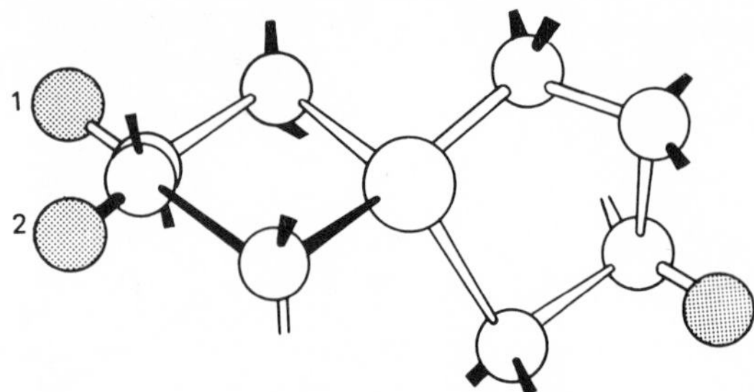

Fig. 1-1 An example of geometrical isomerism in a tetrahedral complex [M(R-pn)$_2$].

OPTICAL ISOMERISM AND DIASTEREOISOMERISM

Optical isomers have structures that are nonsuperimposable mirror images of each other. They are related in the same way as the left hand is to the right. They have been classified as optical isomers* because they are found to be "optically active"; that is, they are able to rotate the plane of polarization of plane-polarized light. It is to be noted that the term "optical activity" is spectroscopic and implies that the rotational strength of at least one of a compound's electronic transitions is different from zero (see Chap. 5).

For a compound to have this kind of isomerism, it must not have a center of symmetry, a plane of symmetry, or any improper axis of higher order. A compound need not lack a proper axis of rotation, however. In fact, in complex chemistry, the majority of compounds that have been resolved into their optical isomers do contain this element of symmetry; for example,

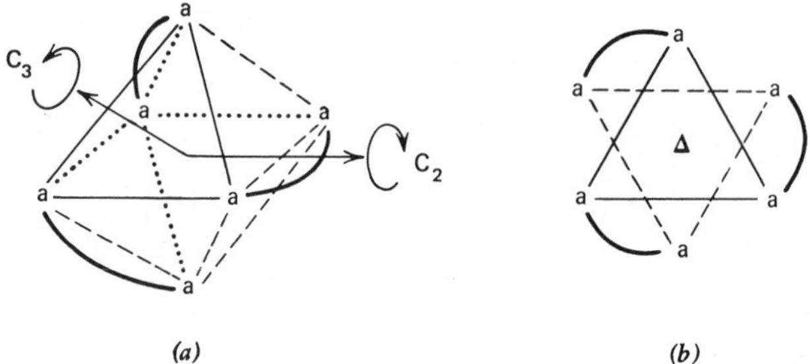

Fig. 1-2 (*a*) Proper axes of rotation in tris(bidentate) complexes; (*b*) view down the C_3 axis.

* Optical isomers are also referred to as optical antipodes (from the Greek ἀντίποδες, those having the feet opposite); enantiomorphs (from the Greek ἐναντίος opposite and μορφή form), originally restricted to crystals but generalized to describe any form related to another as an object is to its mirror image; and finally, enantiomers, derived from the previous term.

although tris(ethylenediamine)cobalt(III) is of the general type $M(aa)_3$, which has D_3 symmetry with a C_3 and three C_2 axes (Fig. 1-2), it is capable of resolution into its optical antipodes, **XXXIII** and **XXXIV**:

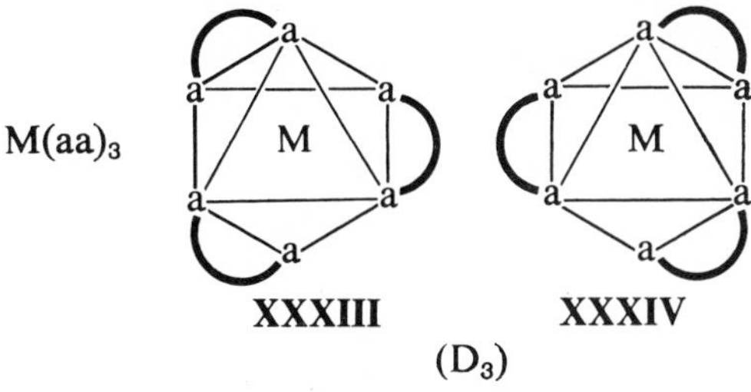

Compounds of this kind are classified as *dissymmetric*, and those having no element of symmetry other than the identity are classified as *asymmetric*. Thus, molecules must belong to either of the point groups C_n or D_n if they are to be nonsuperimposable on their mirror image.

A metal complex can gain dissymmetry through one or more of the following:

1. The distribution of chelate rings about the central metal ion
2. The conformations of the chelate rings
3. The distribution of unidentate ligands about the central metal ion
4. The coordination of unsymmetrical multidentates
5. The coordination of an optically active ligand
6. The coordination of a donor atom that is asymmetric

The first of these has attracted the most study. Structures **XXXIII** and **XXXIV** are examples of complexes gaining dissymmetry from the distribution of chelate rings. Other examples are given below:

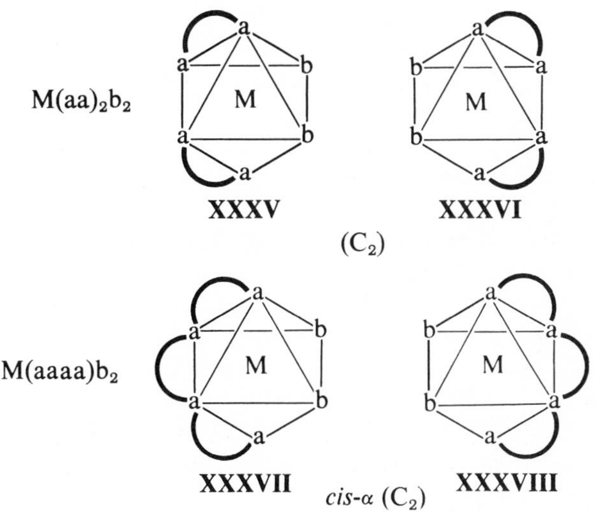

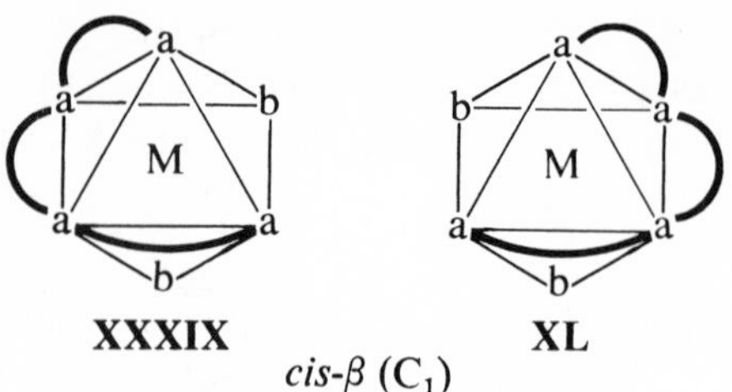

XXXIX **XL**

cis-β (C_1)

Following Theilacker's discovery that chelate rings can be puckered [19], it was realized that the nonplanar nature of the chelates could introduce dissymmetry. In Figs. 1-3 to 1-8 some examples of the conformations of five- and six-membered rings are shown in various elevations (only the atoms in the ring are considered for simplicity).

The third source of dissymmetry is the distribution of unidentate ligands. Little attention has been focused on this aspect. The cis-isomer of $Ma_2b_2c_2$ is asymmetric and exists in two enantiomeric forms, **XLI** and **XLII**:

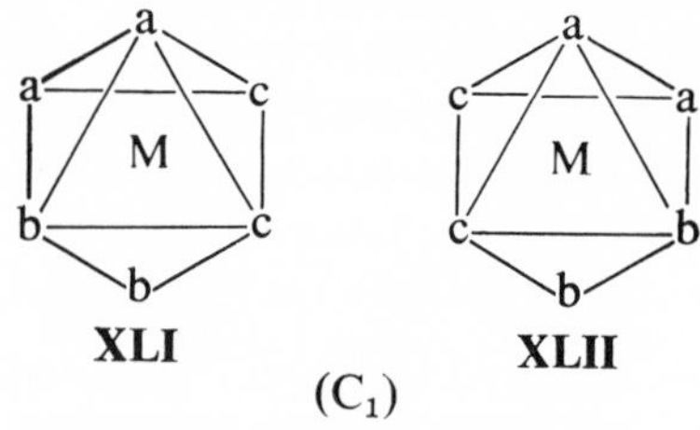

XLI **XLII**

(C_1)

Few compounds of this kind have been resolved, although Russian workers have reported the resolution of $[Pt(NH_3)_2(NO_2)_2Cl_2]$ [7]. The compounds cis-$[Coen(NH_3)_2Cl_2]^+$ [13], cis-$[Pten(NH_3)_2Cl_2]^{2+}$ [4], and two geometrical isomers of cis-$[PtenCl_2(NH_3)py]^{2+}$ [5,6], which belong to this category but have a bidentate instead of two unidentates, have also been resolved. The resolution of *cis*-$[Coox(NO_2)_2(NH_3)_2]^-$ has been reported [17,20], but there is considerable doubt concerning its actual geometrical isomeric form [14].

Some complexes containing multidentate ligands in which, ignoring the individual atoms, the chelate rings are arranged in a nondissymmetric manner, are still potentially resolvable irrespective of the conformations of the rings because of the unsymmetrical nature of the ligand. In the complex $M(aab)_2$, for example, the distribution of chelate rings is symmetrical for the meridional, the *cis*-facial, and the *trans*-facial isomers. The meridional (**XLIII** and **XLIV**) and *cis*-facial (**XLV** and **XLVI**), however, are potentially resolvable because of the unsymmetrical nature of the ligand. Numerous

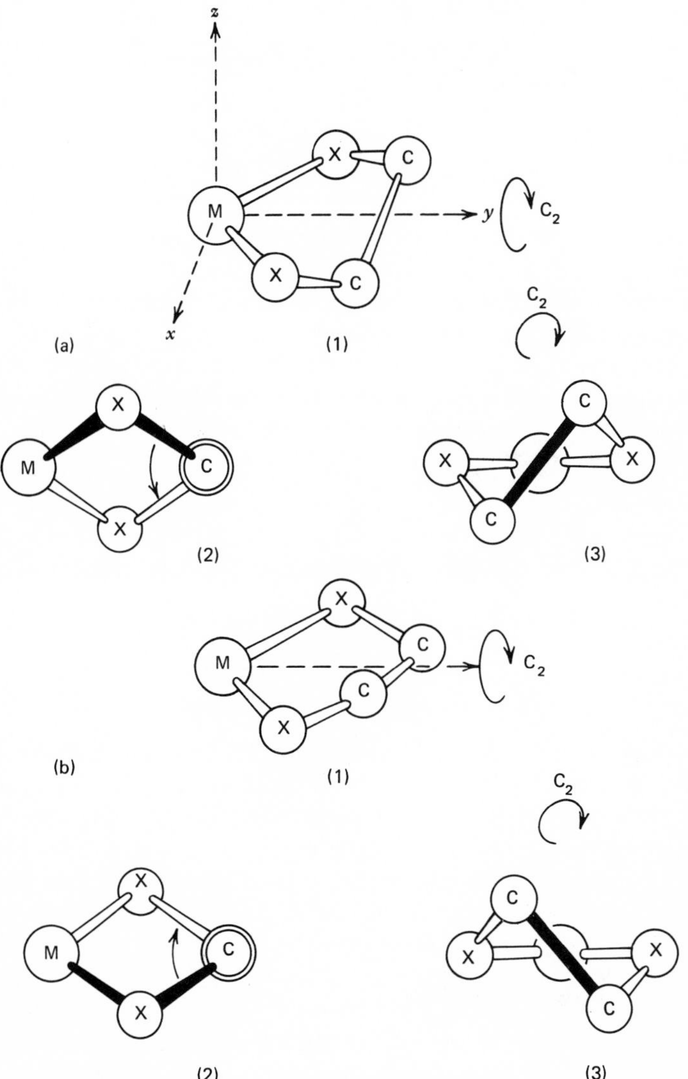

Fig. 1-3 Symmetric skew five-membered chelate ring: (a) and (b) are enantiomeric conformations.

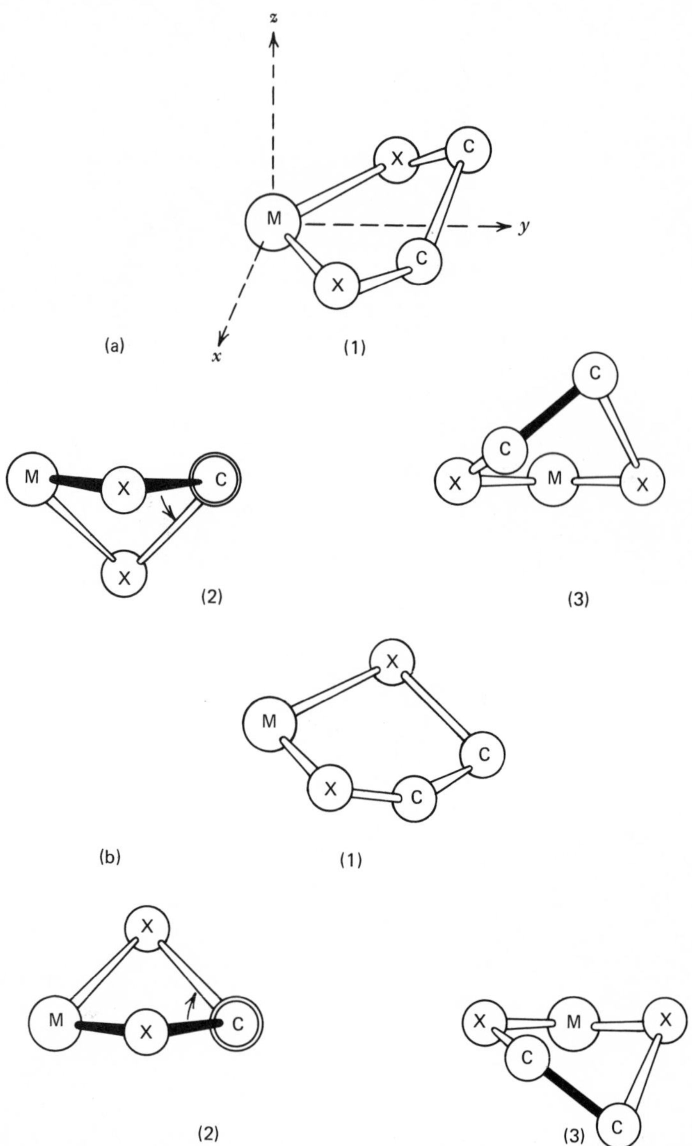

Fig. 1-4 Asymmetric envelope conformation of a five-membered chelate ring: (a) and (b) are enantiomeric conformations.

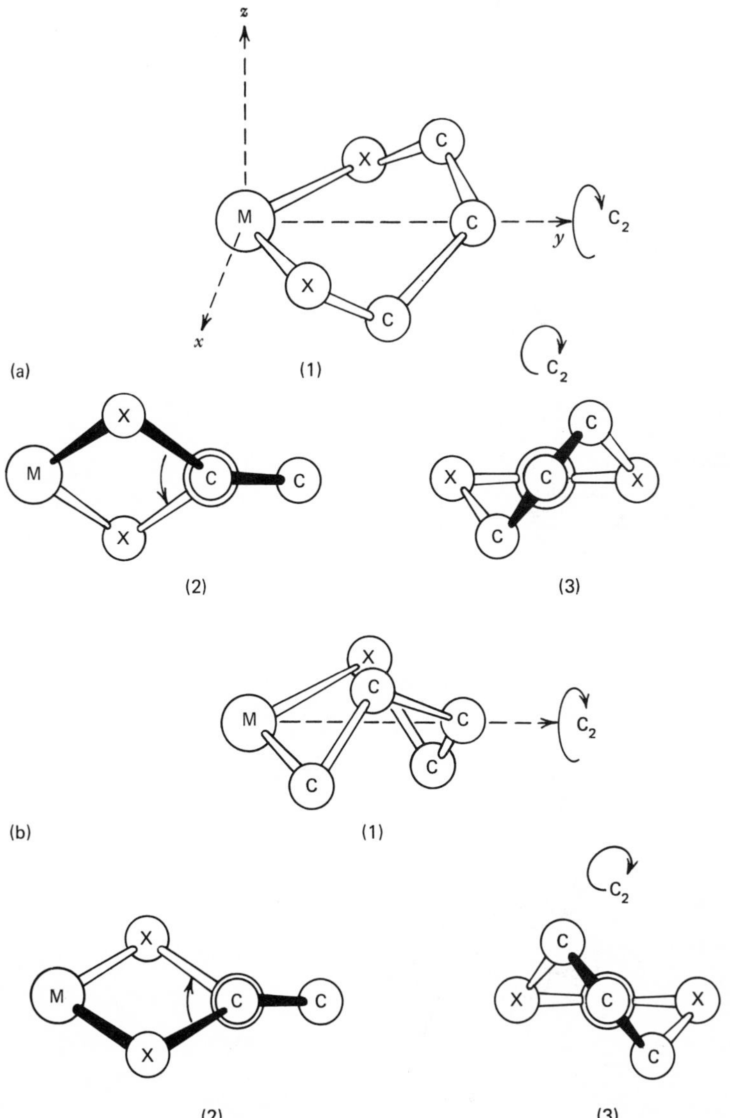

Fig. 1-5 Symmetric skew-boat conformation of a six-membered chelate ring: (a) and (b) are enantiomeric conformations.

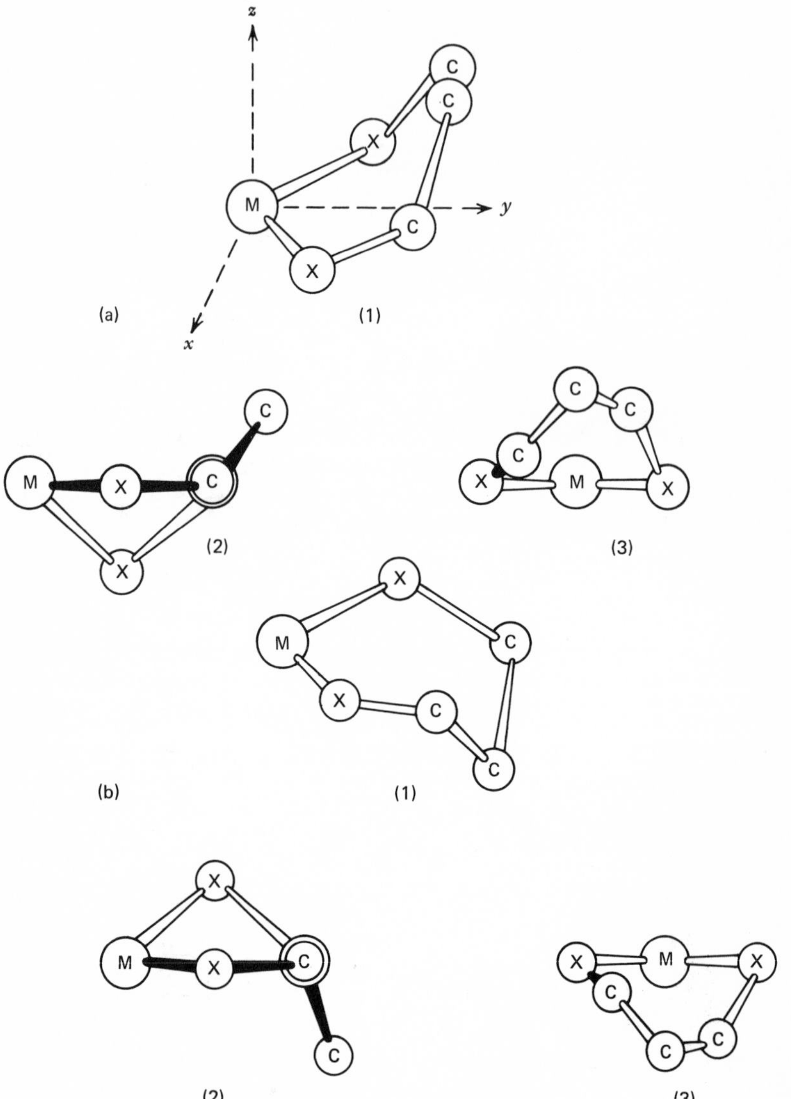

Fig. 1-6 Asymmetric boat conformation of a six-membered chelate ring: (a) and (b) are enantiomeric conformations.

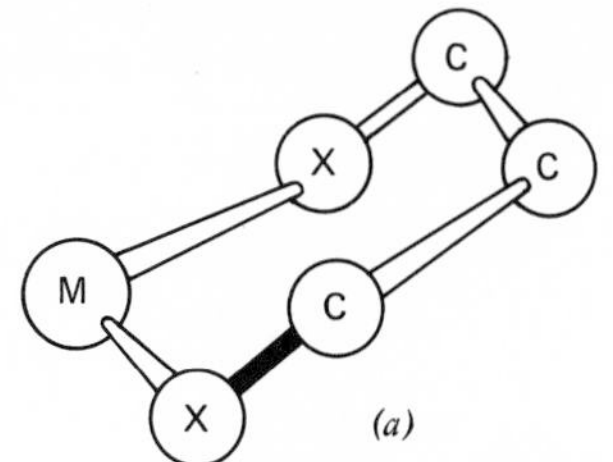

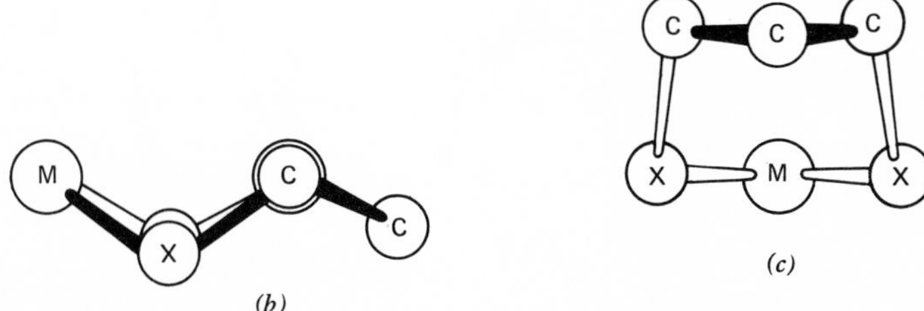

Fig. 1-7 Chair conformation of a six-membered chelate ring.

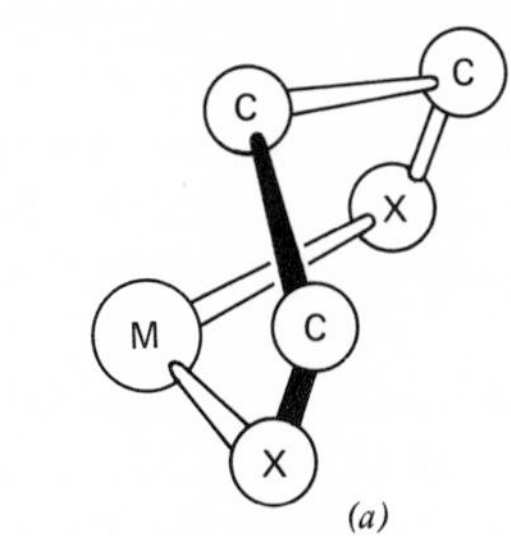

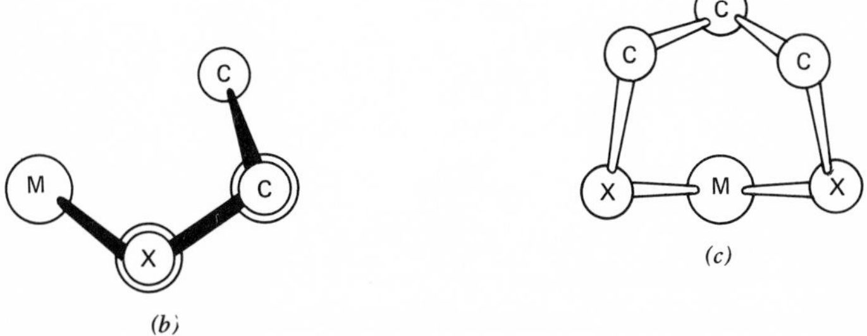

Fig. 1-8 Boat conformation of a six-membered chelate ring.

multidentates are capable of introducing dissymmetry in this way—dipeptides, *N*-(2′-aminoethyl)-2-aminoethanol, for example:

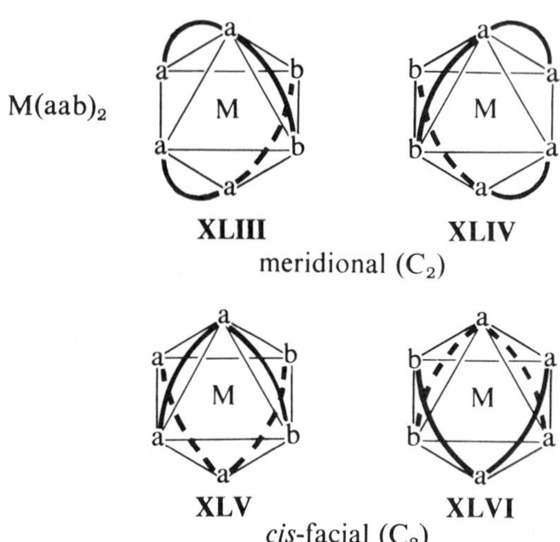

A complex also becomes dissymmetric if it has an asymmetric ligand. For example, $[Co(NH_3)_5\text{-L-phenylalanine}]^{3+}$ is asymmetric because of the coordinated L-phenylalanine. Its optical isomer is $[Co(NH_3)_5\text{-D-phenylalanine}]^{3+}$. It is of interest to note that the asymmetry is transmitted to the central metal ion chromophore, which inherently has a C_{4v} symmetry. This is shown by the measurable rotational strength of the *d–d* transitions [8,11,12,21].

The amino acid anion sarcosinate, $CH_3—NH—CH_2—COO^-$, cannot be resolved because of the lability of the inversion about the nitrogen, but it has been resolved indirectly by the resolution of its complex $[Co(NH_3)_4sarc]^{2+}$ (**XLVII**) [10]. In this complex the asymmetric center is the donor atom itself:

NH_3, H_3N, $NH—CH_3$, CH_2, Co, H_3N, O—C, O, NH_3

XLVII

Similarly an *N*-methyl-1,2-diaminoethane complex has been resolved [3]. For these two complexes, however, there are two sources of dissymmetry: the asymmetric nitrogen enforces one conformation of the chelate ring to be preferred over the other. When prolinate **XLVIII** coordinates as a bidentate, only one configuration of the asymmetric nitrogen can coordinate for a particular configuration of the asymmetric carbon:

H_2C——CH_2
H_2C　　　$CHCOO^-$
N
H

XLVIII

Thus complexes containing optically active proline have asymmetry imposed by the conformation of the chelate ring, from an asymmetric carbon atom, and from an asymmetric donor nitrogen atom. No complex has yet been made with an unidentate ligand having an asymmetric donor atom.

The fact that more than one source of dissymmetry can be present in a complex molecule gives rise to *diastereoisomerism.* Diastereoisomers or diastereomers are stereoisomers that have the same elements of dissymmetry, some but not all of which are enantiomeric.

The phenomenon is perhaps best known when two optical isomers form salts with a particular optically active ion (see Fig. 1-9). The difference in solubility of the salts provides the most useful method of resolution of optical isomers. Diastereoisomerism is often not recognized in its other forms, however. The complexes D-$[Co(R\text{-}pn)_3]^{3+}$ and L-$[Co(R\text{-}pn)_3]^{3+}$, for example, are not optical isomers, as sometimes stated, but diastereoisomers, because the configurations and conformations of the 1,2-diaminopropane chelate rings are identical in the two compounds, not enantiomeric. D-$[Co(S\text{-}pn)_3]^{3+}$ is the correct optical isomer of L-$[Co(R\text{-}pn)_3]^{3+}$.

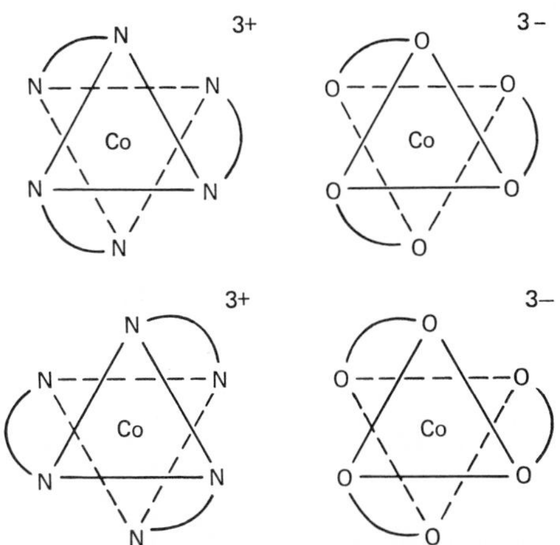

Fig. 1-9 Diastereomers formed between DL-$[Coen_3]^{3+}$ and D-$[Coox_3]^{3-}$.

METHODS OF DETERMINING ABSOLUTE CONFIGURATIONS

Because the conditions for a compound to be optically active are the same as those governing optical isomerism, it is not surprising that the greatest emphasis for the study of absolute configuration has been placed on techniques depending on the measurement of optical activity: optical rotatory dispersion and circular dichroism (see Chap. 5). In most cases, however, these techniques are only able to relate configurations. They rely on other techniques, such as the anomalous X-ray diffraction technique (see Chap. 4), to provide absolute configuration data for model compounds.

Most other techniques for determining absolute configuration can be used to distinguish between diastereomers and between nonenantiomeric conformations, but not between optical antipodes. This is true for the conformational analysis (see Chap. 3), nuclear magnetic resonance (see Chap. 6), and infrared-Raman (see Chap. 7) techniques. Nevertheless they have been extremely useful in the determination of absolute configuration.

CHIRALITY

Because optical isomers are related as an object is to its mirror image, just as the left hand is to the right hand, the terms *handedness* and *chirality* (from

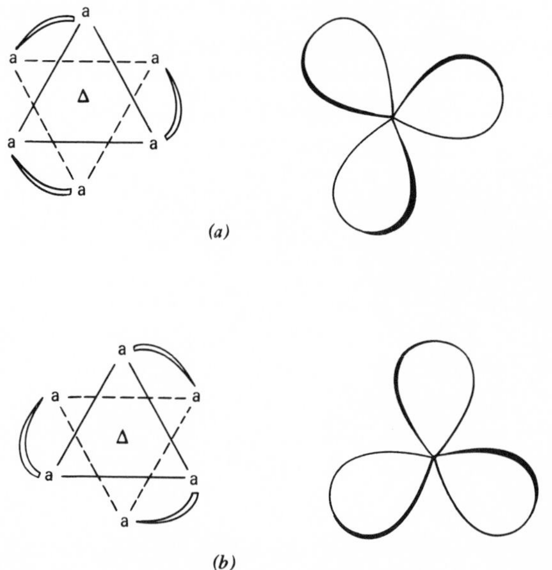

Fig. 1-10 Comparison of the distribution of chelate rings for (a) D- and (b) L-$[M(aa)_3]$ viewed along the C_3 axis with the helical screw of a three-bladed propeller.

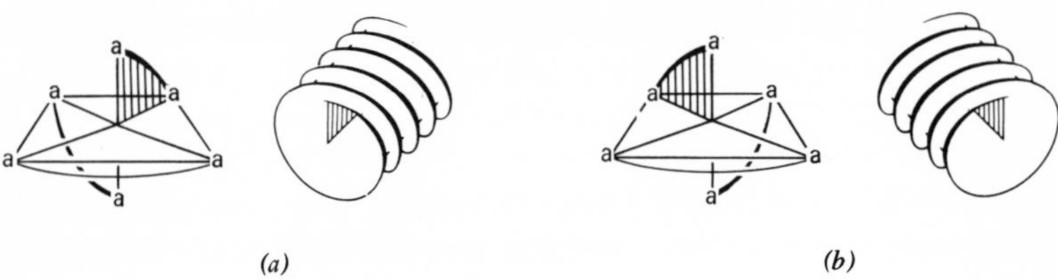

Fig. 1-11 Comparison of the distribution of chelate rings for (*a*) D- and (*b*) L-[M(aa)$_3$] viewed along a C_2 axis with the helical scetion of a screw.

the Greek word for hand, χεὶρ) are often used in connection with the absolute configuration of the optical antipodes. When applied to optical isomers, these terms have no intrinsic meaning and therefore rely on an adopted convention.

If you look down the C_3 axis of a tris-bidentate complex, you see the chelate rings arrayed as if they were the blades of a propeller. The helical screw of the propeller would be opposite for the two optical isomers. One would be twisted as if following the thread of a right-handed screw, the other a left-handed screw. For the former, a clockwise rotation results in a translation away from the observer; for the latter, it results in a movement in the oppo-

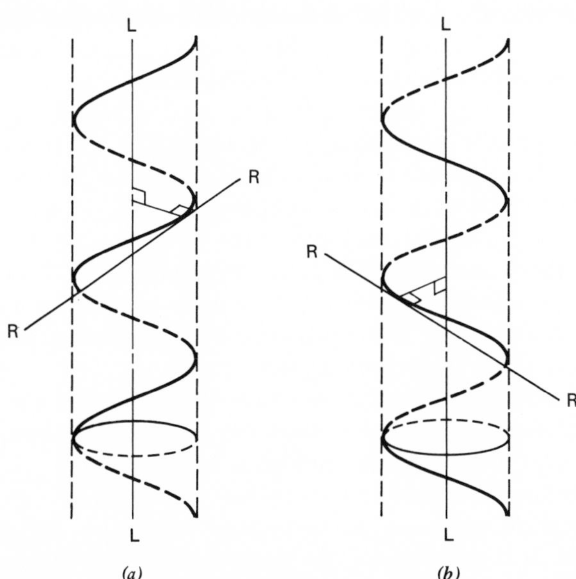

Fig. 1-12 Helicity associated with two skew lines: (*a*) right-handed and (*b*) left-handed. Two skew lines LL and RR uniquely describe a cylindrical helix in which one line (LL) corresponds to the axis, and the other line (RR) is a tangent to the helix whose radius is the normal common to the two lines.

site direction. The distribution of chelate rings is therefore said to be left- (Fig. 1-10*a*) or right-handed (Fig. 1-10*b*) about the C_3 axis.

The chirality of these compounds has some meaning when defined in this way. It is of particular importance, however, that the axis about which the chirality is defined is clearly indicated, because, if the compounds above are viewed along any of the C_2 axes, the opposite handedness is observed (see Fig. 1-11).

The chirality of the distribution of two chelate rings can also be ascertained by considering the rings as lines, one of which is the axis of a cylindrical helix, the other being the tangent to the helix, the radius of which is the normal common to the two lines (see Fig. 1-12). Any two skew lines that are not orthogonal will uniquely define a cylindrical helix in this way.

A sense of handedness can also be associated with conformations. Examples of dissymmetric conformations were shown in Figs. 1-3 to 1-6. Again there are various ways in which this chirality can be viewed. First, if we look down the *y* axis, which in Figs. 1-3 and 1-5 is a C_2 axis, the two C—X bonds show helicity: in Figs. 1-3*a*(3) and 1-5*a*(3), as we move down the *y* axis to the metal ion, that is, into the page, the C—X bonds trace out a clockwise screw motion (right-handed); in Figs. 1-3*b*(3) and 1-5*b*(3) the movement is counterclockwise (left-handed). For the asymmetric envelope conformations, however, the two C—X bonds define motions of opposite helicity. Therefore, this means of viewing the chirality of a conformation is not very satisfactory, although the screws are of different pitch.

If one views the rings according to elevation 2 in Figs. 1-3 to 1-6 along a line that includes the two carbon atoms adjacent to the donor atoms, the C—X bonds are again seen to describe a helical motion. In *a* of these figures the helix is left-handed; in *b* it is right-handed. This is equivalent to considering the two nonorthogonal skew lines X—X and C—C (the carbons adjacent to the X atoms), which define a helix in which C—C is the axis and X—X is a tangent to the helix that has as its radius the normal common to both X—X and C—C. If the two skew lines containing the C—X bonds are chosen, however, the helicity is reversed.

With such a variety of ways in which the chirality of the distribution and the conformations of chelate rings can be viewed, it is imperative that the method used in deriving the chirality is fully specified.

In the literature it is often stated that two compounds have the same absolute configuration, although the compounds may have radically different compositions. It is not unusual, for example, to see the structures **XXXIII** and **XXXV** for tris(bidentate) and *cis*-bis(bidentate) complexes described as having the same absolute configuration. This is a loose usage of the term. What is usually understood, however, is that the distributions of the chelates in the two compounds are of the same chirality.

REFERENCES

1. Bailar, J. C., *J. Inorg. Nucl. Chem.*, **8**:165 (1958).
2. Bailar, J. C., and D. F. Peppard, *J. Am. Chem. Soc.*, **62**:105 (1940).
3. Buckingham, D. A., L. G. Marzilli, and A. M. Sargeson, *J. Am. Chem. Soc.*, **89**:825 (1967).
4. Chernyaev, I. I., T. N. Fedotova, and O. N. Adrianova, *Russ. J. Inorg. Chem.*, **10**:841 (1965).
5. Chernyaev, I. I., T. N. Fedotova, and O. N. Adrianova, *Russ. J. Inorg. Chem.*, **11**:719 (1966).
6. Chernyaev, I. I., T. N. Fedotova, and O. N. Adrianova, *Russ. J. Inorg. Chem.*, **11**:723 (1966).
7. Chernyaev, I. I., L. S. Korablina, and G. S. Muraveiskaya, *Russ. J. Inorg. Chem.*, **10**:567 (1965).
8. Fujita, J., T. Yasui, and Y. Shimura, *Bull. Chem. Soc. Japan*, **38**:654 (1965).
9. Gehman, W. G., Ph.D. thesis, Penn. State Univ., State College, Pa., 1954.
10. Halpern, B., A. M. Sargeson, and K. R. Turnbull, *J. Am. Chem. Soc.*, **88**:4630 (1966).
11. Hawkins, C. J., and P. J. Lawson, *Chem. Commun.*, **1968**:177.
12. Hawkins, C. J., and P. J. Lawson, *Inorg. Chem.*, **9**:6 (1970).
13. Hawkins, C. J., J. Niethe, and C. L. Wong, *Chem. Commun.*, **1968**:427.
14. Ray, B. C., *J. Indian Chem. Soc.*, **14**:440 (1937).
15. Rây, P., and N. K. Dutt, *J. Indian Chem. Soc.*, **20**:81 (1943).
16. Seiden, L., Ph.D. thesis, Northwestern Univ., Evanston, Ill., 1957.
17. Shibata, J., and T. Maruki, *J. Coll. Sci. Imp. Univ. Tokyo*, **41**(2):1 (1917).
18. Sprenger, C. S., and R. E. Sievers, *Inorg. Chem.*, **6**:852 (1967).
19. Theilacker, W., *Z. Anorg. Chem.*, **234**:161 (1937).
20. Thomas, W., *J. Chem. Soc.*, **1923**:617,
21. Yasui, T., J. Hidaka, and Y. Shimura, *Bull. Chem. Soc. Japan*, **39**:2417 (1966).

2

Notation of Absolute Configuration

PREAMBLE

A set of symbols is required to describe the absolute configurations of metal complexes. It is imperative that the system adopted is simple to apply and enables a reader to visualize the detailed structure of the isomer in question readily. Although a system of notation has been successfully used in organic chemistry for some time [1,2], in coordination chemistry there is at present great confusion because of the proliferation of symbols and principles of application.

It would seem that any system sufficiently general to cover all kinds of dissymmetries would be extremely complicated to apply, and the description would not immediately call to one's mind any particular structure. These criticisms could well be leveled at the Cahn, Ingold, and Prelog system [3], which ostensibly is a general one covering all fields of chemistry with the symbols R,S for configuration and P,M for conformation. As they themselves admit, the extension of the system to octahedral metal complexes was made possible only by the introduction of a new set of subrules. The original "self-contained basis of a few general rules" had to be dramatically broadened to cover the octahedral complexes.

To overcome these difficulties, it seems logical to have a separate notation for each source of dissymmetry.

NOTATION FOR THE DISTRIBUTION OF CHELATE RINGS

Most attention has been directed at the problem of how to denote the distribution of chelate rings about a central metal ion. Originally a compound was differentiated from its optical isomer simply by the sense of the rotation of plane-polarized light, usually at the sodium D line. Thus, the compounds were dextro- or levorotatory and were given the symbols *d* or *l*. Often the small capital letters D and L were used to denote the distribution of chelate rings in tris- and *cis*-bis(bidentate) complexes—see [8], for example—but they were also used to describe the sign of rotation. Prior to 1954 no complex had had its absolute configuration determined. The first such determination was by Saito and his coworkers, who studied (d-[$Coen_3$]Cl_3)$_2$NaCl-6H_2O by X-ray analysis [15,18,19]. The distribution was found to be that given by structure **XXXIII** and was denoted D; its enantiomeric structure, **XXXIV**, was denoted L.

Mason recognized that structure **XXXIII** had a chirality that was left-handed about the threefold axis and right-handed about the twofold axes [13]. He recommended that the symbols D,L [14] and later P,M [11,12] be used in conjunction with the relevant axis to denote the chirality, and therefore the absolute configuration, of the various isomers; for example, structure **XXXIII** was to be represented by L(C_3) or D(C_2).

Piper proposed that D,L would lead to confusion following the early usage of these symbols for both rotation and configuration [16]. He, and later Liehr [10], suggested that Δ and Λ be used in their place and that these symbols were to specify the chirality about the axis of highest order. Thus structures **XXXIII** and **XXXIV** were given the symbols Λ and Δ, respectively. According to this system, however, *cis*-bis(bidentate) complexes with the distribution **XXXV**, which is normally related to the distribution **XXXIII**, would be denoted by Δ. Recognizing this problem, Piper [17] and Liehr [10] introduced pseudo-threefold axes of symmetry for *cis*-bis and low-symmetry tris(bidentate) complexes.

These systems were primarily designed for denoting the distribution of chelate rings in tris- and *cis*-bis(bidentate) complexes. They were extendable in a limited number of cases to multidentate complexes by selecting a pair of chelate rings with due regard to the symmetry of the molecule and comparing the chirality of their distribution to that of the equivalent *cis*-bis-(bidentate) complex. Thus, for structure **XXXVII**, the two apical chelate rings would be chosen and their distribution related to structure **XXXV** and given the symbol D, P, or Δ.

In 1965 Hawkins and Larsen introduced a more general system based on an octant sign [7]. A complex is situated in a right-handed coordinate system with the metal ion at the origin. Each of the octants has a sign derived by the multiplication of the signs of the coordinates defining that octant (see Fig. 2-1). In order to determine the position of each chelate ring relative to all other rings, the chelate rings are placed in turn in the xy plane with the donor atoms in the positions $(+x, +y)$ and $(-x, +y)$, and the signs of the octants in which the other rings are situated are determined. The octant

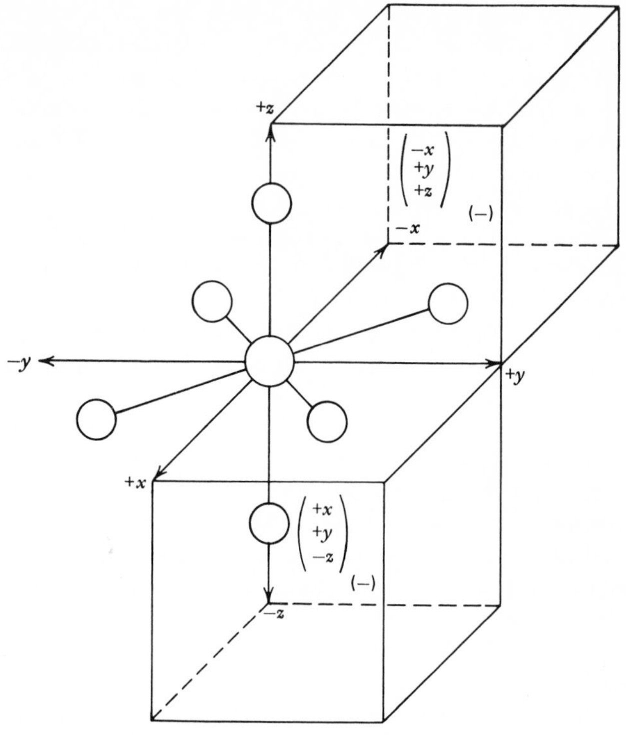

Fig. 2-1 Octahedral complex in a right-handed coordinate system. Octants have a sign given by the products of the signs of the coordinates defining the octant.

sign of the complex is obtained by adding the signs for the individual chelate rings. Some examples are given in Figs. 2-2 and 2-3. The symbols D and L were selected to indicate whether the absolute configuration was defined by a positive or negative octant sign. The structures **XXXIII**, **XXXVII**, and **XLIX** all have the same octant sign (positive) and therefore would be allocated the same symbol. This has some merit, because the related transitions in the various complexes have the same sign for their Cotton effect (see Chap. 5) [7]:

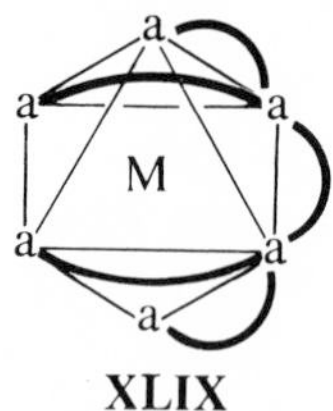

XLIX

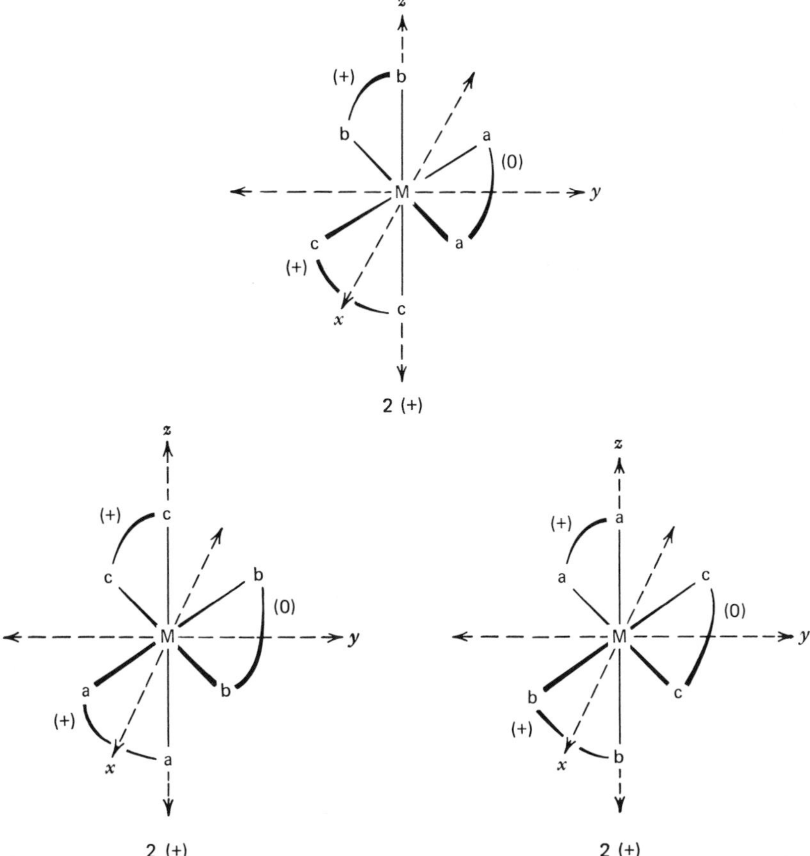

Fig. 2-2 The octant sign for structure **XXXIII**: positive. (Donor atom symbols have been altered to identify individual rings.)

The more recent ring-paired method of Legg and Douglas [9] arrives at the same conclusions as the octant sign method, because the two methods relate the positions of the chelate rings in what is fundamentally the same way. In the ring-paired method the chirality of each dissymmetric combination of two chelate rings is determined about the C_2 axis for the ring combination, and the individual chiralities are added to give the overall handedness of the complex; for example, in structure **XXXIII** there are three sets of two chelate rings in which the rings have a chiral distribution. Each set has a right-handed chirality about its C_2 axis, and therefore the complex was judged to be right-handed and was allocated the symbol Δ. In structure **XLIX** there are also three such combinations, two right-handed and the

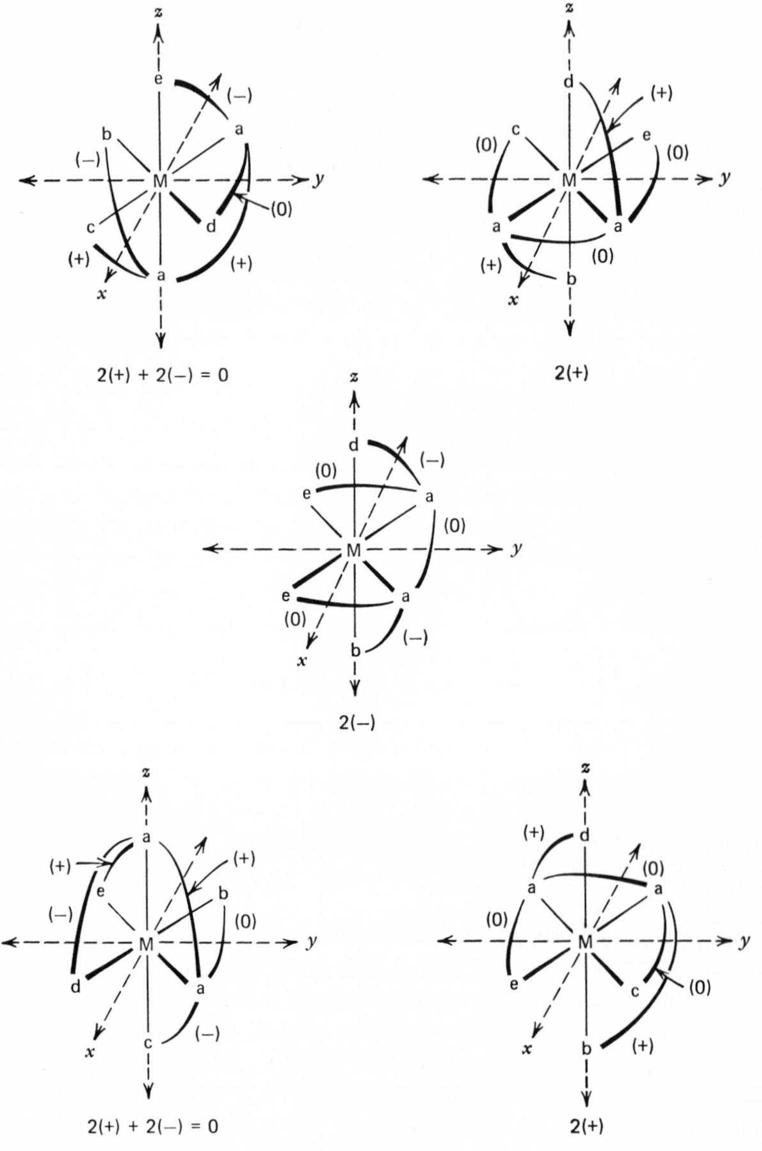

Fig. 2-3 The octant sign for structure **XLIX**: [2(+) + 2(−) + 2(+) = 2(+)] positive. (Donor atom symbols have been altered to identify individual rings.)

other left-handed, as well as five combinations that are not dissymmetric. The complex was therefore given the symbol Δ.

The R,S system of Cahn, Ingold, and Prelog [3] is the most ambitious of all. Unfortunately, the symbol to describe a particular distribution of chelate rings varies, depending on the kinds of ligands present. To overcome this problem the authors introduced a secondary structure notation. This structure is simply the chirality of the chelate distribution and is denoted by sec-P or sec-M, depending on which "helical axis" is chosen. Thus **XXXIII** would be sec-P(C_2).

There are three rules and a number of subrules for the application of the R,S system. The rules have been classified as the octahedral sequence, numbering, and chirality rules. Before they can be applied, however, you must be familiar with the standard subrules for the general sequence rule.

General Sequence Rule

"The ligands associated with an element of chirality are ordered by comparing them at each step in bond-by-bond explorations of them from the element, along the successive bonds of each ligand, and, where the ligands branch, first along branch-paths providing highest precedence to their respective ligands, the explorations being continued to total ordering by the use of the following Standard Sub-rules, each to exhaustion in term, namely:

(0) "Nearer end of axis or side of plane precedes further.

(1) "Higher atomic number precedes lower.

(2) "Higher atomic mass-number precedes lower.

(3) "Seqcis precedes seqtrans. (For a double-bonded unit, abX=Ycd, an atom-pair is called seqcis or seqtrans according as the sequence-rule-preferred ligand bound to each atom of the pair is cis or trans with respect to the other such ligand. Note that "ligand" is used here to mean an atom or group bound to any particular atom).

(4) "Like pair R,R or S,S precedes unlike R,S or S,R; and M,M or P,P precedes M,P or P,M; and R,M or S,P precedes R,P or S,M; M,R or P,S precedes M,S or P,R; . . .

(5) "R precedes S; and M precedes P (Note: 'according as an identified helix is left- or right-handed, it is designated "minus" and denoted by M, or designated "plus" and denoted by P')."

Octahedral Sequence Subrules

The following are successively applied:

(1) "Ligands of higher denticity precede those of lower. (Note: the "denticity" of a ligand is the number of donor atoms coordinated to the central metal ion.)

(2) "The leading ligating (donor) atoms of a ligand being its terminal

ligating atom most preferred by the Standard Subrules, or, in an endless ligand, the ligating atom which is so most preferred, the ligands are ordered, ended before endless, and otherwise as their leading ligating atoms would be ordered by the Standard Subrules."

Octahedral Numbering Subrules

The following are successively applied:

(1) "The numbers 1 and 6 mark trans positions.

(2) "The numbers 2–5 form a cyclic sequence.

(3) "With the number 1 is associated the ligating atom which is most preferred by the Octahedral Sequence Rule; or, in case a choice is open, that most preferred ligating atom which allows the least preferred possible to be associated with the number 6; or, if a choice is still open, that which allows more preferred ligating atoms to be associated with lower numbers in the range 2–5.

(4) "The numbers 2–4 are successively associated with the most preferred of the ligating atoms remaining available to each."

Octahedral Chirality Rule

"The chiral center of the numbered model is specified as R or S, according as the path of the numbers 1, 2, 3 in order appears right- or left-handed from the side of the model remote from 4, 5, 6."

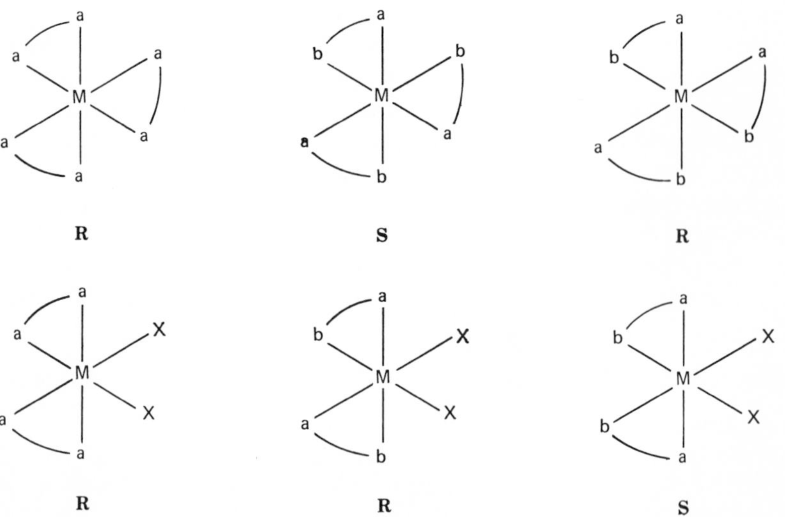

Fig. 2-4 Octahedral metal complexes with absolute configurations assigned symbols, R and S, according to the method of Cahn, Ingold, and Prelog. [3].

These rules are rigorous but they are also frightening to the uninitiated. A few structures with their symbols assigned are given in Fig. 2-4. It remains as an exercise for you to reason how these symbols were assigned.

It would seem from this review that the situation regarding a system of notation is both confused and confusing. On closer examination, however, it is possible to observe considerable agreement among a number of the approaches. The comparison of chirality about common C_2 axes, as proposed by Mason [12] and supported by Schäffer [20], the ring-paired method, and the octant sign method all ascribe a right-handedness to **XXXIII** and **XXXV** and a left-handedness to their enantiomers. This is in agreement with the allocation of the symbol D to **XXXIII** and **XXXV**, which has persisted throughout the present century. It is therefore unfortunate that the IUPAC subcommittee on the nomenclature of absolute configuration of octahedral metal complexes has put forward a preliminary report in which they propose the opposite chirality for these complexes [5].

According to the IUPAC subcommittee proposal each chiral pair of chelate rings can be looked on as two skew lines which are not orthogonal and which describe a cylindrical helix, one line being the axis of the helix, the other a tangent to the helix, which has as its radius the normal common to the two skew lines (see Fig. 1-12). The handedness of the helix for each chiral pair of chelates is determined and summed to give the overall chirality of the complex. This has some aspects in common with the ring-paired method, but the two procedures yield opposite chiralities.

In the absence of a universally accepted system of nomenclature it has been decided to base the system of notation in the present work on the ring-paired and octant sign methods. The symbols D and L have been chosen to denote right- and left-handedness.*

There is at least one compound, **L**, with a chiral distribution of chelates for which these procedures are unable to assign a symbol. It is proposed that, for these complexes, the chirality of the distribution of the terminal chelate rings should be examined. Thus, **L** would be given the symbol D:

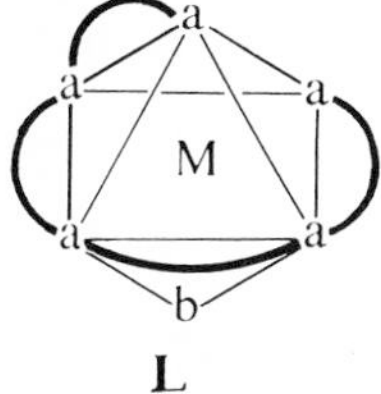

L

* At the time this book was written the IUPAC subcommittee's tentative proposals had not gained any marked acceptance. However, it is apparent at the time of printing that the proposals will be recommended by IUPAC. It should be noted, therefore, that D $\equiv \Lambda$ and L $\equiv \Delta$ and that the two systems of nomenclature have a consistent notation for conformational chirality.

NOTATION FOR THE CONFORMATION OF CHELATE RINGS

Corey and Bailar [4] were the first to denote conformations of chelates by symbols. They chose k and k' to specify the conformations of the ethylenediamine ring corresponding to Fig. 1-3*a* and *b*, respectively.

The octant sign method has also been applied to conformations [7]. The octants in which the various ring atoms lie are determined and the signs added. For the conformation shown in Fig. 1-3*a*(1), three of the ring atoms (M, X, and X) lie in the xy plane and so have no octant sign, whereas the two carbon atoms are to be found in negative octants ($+x$, $+y$, $-z$ and $-x$, $+y$, $+z$). Thus the conformation is said to have a negative octant sign, as does the conformation in Fig. 1-5*a*(1), where four atoms (M, X, X, and C) lie in the xy plane and the remaining two carbons are in the same negative octants as for Fig. 1-3*a*(1). An additional rule has to be introduced for the asymmetric envelope and asymmetric boat conformations in Figs. 1-4 and 1-6. The two critical carbon atoms are in octants with opposite signs: for conformations *a*(1), $+x$, $+y$, $+z$, and $-x$, $+y$, $+z$. In Fig. 1-6*a*(1) the third carbon atom is in the yz plane. The octant sign is determined in these circumstances by the sign of the atom farthest from the xy plane. For conformations 1-4*a*(1) and 1-6*a*(1) the octant sign is therefore negative.

Cahn, Ingold, and Prelog [3] have proposed a method for denoting conformations in which the "molecular conformation is treated in terms of partial conformations about the individual single bonds." Thus, the problem resolves itself into a consideration of a number of 1,2-disubstituted-ethane systems. The two enantiomeric gauche (synclinal) conformations are denoted P or M, depending on whether "the smallest torsional angle between the fiducial groups" is positive or negative, that is, plus or minus. Unfortunately, the subrules introduced to specify which groups are fiducial do not necessitate that, for a ring system, the ring atoms are taken as the reference atoms. As shown in Fig. 2-5, the symbols used to denote a particular conformation depend on the substituents on that ring and do not specify the ring conformation by itself. Further, the rules are specific for tetracovalent atoms. Therefore they cannot be rigorously applied to rings containing octahedral metal ions.

In Chap. 1, various ways were discussed for deducing a chirality associated with conformations. It would seem that the most promising general method is that in which the conformations are viewed according to elevation 2 in Figs. 1-3 to 1-8 along the line including the two carbon atoms bound to the donor groups. The helicity of the distribution of the two C—X bonds defines the chirality of the ring. In the present work the symbols δ and λ are used to specify right- and left-handedness, respectively, the chirality being determined in this way. Thus the conformations *a* in Figs. 1-3 to 1-6

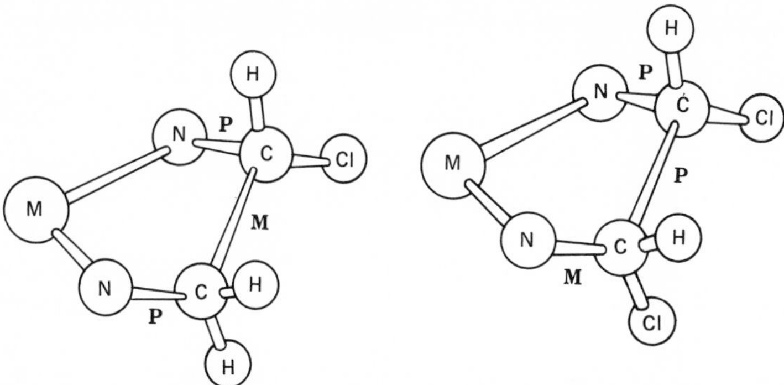

Fig. 2-5 Notation of substituted ethylenediamine chelate ring conformations according to Cahn, Ingold, and Prelog [3]. The hydrogen atoms on the nitrogens and the other groups attached to the metal have been ignored for simplicity.

are λ and the conformations *b* are δ. (If the octant sign procedure is used, positive and negative octant sign conformations should be specified as δ and λ, respectively).

NOTATION FOR THE DISTRIBUTION OF UNIDENTATE LIGANDS

The helicity associated with the distribution of chelate rings and with the conformations of chelate rings is easy to visualize. This is not so for the majority of complexes whose asymmetry comes from the relative positions of unidentates. Perhaps a compound such as *cis*-$Ma_2b_2c_2$ (**V**) can be compared with a tris(bidentate) complex and seen to have a sense of helicity about the pseudo-C_3 axis, but generally this is not possible.

It seems logical to extend the accepted nomenclature for an asymmetric tetracovalent atom to the asymmetric hexacovalent octahedral atom. The R,S system of Cahn, Ingold, and Prelog [3] is not successful when applied to the distribution of chelate rings, because the symbols are not directly related to the distribution. For unidentate systems, however, where there are no established conventions for relating structures, it proves to be very acceptable.

The rules are as given above, but in most cases only a few of them are required. Some examples of the application of the system are given in Fig. 2-6.

NOTATION FOR DISSYMMETRY FROM AN UNSYMMETRICAL CHELATE

Because the dissymmetry in this case is due to differences in the donor groups, it is convenient to use the Cahn, Ingold, and Prelog system [3] again. Examples in which assignments have been made are to be found in Fig. 2-7.

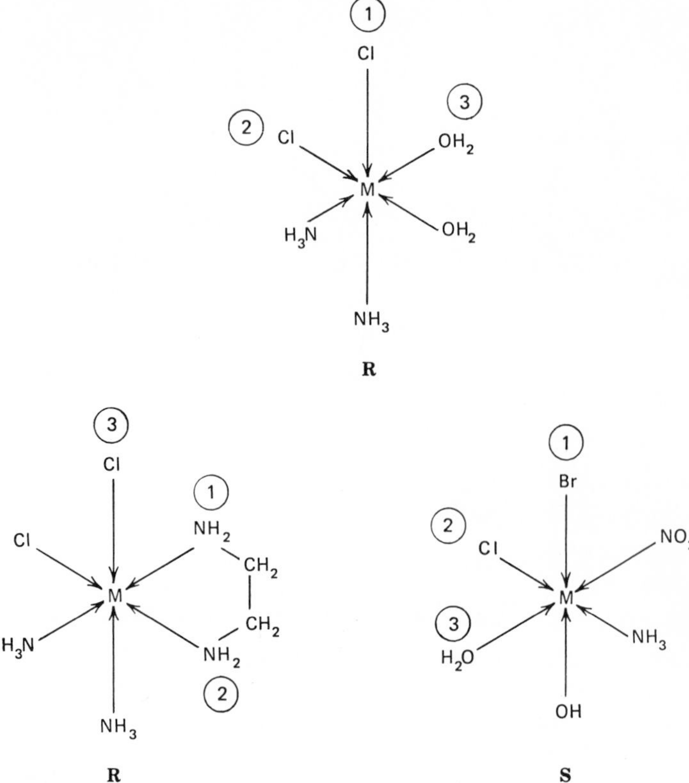

Fig. 2-6 Assignment of symbols to complexes which are asymmetric because of the distribution of unidentates.

NOTATION FOR COMPLEX CONTAINING AN ASYMMETRIC LIGAND

If the only source of dissymmetry in a complex is the coordination of an asymmetric ligand, the normal R,S notation for the absolute configuration of the ligand is sufficient to describe the configuration of the complex. This is correct even if the ligand becomes resolvable only on coordination; for example, the absolute configuration of $[Co(NH_3)_5(\text{S-1-phenylethylamine})]^{3+}$ and $[Co(NH_3)_5(\text{R-sarcosine})]^{3+}$ are completely specified as presented.

Although R and S have been generally accepted for organic structures, the α-amino acids are still more usually classified as D,L. This convention is adopted here.

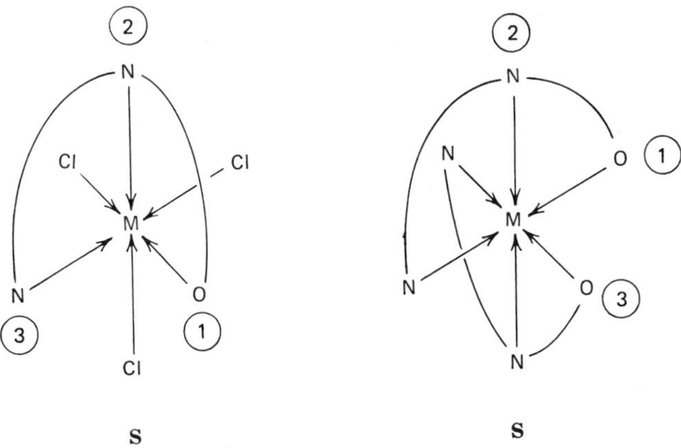

Fig. 2-7 Assignment of symbols to complexes which are dissymmetric because of the unsymmetrical nature of some chelates.

SUMMARY OF ADOPTED SYMBOLISM

The following symbols are used throughout the present work:

D,L to describe the overall absolute configuration of a complex when the dissymmetry is due to the distribution of chelate rings; D and L correspond to positive and negative octant signs, respectively, or to right- and left-handed chirality according to the ring-paired method*

δ,λ to describe the conformations of individual chelate rings; δ and λ correspond to positive and negative octant signs, respectively, or to right- and left-handed chirality according to the helicity of the C—X bonds viewed along a line containing the carbon atoms

R,S to describe the dissymmetry due to the distribution of unidentates, to the unsymmetrical nature of some chelates, and also to describe the absolute configurations of asymmetric ligands according to the method of Cahn, Ingold, and Prelog

d_x,l_x to describe the rotation at a wavelength x

d,l to describe the rotation at the sodium D line

REFERENCES

1. Cahn, R. S , and C. K. Ingold, *J. Chem. Soc.*, **1951**:612.
2. Cahn, R. S., C. K. Ingold, and V. Prelog, *Experientia*, **12**:81 (1956).
3. Cahn, R. S., C. K. Ingold, and V. Prelog, *Angew. Chem. Internat. Edn.*, **5**:385 (1966).

* D,L also specify absolute configurations of α-amino acids.

4. Corey, E. J., and J. C. Bailar, *J. Am. Chem. Soc.*, **81**:2620 (1959).
5. Douglas, B. E., A. M. Sargeson, and C. E. Schäffer, Some Proposals for Nomenclature of Absolute Configurations Concerned with "Octahedral" Chelate Complexes, *I.U.P.A.C. Sub-committee Report*, 1967.
6. Gollogly, J. R., and C. J. Hawkins, *Chem. Commun.*, **1966**:873.
7. Hawkins, C. J., and E. Larsen, *Acta Chem. Scand.*, **19**:185 (1965).
8. Jaeger, F. M., "Optical Activity and High Temperature Measurements," McGraw-Hill, New York, 1930.
9. Legg, J. I., and B. E. Douglas, *J. Am. Chem. Soc.*, **88**:2697 (1966).
10. Liehr, A. D., *J. Phys. Chem.*, **68**:3629 (1964).
11. McCaffery, A. J., S. F. Mason, and R. E. Ballard, *J. Chem. Soc.*, **1965**:2883.
12. McCaffery, A. J., S. F. Mason, and B. J. Norman, *J. Chem. Soc.*, **1965**:5094.
13. Mason, S. F., *Quart. Revs. London*, **17**:20 (1963).
14. Mason, S. F., Communication to I.U.P.A.C. Commission on Symbols, Terminology, and Units, 1964.
15. Nakatsu, K., M. Shiro, Y. Saito, and H. Kuroya, *Bull. Chem. Soc. Japan*, **30**:158 (1957).
16. Piper, T. S., *J. Am. Chem. Soc.*, **83**:3908 (1961).
17. Piper, T. S., Communication to I.U.P.A.C. Commission on Symbols, Terminology, and Units, 1965.
18. Saito, Y., K. Nakatsu, M. Shiro, and H. Kuroya, *Acta. Cryst.*, **7**:636 (1954).
19. Saito, Y., K. Nakatsu, M. Shiro, and H. Kuroya, *Acta Cryst.*, **8**:729 (1955).
20. Schäffer, C. E., Communication to I.U.P.A.C. Commission on Symbols, Terminology, and Units, 1965.

3

Conformational Analysis

INTRODUCTION

Conformational analysis is usually defined as the analysis of the chemical and physical properties of a compound in terms of its conformational* structure. The experimental procedures that have been or could be applied to the study of the conformations of chelate rings are briefly discussed below, and in the respective chapters on the individual techniques.

In theory it should be possible to determine the most preferred conformation of a molecule and the energy differences between this structure and other conformations by way of quantum mechanical computations. In practice, however, this is far from being realized because of the complexity of the calculations. Nevertheless, classical models have been developed that have factorized this problem into the estimation of distinctly different energy contributions that are calculable from relatively simple expressions. This has enabled reliable qualitative predictions to be made of conformational preferences and also, in some cases, of the preferred chirality of the distribution of chelate rings. In the present chapter, each of the energy terms is discussed in considerable detail, and this form of conformational analysis is applied to a number of metal complex systems to determine (a) which conformations are stable, that is, correspond to energy minima, (b) which conformations are the most preferred, and (c) what are the energy differences between the various stable conformations.

Although this approach is well established in the field of organic chemistry, it is still very much in its infancy as far as coordination chemistry is con-

* Conformations, as defined in Chap. 1, are the nonidentical three-dimensional arrangements of atoms in a molecule that can be interconverted by rotation about one or more single bonds.

cerned. Mathieu [94] pioneered its application to metal complexes when he calculated the difference in the nonbonded interactions in the diastereoisomers of *cis*-[Co(*d*-pn)$_2$X$_2$] in an attempt to explain why one isomer was formed preferentially. He restricted himself to the differences in energy due to the interactions of the two methyl groups, and to the dipole–induced dipole interactions between X and CH_3. A period of 15 years elapsed before the second notable paper on the conformational analysis of metal complexes appeared. in 1959 Corey and Bailar published a most important paper detailing the results of their study of a number of metal complex systems [26]. This formed the basis for most of the subsequent work. Very recently, however, a more rigorous approach has been applied [49,50,51,52], and this is followed here.

VAN DER WAALS FORCES

It is currently accepted that interatomic forces exist other than those giving rise to chemical bonding. Molecules come together to form liquids and solids, for example, because of some kind of attractive force, but the individual molecules are held apart in these materials because of the presence of strong repulsive forces that are operative at relatively small interatomic distances. These forces are often very important in determining the conformational and configurational preferences, and, as they are perhaps the most controversial aspect of this work, they are reviewed at length below.

Dispersion Forces

In 1873 van der Waals [131] suggested a hard-sphere model in which deviations from ideal-gas-law behavior, due to the mutual attractive forces between molecules, could be accounted for by the equation

$$\left(p + \frac{a}{V^2}\right)(V - b) = RT \tag{3-1}$$

By 1925, there had been several attempts to explain these attractive forces in terms of the electronic structure of the molecules.

Keesom proposed a dipole–dipole interaction as the origin [74]. This interaction was averaged over all orientations using a Boltzman distribution to account for the preference for the low-energy orientations and led to the expression

$$\mathscr{E} = \frac{-2\mu_A\mu_B r^{-6}}{3kT} \tag{3-2}$$

for $\mu_A\mu_B r^{-3}$ less than kT, and where μ_A and μ_B are the dipoles on the two interacting molecules. This is known as *Keesom's dipole-orientation energy.*

Debye noted that Eq. (3-2) requires the energy to decrease with increasing temperature and that this requirement is not consistent with the observed temperature variation of the empirical van der Waals parameters a and b [28]. He attempted to overcome this deficiency by introducing the interaction between a permanent dipole of moment μ and the dipole it induces in a molecule of polarizability α. Averaging over all mutual orientations,

$$\mathscr{E} = -\alpha\mu^2 r^{-6} \tag{3-3}$$

If both molecules have permanent dipoles μ_A and μ_B and finite polarizabilities α_A and α_B, it follows that

$$\mathscr{E} = -(\alpha_A \mu_B{}^2 + \alpha_B \mu_A{}^2) r^{-6} \tag{3-4}$$

This is the general form of Debye's induction energy.

Neither Keesom's dipole–dipole or Debye's dipole–induced dipole force, however, can account for the presence of van der Waals forces between molecules that lack permanent dipoles—H_2, N_2, rare gases, for example. Both could be extended to cover these cases by considering the interactions of quadrupoles and higher multipoles, but the Keesom force would still vanish with increasing temperature. A more important objection to these two models is that, in contrast to the van der Waals forces, the proposed forces are not additive.*

A significant advance occurred when London noted that electronic motion itself constituted a rapidly fluctuating dipole moment [85,86]. Such instantaneous dipoles are able to induce dipoles of identical phase in other atoms, resulting in a net attractive force that does not depend on the existence of permanent dipoles or higher multipoles. The expression for the energy of interaction was based on Drude's dispersion formula [32]:

$$\alpha_k = \frac{2}{3h} \sum_l \frac{\mu_{kl}{}^2 \nu_{kl}}{\nu_{kl}{}^2 - \nu^2} \tag{3-5}$$

where α_k is the polarizability of the molecule in the state k, ν is the frequency of the applied field, ν_{kl} is the frequency of the vibration of the dipole between the states k and l, μ_{kl} is the corresponding dipole moment, and the sum is taken over all states l that are accessible to a dipole in the state k [37]. If the applied field had an amplitude of F_0, the energy of interaction with the dipole is $-\frac{1}{2}\alpha F_0$. When the applied field derives from another oscillating di-

* A force law is said to be additive if the interaction energy of a system of n molecules is equal to the sum of the energies of interaction of $n!$ pairs. In other words, the interaction law that governs the energy of aggregations of molecules is the same as that governing isolated pairs of molecules.

pole ($\mu_{k'l'}, \nu_{k'l'}$), the total energy due to the interaction between an oscillating dipole in state k and one in state k' can be expressed as

$$\mathscr{E} = \mathscr{E}_{kk'} + \mathscr{E}_{k'k} = \frac{-2}{3h} \sum_{l} \sum_{l'} \frac{\mu_{kl}^2 \mu_{k'l'}^2}{\nu_{kl} + \nu_{k'l'}} \tag{3-6}$$

For many simple gas molecules (e.g., H_2, N_2, O_2, CH_4, rare gases), the empirical dispersion curves can be represented by a one-term expression

$$\alpha_k = \frac{2}{3h} \frac{\mu_k^2}{\nu_k} \tag{3-7}$$

where μ_k corresponds to the ground state vibration of the oscillator and ν_k is its natural frequency. Equation (3-6) can then be rewritten as

$$\mathscr{E} = -\frac{3h}{2} \alpha_k \alpha_{k'} \frac{\nu_k \nu_{k'}}{\nu_k + \nu_{k'}} r^{-6} \tag{3-8}$$

and when the dipoles are identical this reduces to

$$\mathscr{E} = -\frac{3h}{4} \alpha_k^2 \nu_k r^{-6} \tag{3-9}$$

London [85,86] observed that for some simple molecules the values of $h\nu_k$ are very similar to the observed ionization potentials, suggesting that Eq. (3-9) can be approximated to

$$\mathscr{E} = -\frac{3h}{2} \alpha_k \alpha_{k'} \frac{\mathrm{I}_k \mathrm{I}_{k'}}{\mathrm{I}_k + \mathrm{I}_{k'}} \tag{3-10}$$

which has been used for cases where the dispersion curve has not been determined and thus ν_k is not known.

Perturbation theory has been applied to the simplest problem of the interaction of two hydrogen atoms [37]. It was found that the first-order perturbation energy was zero and that the second-order perturbation energy was given by

$$\mathscr{E}'' = -\tfrac{3}{4} h\nu_0 \alpha^2 r^{-6} \text{ kcal mol}^{-1} \tag{3-11}$$

where ν_0 is the *natural frequency* of each of the hydrogen atoms. Equation (3-11) confirmed the form of an earlier equation derived by Wang, who used a second-order perturbation theory approach based on Epstein's analysis [40] of the Stark effect in hydrogen [132]. Wang's original calculation contained a numerical error in the evaluation of the coefficients of r^{-6}, but his approach was reexamined by Pauling and Beach [109] and found to give a good approximate value for this coefficient. Equation (3-11) was later verified by Lennard-Jones [83].

The variation theorem has been applied to the same problem by Hassé [57] and Slater and Kirkwood [122]. The coefficients for r^{-6} determined in this way agreed well with those from perturbation theory. Slater and Kirkwood extended their approach to multielectron systems and, on the basis of their equations, suggested that only the outer shell of electrons contributed to the interaction energy. They derived the simplified expression

$$\mathscr{E} = -1.36N^{1/2}\alpha^{3/2}\mathscr{E}_0 r^{-6} \tag{3-12}$$

where N is the number of outer-shell electrons, α is the polarizability, and $\mathscr{E}_0$ is the zero-point energy of the atom.

Kirkwood [77] further extended the variation treatment to the interaction of two unlike atoms and obtained the formula

$$\mathscr{E} = -\frac{3}{2}\frac{\alpha_A\alpha_B}{[(\alpha_A/N_A)^{1/2} + (\alpha_B/N_B)^{1/2}]} r^{-6} \tag{3-13}$$

where N_A, N_B are the total number of electrons in the atoms A and B. Hellmann [60,61] showed that the major contribution to this came from the outermost electrons, and, following his work, Eq. (3-13) is now used more commonly with N_A, N_B as the number of electrons in the outermost shell. Kirkwood noted that some of the terms that arose from his variation treatment could be related to the diamagnetic susceptibility χ, just as he had related the terms to the polarizabilities [77]. The interaction of spherical atoms led to the equation

$$\mathscr{E} = -6mc^2\frac{\chi_A\chi_B}{\chi_A/\alpha_A + \chi_B/\alpha_B} r^{-6} \tag{3-14}$$

which has been applied by Müller [97]. For helium–helium interactions the r^{-6} coefficient calculated by Müller is much larger than that obtained from Eq. (3-12) or (3-10). The derivation of Eq. (3-14) described by Kirkwood [77] and Müller [97] has been criticized by Hellmann [60,61] on theoretical grounds, and London [86] pointed out that the differences between his and Müller's results become very large for the higher inert gases.

If we consider the interaction of two hydrogen atoms in terms of the classical interaction of two pairs of point charges, the energy may be written in terms of two parallel Cartesian coordinate systems centered on the two nuclei as

$$\begin{aligned}\mathscr{E} = {} & -e^2(2x_Ax_B - y_Ay_B - z_Az_B)r^{-3} \\ & + \tfrac{3}{2}e^2[r_A{}^2x_B - x_Ar_B{}^2 + (x_A - x_B)(2y_Ay_B + 2z_Az_B - 3x_Ax_B)]r^{-4} \\ & + \tfrac{3}{4}e^2[r_A{}^2r_B{}^2 - 5r_B{}^2x_A{}^2 - 5r_A{}^2x_B{}^2 - 15x_A{}^2x_B{}^2 \\ & \qquad + 2(4x_Ax_B + y_Ay_B + z_Az_B)^2]r^{-5}\end{aligned} \tag{3-15}$$

where $r_A{}^2 = x_A{}^2 + y_A{}^2 + z_A{}^2$ and $r_B{}^2 = x_B{}^2 + y_B{}^2 + z_B{}^2$. The r^{-3} term may be associated with a dipole–dipole interaction. The remaining terms may be associated with dipole–quadrupole (r^{-4}) and quadrupole–quadrupole (r^{-5}) interactions, and so on. The quantum-mechanical treatments described so far considered the interaction as a perturbation based on the r^{-3} term alone, neglecting the higher multipole interactions. These methods yielded an interaction energy of the form $-Ar^{-6}$. Margenau performed a calculation of the second-order perturbation energy in which he included these higher multipole terms [91]. In this case the interaction energy has the form

$$\mathscr{E} = -Ar^{-6} - \mathrm{A}'r^{-8} - A''r^{-10} - \cdots \tag{3-16}$$

Pauling and Beach performed a similar calculation using the variation method [109]. The values of the coefficients A, A', and A'' for the H···H interaction were found to be 89, 477, and 1247 to be compared with 82, 518, and 1521 determined by Margenau [91]. Hornig and Hirschfelder have expressed these coefficients in forms comparable to Eq. (3-8) of London [68].

It is clear from inspection of the equations derived by Margenau [91] and Pauling and Beach [109] that, for moderate to large interatomic distances, the contribution from these higher-multipole interactions to the interaction energy is quite small.

Coulombic Interactions

If point charges, permanent dipoles, or higher multipoles exist in a molecule, their coulombic interactions must be taken into account. Once the distribution of charge within a molecule is known, the energy of such interactions follows directly from Coulomb's law. Several authors—see Ref. 92, for example—have made use of the method of Eyring and his coworkers [123], where charge was distributed on the basis of (a) polarizabilities, (b) screening constants, (c) covalent radii, and (d) electric dipole moments. For molecules where such detailed information is not available, a rough idea of the charge distribution may be obtained by dividing the dipole moment of each bond by the bond length. Because most papers on the various aspects of the conformational analysis of organic molecules have been concerned with nonpolar systems, however, there has been little need to perform such calculations. In metal complexes, such Coulombic interactions must exist. Nevertheless, it should be realized that even in somewhat polar molecules simple Coulombic interactions between charges are rarely important in determining conformational preferences because of the relatively slow variation of terms in r^{-1}. Where a higher order polarity is involved, this becomes less true. In fact some authors have suggested that dipole and quadrupole interactions may be important in determining torsional barriers [80,81,104].

Repulsive Forces

In addition to the Coulombic and attractive dispersion forces discussed previously, it is necessary to take into account the fact that, at small interatomic distances, the electron clouds of the atoms begin to repel one another. This may be done either by assuming the existence of a separate repulsive term (or terms) in addition to the above or by determining a general formula that includes all these terms implicitly. Some of the proposed repulsive terms are discussed below.

Exponential repulsive term $Be^{-\beta r}$. Heitler and London's theory of valence considered interatomic repulsive forces in terms of an interference effect of the atomic wave functions, which depends on the degree to which the wave functions overlap [59]. If the form of the wave function is exponential, that is, $\psi = Ne^{-r}$, where N is a normalizing constant, it is logical that the interference (or interpenetration) energy should vary exponentially. This term is most commonly used in what we call the *exponential 6* form of the total van der Waals equation:

$$\mathscr{E} = Be^{-\beta r} - Ar^{-6} \tag{3-17}$$

Inverse-power repulsive term Br^{-b}. Lennard-Jones [82] suggested on purely empirical grounds that the interaction energy could be represented by the formula

$$\mathscr{E} = Br^{-b} - Ar^{-a} \tag{3-18}$$

This general equation has been used in a number of special forms, two of which are

$$\mathscr{E} = \varepsilon\left(\frac{r^*}{r}\right)^{12} - 2\varepsilon\left(\frac{r^*}{r}\right)^{6} \tag{3-19}$$

due to Hill [65], where $\mathscr{E}$ is a minimum when $r = r^*$ (the sum of the van der Waals radii), and

$$\mathscr{E} = 4\varepsilon\left[\left(\frac{\sigma}{r}\right)^{12} - \left(\frac{\sigma}{r}\right)^{6}\right] \tag{3-20}$$

due to Hirschfelder, Curtiss, and Bird [66], where $\mathscr{E}$ is zero for $r = \sigma$ and is a minimum for $r = 2^{1/6}\sigma$. Equations (3-18) to (3-20) are commonly known as *Lennard-Jones potential functions.*

It is theoretically possible to calculate the exponential repulsive force by applying the method of Heitler and London [59] to a known wave function and then by choosing B and β so that the results may be represented by an exponential term. This has been done by Slater, who obtained an approximate term for helium–helium interactions [121]. In practice, such calculations using Heitler and London's method are both difficult and tedious, and the exponential 6 and Lennard-Jones equations are more usually determined empirically.

By an appropriate choice of parameters Eqs. (3-17) and (3-18) may be made equivalent over a particular range of interatomic distances; such conversions are sometimes used to simplify mathematical manipulations [65].

Nonbonded repulsive states. Valence-bond theory applied to interactions between hydrogen atoms yields the following expression for the H—H interaction energy:

$$\mathscr{E}_{\mathrm{HH}} = Q_{\mathrm{HH}} - \tfrac{1}{2}J_{\mathrm{HH}} \tag{3-21}$$

where Q_{HH} is the Coulomb integral for the two electron orbitals and J_{HH} is the corresponding exchange integral. The two states arising from this interaction, $^1\Sigma$ and $^3\Sigma$, have the energies (see Fig. 3-1)

$$\mathscr{E}(^1\Sigma) = Q_{\mathrm{HH}} + J_{\mathrm{HH}} \tag{3-22}$$

$$\mathscr{E}(^3\Sigma) = Q_{\mathrm{HH}} - J_{\mathrm{HH}} \tag{3-23}$$

Eyring [41] calculated values for the interaction energy from Eq. (3-21) using values of the integrals evaluated by Sugiura [127].

Attempts have also been made to relate the interaction energy to the energies of the singlet and triplet states. If Q_{HH} is negligible, the interaction energy is equal to half the energy of the triplet state. Pritchard and Sumner [118] used this relationship to calculate $\mathscr{E}_{\mathrm{HH}}$ using the equation for the

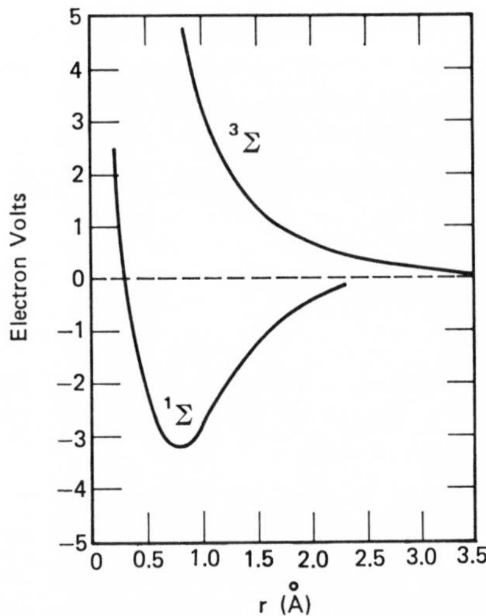

Fig. 3-1 Potential energy curves for the $^3\Sigma$ and $^1\Sigma$ states of hydrogen.

triplet state proposed by Hirschfelder and Linnett [67]. Mason and Kreevoy [92] and Howlett [69] used the general form

$$\mathscr{E}_{\mathrm{HH}} = K\mathscr{E}(^3\Sigma) \tag{3-24}$$

with K equal to 1.0. The data of Hirschfelder and Linnett were again used for the triplet-state energy. Pauncz and Ginsburg [110] have proposed an equation in which the coefficient 0.5, suggested by Pritchard and Sumner [118], is used with Buckingham's approximate formula [22]

$$\mathscr{E}(^3\Sigma) = (166r^{-1} + 366 + 23.1r + 580r^2 + 1828r^3)e^{-4.35r} \tag{3-25}$$

Because of the arbitrary nature of the derivation of these equations, their usefulness lies in their quantitative predictions in conformational calculations. All but the last have been applied to the problem of the torsional barrier in ethane, where the significance of their estimations is difficult to judge.

Morse curve. In 1929 Morse [96] showed that the Schrödinger equation for a diatomic molecule could be solved, and energy levels calculated, using a potential function

$$\mathscr{E} = D[\exp(-2a(r - r_e)) - 2\exp(-a(r - r_e))] \tag{3-26}$$

where D is the bond strength and r_e is the equilibrium bond length. The Morse curve is qualitatively similar to the expected form of the nonbonded interaction curve (small attractive forces at large interatomic distances and large repulsive forces at small distances), although its energy minimum is at a much smaller interatomic distance. This similarity suggested its use as an empirical nonbonded potential energy curve similar to the exponential 6 and Lennard-Jones functions. Wiberg has used a Morse function in this way, choosing parameters to fit certain experimental data [135].

The Morse curve for the hydrogen molecule may also be related to the graph of potential energy against distance for the $^1\Sigma$ (bonding) state of two hydrogen atoms. Magnasco has used the inverse of the experimental Morse curve for molecular hydrogen as an approximate H···H interaction curve [90].

Evaluation of Parameters

The van der Waals equations that have been used for conformational analysis are summarized in Table 3-1. Some of the methods that have been applied to the evaluation of the parameters for these equations are discussed below.

Second-order virial coefficients. Deviations from ideal gas laws arise from interatomic interactions. The effects of these deviations on the equation of state are accounted for by equations such as that of Kammerlingh Onnes [70]:

$$pV = A_V + B_V V^{-1} + C_V V^{-2} + \cdots \tag{3-27}$$

Table 3-1 Nonbonded interaction equations

Derivation	Equation	Parameters[a]	Derivation of parameters	r_0, Å[b]	r_{min}, Å[c]
H···H interactions					
Aston et al. [10]	Br^{-b}	$B = 499$ $b = 5$	Repulsive term Br^{-b} fitted to torsional barriers of ethane and tetramethylsilane	6.30	
Hill [65]	$-2.25\varepsilon(r^*/r)^6 + 8.28 \times 10^5\varepsilon \exp(-r/0.0736r^*)$	$\varepsilon = 0.042$ $r^* = 2.4$ Å	Exp 6 equation derived from Lennard-Jones' 12-6 function for H_2	2.15	2.4
Mason and Kreevoy [92]	$Be^{-\beta r} - Ar^{-6}$	$A = 89.52$ $B = 3.7164 \times 10^3$ $\beta = 3.0708$	Exp 6 equation fitted to Hirschfelder-Linnett [67] values of $\mathscr{E}(^3\Sigma)$ for two H atoms at different interatomic distances ($r < 1.8$)	3.85	>4.2
Pitzer and Catalano [115]	$-Ar^{-6}$	$A = 49.2$	Calculated from Slater-Kirkwood equation (3-13)	3.15	
Pauncz and Ginsburg [110]	$(166r^{-1} + 366 + 23.1r + 582r^2 + 1{,}828r^3)e^{-4.35r}$		Derived from formula $\mathscr{E} = K\mathscr{E}(^3\Sigma)$, with $K = 0.5$ and $\mathscr{E}(^3\Sigma)$ calculated by Buckingham [22]		
Bartell [11]	$Be^{-\beta r} - Ar^{-6}$	$A = 49.2$ $B = 6.6 \times 10^3$ $\beta = 4.08$	A from Pitzer and Catalano [115], B and β chosen to fit values of $\partial\mathscr{E}/\partial r$ at 1.36 Å (Mulliken [98]) and $\mathscr{E}$ at $r > 2.6$ Å (Boer [16])	2.6	3.0
Hendrickson [64]	$Be^{-\beta r} - Ar^{-6}$	$A = 49.2$ $B = 1 \times 10^4$ $\beta = 4.6$	A from Pitzer and Catalano [115], β chosen from Amdur's scattering data [7] for He, B chosen so that $r_{min} = 2.5$ Å	2.15	2.5
Hendrickson [64]	$Be^{-\beta r} - Ar^{-6}$	$A = 49.2$ $B = 1.36 \times 10^4$ $\beta = 4.72$	A, B, and β calculated using Bartell's criteria [11], with additional requirement that $r_{min} = 2.5$ Å	2.15	2.5

Amdur and Mason [9]	Br^{-b}	$B = 33.2$ $b = 6.18$	Derived by analysis of $CH_4 \cdots CH_4$ interaction curves determined from scattering data	2.85	
Scott and Scheraga [120]	$Be^{-\beta r} - Ar^{-6}$	$A = 45.2$ $B = 9170$ $\beta = 4.54$	A calculated from Slater-Kirkwood equation, replacing N by N_{eff} [114]; β estimated by extrapolating graph of β v's Z for inert gases to $Z = 1$; B determined by requiring that $r_{min} = 2.8$ Å	2.25	2.8
Abe et al. [4]	$Be^{-\beta r} - Ar^{-6}$	$A = 45.2$ $B = 9{,}950$ $\beta = 4.54$	As for Scott and Scheraga [120] except that B chosen to reproduce torsional barriers in the lower n-alkanes	2.3	2.6
McCullough and McMahon [87]	$\varepsilon(r^*/r)^{12} - 2\varepsilon(r^*/r)^6$	$\varepsilon = 0.0645$ $r^* = 3.06$ Å	r^* and ε chosen to fit gas viscosity data for H_2 and experimental torsional barriers of ethane and propane	3.06	3.45
Wiberg [135]	$D\,[\exp(-2a(r - r_e)) - 2\exp(-a(r - r_e))]$	$D = 0.35$ $a = 3.5$ $r_e = 2.3$ Å	Morse function fitted to energies of Pitzer and Catalano Ar^{-6} term at 2.3 Å and 3.0 Å and to condition that $\mathscr{E} = 3.0$ kcal mol^{-1} at 1.9 Å, latter being chosen from destabilization energy of cyclodecane	2.15	2.3
Allinger et al [5]	$-2.25\varepsilon(r^*/r)^6 + 8.28 \times 10^5\varepsilon \exp(-r/0.0736r^*)$	$\varepsilon = 0.049$ $r^* = 3.0$ Å	Hill's parameters r^* and ε empirically chosen to account for (1) heat of sublimation and crystal structure of n-hexane, (2) free energy difference of 1.9 kcal mol^{-1} between axial and equatorial conformations of methylcyclohexane, and (3) torsional barriers of substituted ethanes	2.75	3.0
Gollogy and Hawkins [52]	$-2.25\varepsilon(r^*/d)^6 + 8.28 \times 10^5\varepsilon \exp(-d/0.0736r^*)$	$\varepsilon = 0.042$ $r^* = 2.4$ Å $d = r - (r_{min} - r^*)$ $r_{min} = 2.50$ Å	Curve with same shape as Hill's curve fitted to experimental energy differences between configurations of $[Coen_3]^{3+}$	2.25	2.50

Derivation	Equation	Parameters[a]	Derivation of parameters	r_0, Å[b]	r_{min}, Å[c]
C···C interactions					
Pitzer and Catalano [115]	$-Ar^{-6}$	$A = 325$	Calculated from Slater-Kirkwood equation (3-13)	4.30	
Bartell [11]	$Br^{-12} - Ar^{-6}$	$A = 325$ $B = 3 \times 10^5$	A from Pitzer and Catalano [115], B chosen to minimize $\mathscr{E}$ at 3.5 Å	3.15	3.5
Hendrickson [64]	$Be^{-\beta r} - Ar^{-6}$	$A = 325$ $B = 1.66 \times 10^4$ $\beta = 3.63$	A from Pitzer and Catalano [115]; B and β chosen by curve-fitting Amdur's scattering data for neon [8], which has similar van der Waals radius as carbon, and by requiring also that $r_{min} = 3.2$ Å	2.75	3.2
Allinger and Szkrybalo	$-2.25\varepsilon(r^*/r)^6 + 8.28 \times 10^5\varepsilon \exp(-r/0.0736r^*)$	$r^* = 3.4$ Å $\varepsilon = 0.107$	r^* and ε chosen by comparison with similar functions involving F, O, and N atoms	3.05	3.4
Abe et al. [4]	$Be^{-\beta r} - Ar^{-6}$	$A = 363$ $B = 9.086 \times 10^5$ $\beta = 4.59$	A from Slater and Kirkwood [77,122] and Pitzer [114], β estimated by comparison with scattering data for inert gases, B chosen to fit torsional barriers of lower n-alkanes	3.25	3.6
Allinger et al. [5]	$-2.25\varepsilon(r^*/r)^6 + 8.28 \times 10^5\varepsilon \exp(-r/0.0736r^*)$	$\varepsilon = 0.65$ $r^* = 2.2$ Å	r^* and ε chosen in conjunction with H···H function described previously		2.2
C···H interactions					
Pitzer and Catalano [115]	$-Ar^{-6}$	$A = 125$	Calculated from Slater-Kirkwood equation (3-13)	3.70	
Bartell [11]	$B[e^{-\beta r} - 1]r^{-6}$	$B = 44{,}620$ $\beta = 2.04$	Calculated as geometric mean of Bartell's H···H and C···C repulsive functions	2.9	3.2

Hendrickson [64]	$Be^{-\beta r} - Ar^{-6}$	$A = 125$ $B = 1.29 \times 10^4$ $\beta = 4.12$	A from Pitzer and Catalano [115], B and β determined from geometric mean of repulsive terms in Hendrickson's H···H and C···C interaction equations	2.5	2.8
Allinger and Szkrybalo [6]	$-2.25\varepsilon(r^*/r)^6 + 8.28 \times 10^5\varepsilon \exp(-r/0.0736r^*)$	$r^* = 2.9$ Å $\varepsilon = 0.067$	r^* and ε chosen as arithmetic and geometric means, respectively, of values of r^* and ε calculated by Hill [65] for H···H interactions and by Allinger and Szkrybalo [6] for C···C interactions	2.6	2.9
Abe et al. [4]	$Be^{-\beta r} - Ar^{-6}$	$A = 127$ $B = 8.61 \times 10^4$ $\beta = 4.57$	A calculated from Slater-Kirkwood equation, replacing N by N_{eff}; β chosen to be average of β values for previously derived H···H and C···C interaction equations; B chosen to reproduce torsional barriers in lower n-alkanes	2.75	3.1
Allinger et al. [5]	$-2.25\varepsilon(r^*/r)^6 + 8.28 \times 10^5\varepsilon \exp(-r/0.0736r^*)$	$\varepsilon = 0.179$ $r^* = 2.6$ Å	r^* chosen as arithmetic mean of r^* for corresponding H···H and C···C interaction equations, ε chosen as geometric mean of values of ε for H···H and C···C interactions	2.35	2.6
Gollogly and Hawkins [52]	$-2.25\varepsilon(r^*/d)^6 + 8.28 \times 10^5\varepsilon \exp(-d/0.0736r^*)$ with $d = r - (r_{min} - r^*)$	$\varepsilon = 0.067$ $r^* = 2.9$ Å $r_{min} = 3.1$ Å	Curve with same shape as Hill curve of Allinger and Szkrybalo [6] fitted to experimental energy differences between configurations of $[Coen_3]^{3+}$ in conjunction with H···H interaction equation	2.8	3.1

[a] In kcal.

[b] r_o is value of r at which $\mathscr{E}_v = 0$.

[c] r_{min} is value of r at which $\mathscr{E}_v$ is a minimum.

Equations and parameters for other interaction systems have been published (e.g., N···H, Cl···H). Some of these are given in the following papers: Scott and Scheraga [120], Hill [1,65], Mason and Kreevoy [92].

where A_V is the first-order virial coefficient, B_V is the second-order coefficient, and so on. This equation can usually be limited to two terms $(A_V + B_V V^{-1})$.

Lennard-Jones interpreted the second-order virial coefficient in terms of Eq. (3-18) and showed that, at least in principle, the four parameters of Eq. (3-18)—a, b, A, and B—could be calculated from the experimentally determined values of A_V and B_V for different temperatures [82]. In practice, however, slight experimental errors make it difficult to calculate a unique set of the four parameters, and it is usually necessary to assume values for a and b before A and B can be determined.

Lennard-Jones assumed values of $a = 6$ and $b = 12$ and calculated A and B for a number of gases—rare gases, H_2, N_2, O_2, for example. His data have been used by Hill [65] to evaluate the parameters in Eq. (3-19). The values of r^* and ε that may be obtained for diatomic molecules are quite different from those which may be associated with interactions between individual atoms. Hill has discussed a simple empirical method of relating the two sets of parameters. Allinger and Szkrybalo [6] have estimated r^* and ε for the carbon atom by comparison with values for atoms adjacent to it in the periodic table—N, O, F, for example.

For interactions between non identical atoms Hill suggested using the formulas,

$$r^*_{AB} = \frac{r^*_{AA} + r^*_{BB}}{2} \tag{3-28}$$

$$\varepsilon_{AB} = (\varepsilon_{AA}\varepsilon_{BB})^{1/2} \tag{3-29}$$

Values of r^* and ε derived in this way for a series of atoms have been calculated by Eliel and coworkers [1].

Mason and Rice [93] have performed similar calculations to the above with second-order virial coefficients for rare gases using a potential function of the exponential 6 type. They related the resultant equations to the interaction of halogen atoms at moderate to large interatomic distances.

Scattering of high-energy neutral particles. Interatomic repulsive energies for small interatomic separations may be determined using data from the scattering of high-energy neutral particles [7,8,9]. The results obtained from such investigations have been conveniently represented for a number of rare gas atoms by exponential repulsive terms.

Mason and Kreevoy have used these repulsive terms to represent the repulsion between halogen atoms, which lie adjacent to the various rare gases in the periodic table [92]. This presumes that the effects on the interactions arising from the smaller nuclear charge on the halogen are balanced by those due to the smaller electron cloud.

X-ray structures. Kitaigorodsky has attempted to derive an interaction equation directly from the distortions (due to the nonbonded repulsions) that have been observed in the X-ray structure of various compounds [79]. He has assumed that these structural distortions arise from an effort by the molecule to minimize the sum of two energy terms associated with nonbonded interactions and angular distortion:

$$\mathscr{E} = \sum_i f_i \left(\frac{\Delta r}{r^*}\right)_i + \sum_j \tfrac{1}{2}k_\theta(\Delta\theta_j)^2 \qquad \text{(3-30)}$$

where $f_i(\Delta r/r^*)_i$ is the nonbonded interaction energy of each pair of atoms at a distance r; r^* is the distance at which this energy is a minimum; $\Delta r = r - r^*$; and $\frac{1}{2}k_\theta(\Delta\theta_j)^2$ is the energy required to distort an angle by $\Delta\theta_j$ from its natural angle. The last energy term excludes the effect of angular distortions on the 1,3 nonbonded interactions, which are considered in the first term in Eq. (3-30).

For the interaction systems that are observed,

$$\frac{\partial\mathscr{E}}{\partial\varphi_k} = 0 \qquad \text{(3-31)}$$

where φ_k is one of the geometrical parameters by which the structure may be specified and varied. Since the values of φ_k are known from the X-ray structure, they may be inserted into the differential equations (3-31), which may then be solved for values of

$$\eta = \frac{f_i'(\Delta r/r^*)_i}{k_\theta} \qquad \text{(3-32)}$$

associated with each pair of interacting atoms in the structure. The parameter η is known as the *relative force of interaction between pairs of atoms.*

By repeating this procedure for a series of compounds, sufficient information may be obtained to plot η against the corresponding values of $\Delta r/r^*$ for each kind of interaction. Kitaigorodsky found that the points corresponding to H···H, C···H, and C···C interactions all fell on the same curve, and he used this curve to predict distortions to a limited number of other structures [79].

Dispersion terms. Equation (3-13) has been used mainly to evaluate the dispersion energy $-Ar^{-6}$. Pitzer and Catalano [115] applied the polarizabilities of Ketalaar [76] for H···H, C···H, and C···C interactions. The number of electrons in the outermost shell (N) was taken to be 1 and 4 for hydrogen and carbon, respectively. However, when these energies were compared with results from second virial coefficient and viscosity data, it was found that, although there was quite good agreement for helium, large differences appeared with increasing atomic number [114]. These discrepancies were thought to be due to contributions from the inner shells [114] and it was proposed that this could be overcome if atoms were assigned an *effective* number of outer-shell electrons [120]. A graph of N_{eff} against atomic number was plotted using Pitzer's data for the rare gases, and values of N_{eff} for hydrogen and carbon were determined. Slightly different values for the coefficient A were obtained by substituting these values into Eq. (3-13), with Ketelaar's polarizabilities. The calculations of Pitzer and Catalano [115] and Scott and Scheraga [120] have provided the basis for a number of semi-empirical interaction equations discussed later.

Lennard-Jones potential functions. Lennard-Jones derived values for A and B for helium, neon, argon, hydrogen, and nitrogen from the second-order virial coefficients of these gases [82]. Hill has converted these results into the parameters r^* and ε in Eq. (3-19) [65]. For the interactions between hydrogen atoms, Barton calculated A from London's formula and determined B from the condition that $\partial\mathscr{E}/\partial r = 0$ when r is equal to the sum of the van der Waals radii [12]. Two values of B were evaluated, corresponding to two different van der Waals radii, 1.2 Å [108] and 1.3 Å [89]. McCullough and McMahon have used minimum-energy distances from gas viscosity data and second-order virial coefficients in Eq. (3-19) and determined a value of ε such that the equation accounted for the torsional barriers in substituted ethanes [87].

Exponential 6 equations. The major difficulty with this kind of equation (3-17) is the evaluation of β for a particular pair of interacting atoms. It is usually assumed that the interaction curve will have the same shape as that for a similar pair of atoms for which a value of β is available from other sources. The difference between the two interacting pairs is then accounted for by the choice of B.

Dostrovsky, Hughes, and Ingold [30] chose $\beta = 0.345$ for H···Br interactions by comparison with the value obtained by Born and Mayer [18] from their study of the crystal structure of a range of alkali halides. The constant A was calculated from London's formula and B determined from the condition that $\mathscr{E}$ is a minimum at $r = 3.55$ Å. Barton chose $\beta = 4.6$ for H···H interactions and evaluated A from London's formula [12]. He calculated two values of B, corresponding to the minimum-energy distances $r = 2.4, 2.6$ Å suggested by Pauling [108] and Mack [89].

Bartell [11] has calculated potential functions for H···H, C···H, and C···C interactions, using Pitzer and Catalano's values for the coefficient of r^{-6} [115]. For the repulsive part of the H···H interactions, an exponential 6 form was fitted to an analytical expression for H···H interactions at $r > 2.6$ Å from de Boer's Heitler-London calculations [16] and to Mulliken's estimation of $\partial\mathscr{E}/\partial r$ at 1.36 Å [98]. For C···C interactions, a general Lennard-

Jones function was used with B chosen so that $r = 3.5$ Å at minimum energy. The repulsive part of the C···H potential was arbitrarily assumed to be the geometric mean of the H···H and C···C repulsive terms. Hendrickson [64] has also used Pitzer and Catalano's values of A for these interactions and chosen β for the H···H and C···C interactions to be 4.55, and 3.63, respectively, by comparison with the values calculated by Amdur and his coworkers [7,8] on the basis of the high-velocity scattering of helium and neon. The parameter B was determined by minimizing the interaction energies at $r = 2.5$ and 3.2 Å. Once again the repulsive part of the C···H interactions was taken to be the geometric mean of the H···H and C···C repulsive terms.

Scott and Scheraga [120] plotted the experimentally determined values of β for the rare gases [2,7,8] against their atomic numbers and from this graph determined values of β that corresponded to the atomic numbers of hydrogen, oxygen, halogens, and so on. Using these values of β, they calculated A from Eq. (3-17). For H···H interactions Scott and Scheraga minimized $\mathscr{E}$ at $r = 2.8$ Å to find B. These authors wrongly attribute their minimum-energy positions to the van der Waals radii recommended by Bondi [17], whose values, in fact, lead to a distance of 2.4 Å. Abe and coworkers [4] followed Scott and Scheraga's method for hydrogen and carbon, except that they chose B to fit the observed torsional barriers of alkanes.

Other equations of the exponential 6 kind have been proposed. Mason and Kreevoy fitted an exponential 6 function to Hirschfelder and Linnett's calculated values of $\mathscr{E}(^3\Sigma)$ for two hydrogen atoms at a range of interatomic distances [92]. Dows adopted the simple exponential repulsion term calculated by de Boer [16] and has used it in conjunction with Pitzer and Catalano's attractive r^{-6} term [31].

Hill proposed a special form of the exponential 6 equation

$$\mathscr{E} = -2.25\varepsilon\left(\frac{r^*}{r}\right)^6 + 8.28 \times 10^5 \varepsilon \exp\left(\frac{-r}{0.0736r^*}\right) \tag{3-33}$$

which, by his choice of the parameters r^* and ε, has been made equivalent to Eq. (3-19) for certain ranges of r [65]. Allinger and coworkers [57] have accepted this form of the equation for H···H interactions and used trial values of r^* and ε to find a set that could accurately account for the following experimentally observed quantities: (a) the heat of sublimation of n-hexane; (b) the spacings of the molecular chains in crystals of n-hexane; and (c) the energy difference between the axial and equatorial forms of methylcyclohexane. They chose one of several suitable sets of r^* and ε that satisfied these conditions.

Morse curves. Wiberg has noted that cyclodecane, whose X-ray structure shows considerable angular distortion with two pairs of hydrogen atoms

1.9 Å apart, is destabilized with respect to the chair form of cyclohexane by about 13 kcal mol^{-1} [135]. He made the simplifying assumption that this large destabilization derives equally from structural distortion and H···H interactions and chose a Morse function for the H···H interaction energy that satisfied the following conditions: (a) $\mathscr{E} = 3$ kcal mol^{-1} at $r = 1.9$ Å; (b) its form corresponds approximately to that of Pitzer and Catalano's r^{-6} term for $r \geqslant 2.3$ Å. In the molecular structures that Wiberg investigated, the C···H and C···C interaction distances were sufficiently large so that only the attractive Pitzer and Catalano terms were required.

Equations Applied to Chelate Ring Systems

Only a few of the available van der Waals equations have been applied to chelate ring systems. Corey and Bailar [26] have used the Mason and Kreevoy equation (3-24) [92] as a general nonbonded interaction equation, although it was considered by its proponents to be effective only in the interaction region $r \leqslant 1.8$ Å. Sargeson and his coworkers [19,20,21] have applied Hill's equation (3-33) to a number of systems. These two sets of interaction equations are representative of the extremes of the van der Waals equations. It can be seen from Fig. 3-2 that, for the nonbonded distances of most interest to the study, 2.0 Å $\leqslant r \leqslant$ 3.0 Å, Hill's equation yields very small interaction

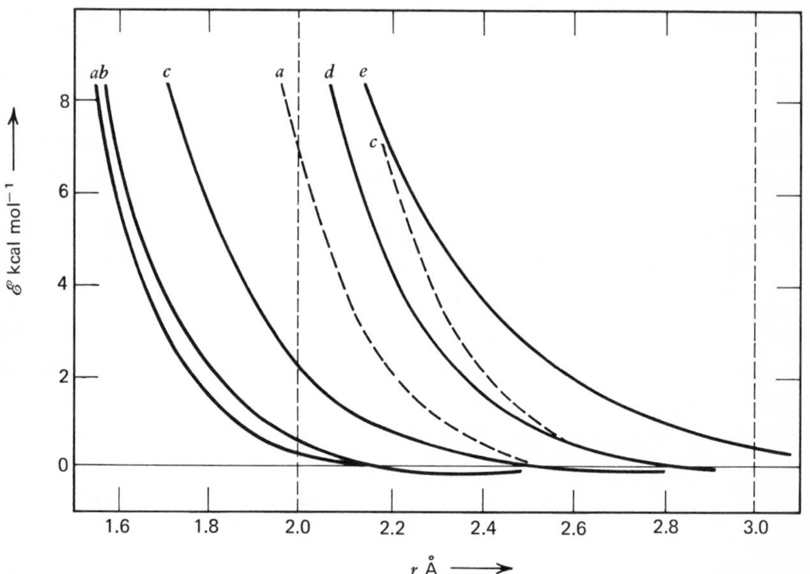

Fig. 3-2 Van der Waals energy curves: (*a*) Hill, [65]; (*b*) Hendrickson, [64]; (*c*) Bartell, [11]; (*d*) Allinger and coworkers [5]; and (*e*) Mason and Kreevoy [92]. A full line represents H···H interactions and a broken line represents C···H interactions.

energies, whereas the Mason and Kreevoy equation gives rise to relatively large energies. For the first series of chelate ring systems to be discussed in the present chapter, both the Hill and the Mason and Kreevoy equations, as well as Bartell's equation [11], which is representative of the moderate-interaction-energy expressions, have been applied in order to see how the general conclusions regarding the conformations of simple ring systems depend on the kind of van der Waals energy equations used.

Considerable experimental evidence has been obtained that suggests that both the low- and high-energy equations are unsatisfactory. For example, the D($\delta\delta\delta$) configuration of $[Coen_3]^{3+}$ is known to be preferred over the D($\lambda\lambda\lambda$) by about 1.6 kcal mol^{-1} (see below). A detailed conformational analysis using Hill's equation predicted no energy difference between the two configurations, whereas the Mason and Kreevoy equation favored the D($\lambda\lambda\lambda$) configuration by 2.4 kcal mol^{-1} [51] (see below). An empirical set of van der Waals curves has been determined to satisfactorily account for the observed energy differences among the D($\delta\delta\delta$), D($\delta\delta\lambda$), D($\delta\lambda\lambda$), and D($\lambda\lambda\lambda$) configurations of $[Coen_3]^{3+}$ [51,52]. The curves were assumed to be the same shape as Hill's interaction curves, and a further condition was applied that the accepted van der Waals interaction distances had to lie between the (attractive) minimum of the nonbonded curves and the distance at which the interactions become significantly repulsive. The curves that were found to give a good fit to the experimental data are given by

$$\mathscr{E} = -2.25\varepsilon\left(\frac{r^*}{d}\right)^6 + 8.28 \times 10^5 \varepsilon \exp\left(\frac{-d}{0.0736r^*}\right) \qquad (3\text{-}34)$$

where ε is the usual Hill parameter $d = [r - (r_{\min} - r^*)]$, and $r_{\min}$ is the distance at which the energy is a minimum. For the H$\cdots$H interactions $r_{\min} = 2.5$ Å; for C$\cdots$H, $r_{\min} = 3.1$ Å. The equation was further tested by its application to the energy difference between the axial and equatorial orientations of the methyl group in methylcyclohexane. The calculated free energy difference of 1.5 kcal mol^{-1}, which includes an entropy contribution due to the difference in the vibrational freedom of the two conformations, [52] is of the right order of magnitude, because the energy has been found experimentally to be approximately 1.7 kcal mol^{-1} [38].

The empirical nonbonded interaction curves, which are not far removed from those of Bartell (Fig. 3-2), are applied later in the present chapter to a number of chelate ring systems. Presently, the interactions for which the curves have been empirically derived in this way are limited to H$\cdots$H, and C$\cdots$H. Approximate curves for N$\cdots$H, O$\cdots$H, Cl$\cdots$N, N$\cdots$C, O$\cdots$C, C$\cdots$C, N$\cdots$N, and so on, have been derived by moving the Hill's curves for those interactions until r^* approximates to r at which $\mathscr{E} = 0$ [52].

For the chelate ring systems it is desirable to evaluate the energies of the

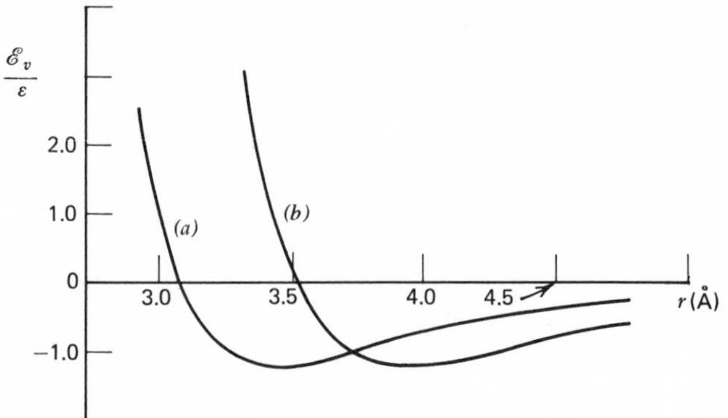

Fig. 3-3 Variation of $\mathscr{E}_v/\varepsilon$ with interatomic distance r for interactions of the central cobalt atom with (*a*) hydrogen and (*b*) carbon.

nonbonded interactions involving the central metal ion. This has not been accomplished because reliable values of r^* and ε are not available; they cannot be easily determined in the same way as for the other interactions. It is possible, however, to estimate qualitatively the importance of these interactions. For example, consider cobalt(III). Pauling has postulated that the covalent radius of cobalt(III) is 1.2 Å [108]. He has noted that for many atoms the covalent radius is about 0.8 Å less than the van der Waals radius. If this condition also applies to cobalt(III), the van der Waals radius of the ion is 2.0 Å. Assuming a curve of the Hill kind and that $\mathscr{E} = 0$ at about $r = r^*$, a graph of $\mathscr{E}_v/\varepsilon$ against r can be plotted (see Fig. 3-3). From this graph a qualitative estimate of the effect of the metal ion can be made [49].

TORSIONAL BARRIERS

One of the major problems of conformational analysis concerns the origin of torsional barriers, such as those observed in ethane. It is well established that the staggered ethane conformation is preferred over the eclipsed by about 3 kcal mol^{-1}. Since the restrictions to free rotation about the C—C bond were first observed, however, many theoretical interpretations of the barrier have been put forward using various bonding and interaction schemes.

In a valence-bond model involving sp^3 hybridization of the carbon orbitals for the bonding in ethane, it has been assumed that the C—H bond orbitals are effectively delocalized about the C—C axis, resulting in a cone-shaped electron cloud that contains the hydrogens (see, for example, [73,112]). As noted by Penney [112], the cylindrically symmetrical distribution of charge implies that only hydrogen–hydrogen interactions contribute to any barrier to rotation. When the effect of the hydrogen atoms on the shape of the electron cloud is taken into account, in the formation of localized bond orbitals, other significant interactions may occur.

Eyring evaluated the H···H interactions on the basis of Sugiura's calculations for the exchange interactions of two hydrogen atoms [126] and predicted a barrier of 0.36 kcal mol^{-1} that favored the staggered conformation [41].* In a later paper, Eyring and his coworkers proposed a bonding scheme in which some *d* atomic orbital character was mixed into the carbon–hydrogen bond orbitals [53]. This led to a charge distribution on each of the carbon atoms that was not cylindrically symmetrical with respect to the C—C axis. The interactions of the charge distributions would then contribute to the barrier. Their calculations, which evaluated the three kinds of interactions involving the hydrogen atoms and the hybridized carbon orbitals, yielded a barrier of up to 0.6 kcal mol^{-1} with the eclipsed form preferred. The effect of mixing some 4*f* character into the carbon–carbon bond was also considered. This contributed to the removal of the cylindrical symmetry of the charge distribution about the C—C axis and, consequently, led to some resistance to free rotation. This procedure is really only a rough method of allowing for the effect of localized carbon–hydrogen bonds on the "shape" of the carbon–carbon bond. Calculations showed, however, that because of the high energy of the 4*f* orbitals their contribution to the bond is small and can account for only a small part of the barrier.

Pauling mixed both *d* and *f* character into the C—H bonds, estimating the contribution of each by comparing the increase in bond strength with the energy necessary to include these orbitals [107]. Pauling evaluated the energy of interaction between the carbon–hydrogen orbitals alone and found that the staggered form was preferred by about 3 kcal mol^{-1}. This method, however, requires a large number of assumptions, and the value above can only be regarded as an indication that this method calculates a barrier of the right order of magnitude.

Clinton pointed out that the assumption of a cylindrically symmetrical charge distribution violates the quantum-mechanical virial theorem because it implies that when there is a change in the electrostatic potential energy (through rotation), there is no accompanying change in the kinetic energy [24]. It follows that any proposed model for torsional barriers must allow for the variation of electron density with rotation before it can describe the physical source of the barriers with any accuracy.

Eyring and his coworkers noted that the staggered conformation of ethane was the one which permitted the greatest delocalization of electrons on the H—C—C—H fragment and suggested that this delocalization might account

* In the literature it is often erroneously stated that in this paper Eyring found the eclipsed conformation to be the most preferred (e.g. [27]).

for the torsional stabilization of this form [43]. A molecular-orbital treatment was applied to the H—C—C—H fragment with six bonding electrons, and the H···H interactions were evaluated using integrals that had been weighted with appropriate bond orders to account for the delocalization [58]. The bond orders were determined on the basis of approximate coupling constants that had been assumed for the vicinal protons. They obtained a barrier of 2.3 kcal mol^{-1}.

Harris and Harris have performed valence-bond treatments of the H—C—C—H fragments of ethane in which the eclipsed and staggered forms of the five, covalent, valence-bond structures involving six bonding electrons have been investigated [56]. A torsional barrier of the order of 7.5 kcal mol^{-1} was estimated. A similar calculation has been attempted by Karplus [72]. Using different assumptions in evaluating the exchange integrals, he obtained a barrier of 0.02 kcal mol^{-1}.

The 10-electron CH_3—CH fragment of ethane has been investigated using s and p orbitals to build up the 42 resonance structures [42]. In the absence of contributions from other integrals, the torsional barrier can be attributed simply to the H···H interactions. This result, however, depends on the value of the other exchange integrals, which are not easily evaluated yet.

In 1963 Pitzer and Lipscomb performed an LCAO-SCF calculation for ethane in which they evaluated the total energy of the eclipsed and staggered conformations and determined the torsional barrier as the energy between them [117]. Although the calculated values of the total energies differed greatly from experimentally determined values, the torsional barrier obtained (3.3 kcal mol^{-1}) was of the correct order of magnitude. Pitzer and Lipscomb used a minimum set of Slater atomic orbitals to build up the molecular orbitals. Later authors have performed similar calculations with Gaussian basis sets and have obtained similar values for the barrier [23,44,111]. Pitzer has recalculated the torsional barrier with a set of exponential atomic orbitals chosen to correspond to a minimum energy in methane and has discussed the rather dramatic canceling out of the errors that appear in the estimates of the total energy [116].

The torsional barrier has also been attributed to electrostatic interactions, such as those between the charge distribution on the C—H bonds [78,80,81].* It would appear, however, that bond dipole interactions are not very important, because the torsional barriers of fluoroethanes are not greatly affected by the substitution of less polar groups [137].

Clinton observed that the torsional barrier was of the same order of magnitude as the difference between the proton–proton repulsive energies in the eclipsed and staggered forms [24]. From a model for these interactions, in

* Subsequent calculations have indicated that the quadrupole moments found by Lassetre and Dean [80] were unreasonably high [104,124].

which each hydrogen was given a unit charge, barriers of 4.77 and 2.6 kcal mol^{-1} were obtained by using slightly different geometrical structures for ethane [24,73]. Wyatt and Parr have recently discussed a more rigorous interpretation of this model that allowed for the redistribution of charge during rotation as required by the quantum-mechanical virial theorem [139].

In many of the theoretical approaches discussed so far the interactions between nonbonded atoms have been evaluated either explicitly as exchange integrals or implicitly in other calculations. However, a number of authors have used interatomic-interaction equations derived from other sources to investigate the barrier.

In 1944 Aston and coworkers made the assumption that the barrier in ethane molecules was due entirely to the H···H interactions and fitted an empirical interaction energy function kr^{-n} to the known barriers of ethane and 2,2-dimethylpropane [10]. When applied to other similar systems, this equation predicted rotational barriers that agreed quite well with experimentally determined values; this agreement is, of course, not unexpected in view of the empirical derivation of the equation. A basically similar approach has been used by French and Rasmussen [47].

Using a nonbonded potential function derived from the Hirschfelder–Linnett equation for the $^3\Sigma$ (repulsive) state of H_2, Mason and Kreevoy [92] predicted a barrier of 1.77 kcal mol^{-1} favoring the staggered conformation. A similar calculation was performed by Howlett [69]. In 1962 Magnasco calculated a barrier of 2.63 kcal mol^{-1} with an equation derived from the Morse curve of H_2 [90]. Energy barriers have also been calculated for a wide range of molecules using Pauling's method to estimate the exchange interactions of the opposing bonds and an exponential 6 equation derived from scattering data of Amdur [7]. The barrier found for ethane was 3.29 kcal mol^{-1} [120].

In 1966 Sovers and Karplus pointed out that nearly all the calculations involving nonbonded interactions had assumed a constant set of bond lengths and bond angles for ethane, and the workers had compared staggered and eclipsed conformations simply by a 60° rotation about the carbon–carbon bond [125]. If, as these calculations suggested, the interatomic interactions were very strong, it is reasonable to expect the ethane molecule to undergo distortion in order to minimize these interactions. This is particularly obvious when one considers the steepness of the postulated interaction curves; small bond-angle and bond-length distortions should produce considerable changes in the interaction energy. Sovers and Karplus performed calculations that suggested that distortions could reduce the barriers predicted on the basis of these high-energy equations by 50% or more. This implies that a simple interatomic-interaction model for torsional barriers is inadequate.

The implications of such distortions in other models are less clear, although Fink and Allen [44] have suggested that the barriers predicted on the basis of a molecular-orbital approach will be fairly insensitive to structural distortion.

Shape of the Torsional Barrier

It has been shown that the thermodynamic properties of entropy, heat capacity, and so on, may be related to a molecule's torsional characteristics by a statistical-mechanical treatment of the problem of hindered rotation. Provided that some assumptions can be made about the shape of the barrier, it is possible to calculate the size of the barrier from the observed thermodynamic data. In 1937, Kemp and Pitzer assumed a torsional function

$$\mathscr{E}_t = \frac{V_0}{2}(1 + \cos 3t) \tag{3-35}$$

for the threefold barrier in ethane, where V_0 is the barrier to free rotation and t is the dihedral angle between the H—C—C and C—C—H planes, and calculated a barrier of 3.15 kcal mol^{-1} using entropy and heat-capacity data [75].

Since the early work of Pitzer and others using thermodynamic data, several other methods have been developed for determining torsional barriers; these have been reviewed by Wilson [137]. Most of the experimentally determined barriers, including those obtained from microwave spectroscopy [84], have been calculated on the basis of the cosine form and have been applied in this form to conformational energy calculations.

Blade and Kimball [15] have discussed the use of a different torsional function consisting of two parabolas associated with the maximum and minimum, respectively; in this way, the shape of the minimum, which determines the spacing of the lowest rotational energy levels, and the height of the barrier can be varied independently. They calculated the torsional barrier from heat-capacity data using a function of this kind. Pitzer [113], however, has criticized their use of this data and has suggested that the two-parabola form is not a significant improvement over the cosine form.

Although other torsional functions have been proposed [54,55], they have had only limited use. The cosine form of the torsional energy has been used for all the calculations discussed in the present chapter. Because the contributions of the various energy terms to the torsional barrier are still uncertain, the barriers associated with the bonds in chelate ring systems have been estimated by comparison with the experimentally determined barriers for organic molecules containing similar bonds. Torsional barriers determined from microwave spectra and thermodynamic data for a number of molecules containing C—C and C—N bonds with a threefold (or pseudo threefold, as in CH_3NH_2) distribution of substituents are listed in Table 3.2.

Table 3.2 Experimental torsional barriers[a]

Molecule	Microwave	Thermodynamic studies
CH_3—CH_3	3.0	2.9
CH_3—CH_2F	3.30	
CH_3—CHF_2	3.18	
CH_3—CF_3	...	3.45
CH_3—CH_2Cl	3.56	2.7
CH_3—CH_2Br	2.80	
CH_3—CH_2—CH_3	...	3.3
$(CH_3)_2CH$—CH_3	...	3.87
CH_3—NH_2	1.98	
CH_3—CHO	1.15	
CH_3—CO—F	1.08	
CH_3—CO—Cl	1.35	
CH_3—CO—CN	1.27	
CH_3—$COOCH_3$	1.17	

[a] In kcal mol^{-1} from [1], [84], [130], and [137].

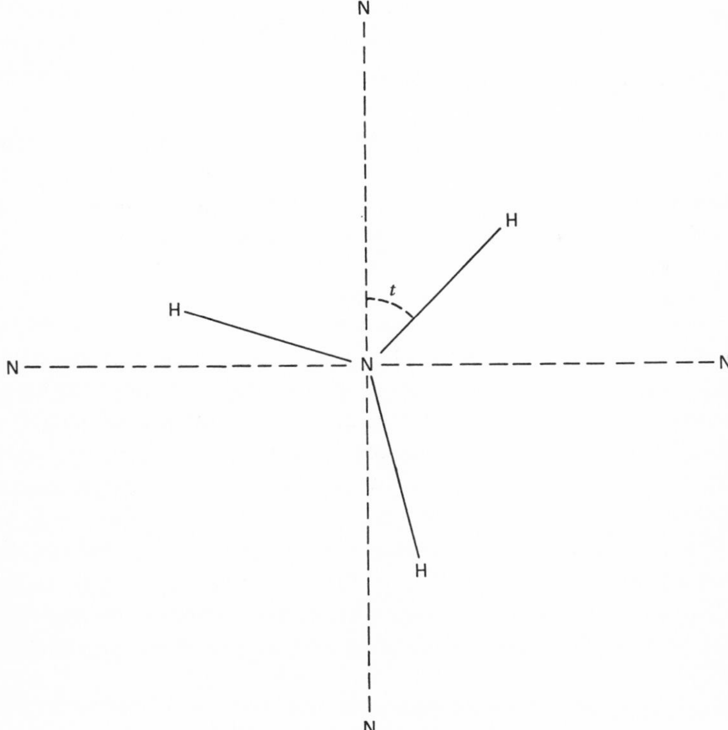

Fig. 3-4 Torsional arrangement about a Co—N bond.

It can be seen that the size of these barriers about C—C bonds is fairly insensitive to the substituents, leading to barriers of the order of 3 to 4 kcal mol^{-1}. In most structures, the preferred conformations are determined by the shape of the torsional barrier in the region of the torsional-energy minimum and the approximate height of the barrier. Consequently, values of 3.0 and 2.0 kcal mol^{-1} have been chosen as being of the right order of magnitude for C—C and C—N bonds in the chelate ring systems.

Unfortunately there are no experimental data available relating to the torsional barriers about the metal-ligand coordinate bonds. For octahedral complexes with ammonia, amines, and related ligands the assumption has been made that if Eq. (3-35) is used to describe the torsional behavior of one coordinate bond with respect to the three substituents of the donor atom the torsional energy about the coordinate bond (see Fig. 3-4), which derives from four such terms, could be given by [50]

$$\mathscr{E}_t = \sum_{i=1}^{4} \tfrac{1}{2}V_{0_i}[1 + \cos 3(t + (i - 1)90)] \qquad (3\text{-}36)$$

For these complexes the barriers V_{0_i} would be of the same order of magnitude and therefore the summation would approach zero.

ANGLE-BENDING ENERGY

A number of analytical expressions have been proposed to account for the energies of the angle-deformation modes in a molecule's infrared spectrum. The simplest of these is the Hooke's law relationship in which, for a tetrahedral molecule AB_4, the angle-bending energy is given by

$$\mathscr{E}_\theta = \frac{k_\theta}{2}(\Delta\theta)^2 \qquad (3\text{-}37)$$

where $\Delta\theta$ is the deviation of the BAB angle from its unstrained value, and k_θ is the angle-bending force constant. The infrared spectra of a wide range of compounds have been interpreted, and force constants calculated, using a function of this kind. Such force constants have been used with apparent success in a number of a priori calculations of conformational energies [5,64,135].

In 1931 Urey and Bradley proposed a more complex angle-bending energy term in which the energy of distortion was assumed to derive from two Hooke's law terms, one associated with valence deformation and another associated with nonbonded interactions between the atoms terminating the angle [129]:

$$\mathscr{E}_\theta = \frac{H}{2}(\Delta\theta)^2 + \frac{F}{2}(\Delta d)^2 \qquad (3\text{-}38)$$

where H is the angle-bending force constant, F is the nonbonded interaction force constant, and d is the distance between the interacting atoms, Δd being the change in this distance. With an additional term to describe bond-length distortions, this equation forms the basis for the Urey-Bradley force field. Using this or a similar force field—for example the modified Urey-Bradley force field or the general force field—it is possible to choose a set of force constants that satisfactorily account for experimental spectra. Angle-bending force constants calculated in this way, however, are very sensitive to the assumptions made in their derivation and to the magnitude of the other force constants in the system. Consequently, there is some question as to the physical significance of such force constants in calculations of this kind.

The two Urey-Bradley force constants for a particular angle may be used to derive a single composite force constant of the kind used in Eq. (3-37), H and F being combined in such a way as to reproduce the variation of $\mathscr{E}_\theta$—Eq. (3-38)—over the important range of angular distortions. In view of this, a simple Hooke's law term is generally used to represent the energy of the angular distortion. The relevant force constants are listed in Table 3-3, together with the *unstrained* values of these angles necessary to calcu-

Table 3-3 Angle-bending force constants and unstrained angles

Bond angle	k_θ[a]	θ_0
NCoN	60	90°
NCoCl	40	90°
NCoO	50	90°
CoNC	40	109.5°
CoNH	40	109.5°
CoOC	40	109.5°
CCC	100	109.5°
CCN	100	109.5°
CCH	100	109.5°
CNH	100	109.5°
CCO (C, C=O, O)	50	120°

[a] In kcal mol^{-1} rad^{-2}. The criteria used in choosing these force constants have been discussed by Gollogly and Hawkins [50,51,52]. It may be necessary to modify some as calculations become more refined.

late $\Delta\theta$. These force constants represent only the order of magnitude of the individual constants. Greater precision is not possible at this time and is not warranted by the calculations.

Hendrickson has pointed out that, when one of the angles in a tetrahedrally bound atom is distorted, the resultant strain in the other three angles may be easily relieved by small distortions of these other angles [64]. Similar considerations should apply to octahedrally bound atoms. Therefore it is convenient to assume that, for such atoms, the angle-strain energy arising from the distortion of one angle is similar to the angle-bending energy of the angle alone.

BOND-LENGTH DISTORTION ENERGY

When the distance between two bound atoms is distorted away from its normal unstrained value by an amount equal to Δl, the energy of the molecule is increased by

$$\mathscr{E}_l = \tfrac{1}{2}k_l(\Delta l)^2 \tag{3-39}$$

where k_l is the stretching force constant for the particular bond. The work of Condrate and Nakamoto [25] and Nakagawa [100] suggests that chelation does not alter the value of the bond-stretching force constants of a ligand. The stretching force constants given in Table 3-4 were obtained from their published work.

For the complexes that are to be discussed in the present chapter it has been found that unfavorable nonbonded interactions can be alleviated with a lower expenditure of energy by the distortion of bond angles than by the distortion of bond lengths because of the large differences in their respective force constants. Thus, throughout the present chapter, bond-length

Table 3-4 Bond-stretching force constants and unstrained bond lengths

Bond	k_l[a]	l, Å	Ref.
Co—N	160	2.0	100
Co—Cl	120	2.3	100
H—N	820	1.00	25, 100
H—C	610	1.10	25, 100
C—N	470	1.47	25, 100
C—C	360	1.54	25, 100
C—O	900	1.25	25
C=O	1100	1.29	25

[a] In kcal mol^{-1} Å^{-2}.

distortion is not considered further. It must be remembered, however, that for some other system this mode of distortion might be energetically economical, and therefore it must always be taken into account.

COMPUTATIONAL METHODS FOR CONFORMATIONAL ENERGIES

The total conformational energy of a structure can be expressed as a function of the various (n) parameters that define the geometry of the molecule:

$$\mathscr{E} = f(\xi_i) \qquad i = n \tag{3-40}$$

This general equation defines an n-dimensional conformational energy surface whose n-dimensional minima represent the various potentially stable conformations, and whose maxima correspond to the activated complexes for the interconversion between the minimum-energy conformations. For all but the simplest structures, Eq. (3-40) is an exceedingly complex function and cannot be solved directly. However, since the conformational energy may be evaluated for any chosen set of parameters $\sum_i \xi_i$, each of the minima and maxima may be located by an indirect method. This usually involves choosing a set of trial structures of the type $\sum_i \xi_i^t$ and minimizing their energy by analytical or iterative procedures.

Many minimization methods utilize the gradients of the potential energy surface in the region of an assumed trial structure as a basis for locating the nearest energy minimum. These gradients are defined by the set of partial derivatives $\sum_i \partial\mathscr{E}/\partial\xi_i$. If the parameters which determine the energy terms, $\mathscr{E}_v$ (the van der Waals energy), $\mathscr{E}_t$, $\mathscr{E}_\theta$, and $\mathscr{E}_l$—interatomic distances, torsional angles, bond angles and bond lengths—can be expressed as simple functions of the parameters $\sum_i \xi_i$, used to define the geometry, the energy gradient can be determined by direct differentiation. This is only possible for simple systems such as ethane, where the variable geometry of the molecule can be defined directly in terms of one torsional angle. In most systems, however, the energy parameters are interdependent and direct differentiation can only be performed if simplifying assumptions are made regarding the relationship between the geometrical parameters and the energy parameters. Approximations of this type will usually be valid only for a limited range of structures centered on the trial structure.

The gradients at any point in the energy surface may be also evaluated by a more general type of approximation. If the calculated conformational energy changes by $\Delta\mathscr{E}$ when each parameter ξ_i is incremented by $\Delta\xi_i$, the approximate partial derivative is given by

$$\frac{\partial\mathscr{E}}{\partial\xi_i} \simeq \frac{\Delta\mathscr{E}}{\Delta\xi_i} \tag{3-41}$$

This method requires no prior approximations and the precision of the gradient depends only on the size of the parameter increment $\Delta\xi_i$.

An iterative minimization method utilizing gradients calculated by Eq. (3-41) has been proposed by Wiberg [135], who used the sign and size of $\partial\mathscr{E}/\partial\xi_i$ to direct further incrementation toward the minimum-energy structure: initially a trial structure $\sum_i \xi_i{}^t$ is defined and each of the parameters incremented by $\Delta\xi_i$ to determine n partial derivatives. The energy is then recalculated using successive sets of parameters defined by

$$\sum_i \xi_i{}^1 = \sum_i (\xi_i{}^t + k\,\Delta\xi_i) \tag{3-42}$$

where k is varied until the energy stops decreasing. Using the final value of k, Eq. (3-42) defines a new set of trial parameters $\sum_i \xi_i{}^t$ for which the calculations may be repeated. This procedure is continued until an energy minimum is reached, and $k = 0$. This minimization technique, which uses the gradient to define the magnitude and direction of incrementation toward the minimum, corresponds to a "method of steepest descent".

Most of the calculations that are discussed below have been based on a systematic incrementation procedure which initially scans the conformational energy surface by simultaneously incrementing each of the geometrical variables over a range of values. Approximate minima and maxima are selected directly from the resultant profile of the energy surface and used as trial structures for a general minimization procedure. The scanning procedure has the advantage that it automatically provides a picture of the shape of the energy surface, from which the population and the entropy of each conformation can be estimated.

The effectiveness of the incrementation procedure is limited by the number of geometrical parameters that are effective in minimizing the conformational energy. If the geometry is defined simply by the Cartesian coordinates of the N atoms in the molecule, there are $3N$ geometrical variables which may be required for minimization. Wiberg [135] and others have used the Cartesian coordinates as variables for investigations where only minor distortions of the trial structure were required. In the calculations discussed below, however, the trial structures have been generated from a set of internal coordinates. This reduces the maximum number of variables from $3N$ to $3N-6$ and also allows a large range of conformations to be examined as part of a single investigation. In addition, the internal coordinates may be selected so that as many as possible are invariant or, at least, limited to a very narrow range of values.

The initial scanning procedure is potentially rather tedious for systems containing many significant variables and has only been rigorously applied to relatively simple systems—for example, the isolated ethylenediamine and 1,3-diaminopropane chelate rings [48]. Once the basic properties of an isolated chelate ring are known, however, many simplifications may be made in the investigation of its octahedral complexes since scanning the energy surface reveals the likely modes of distortion of the ring. In particular, the energy surface of the isolated ring may be used to select trial structures and parameter variables for minimizing the energy of its mono, bis and tris complexes.

ENTROPY OF RING CONFORMATIONS

The energies discussed above are enthalpies. Each conformation also possesses an entropy associated with the vibrational freedom of the ring. The simplest model for calculating this entropy is a one-dimensional Hooke's law oscillator that has its vibrational energy (enthalpy) defined by a parameter ξ according to the equation

$$\mathscr{E} = \tfrac{1}{2}k_\xi(\Delta\xi)^2 \tag{3-43}$$

A simple quantum-mechanical treatment of this vibration in terms of the Schrödinger equation gives $\Delta\mathscr{E}$ the spacing of the vibrational energy levels,

$$\Delta\mathscr{E} = \frac{1}{2\pi c}\sqrt{\frac{k_\xi}{\mu}} \tag{3-44}$$

and the entropy S° associated with this vibration,

$$S_\xi^\circ = \frac{N\,\Delta\mathscr{E}}{T(e^{\Delta\mathscr{E}/\mathbf{k}T}-1)} + R\ln\frac{1}{1-e^{-\Delta\mathscr{E}/\mathbf{k}T}} \tag{3-45}$$

where k_ξ is the force constant for the vibration, ξ is the deformation parameter, c is the velocity of light, μ is the effective mass of the vibrating system, N is Avogadro's number, and $\mathbf{k}$ is the Boltzmann constant.

FIVE-MEMBERED DIAMINE RINGS

Complexes with ethylenediamine and its derivatives, such as propylenediamine, have been by far the most important in the development of the fields of inorganic stereoisomerism, and, more particularly, inorganic conformational analysis. Rosenblatt and Schleede [119] were the first to propose a nonplanar stereochemistry for a chelate ring. A little later Theilacker [127] suggested that ethylenediamine could have the skew form shown in Fig. 1-3. Mathieu [94] used these conformations when he investigated the conformational energies of the $[Co(d\text{-pn})_2X_2]^+$ system. Corey and Bailar [26] carried out a more detailed conformational analysis of the metal–ethylenediamine ring system and concluded that the ring existed in the enantiomeric skew conformations of C_2 symmetry shown in Fig. 1-3. The latter authors, as well as most subsequent workers in the field, have assumed that the conformations of substituted ethylenediamine rings are identical to these irrespective of whether the substituents are axial or equatorial. Any large nonbonded interaction that is present in the complex has not been alleviated by small distortions to the ring system. This has led to unrealistically high energy differences being published for some systems.

The Corey and Bailar method involved inspection of the torsional and angle-bending energy terms to determine the possible ring conformations, followed by a routine calculation of the van der Waals term based on these conformations. Such an approach is appropriate only in situations where the van der Waals energy from interactions between substituents on the chelate ring and other atoms in the complex molecule is relatively unimportant, because it does not allow for any effect of the van der Waals interactions on the actual ring conformations. Gollogly and Hawkins [50] have varied the molecular geometry in such a way as to minimize the sum of the various conformational energy terms. This approach is followed here for a number of octahedral and square planar complexes involving five-membered diamine chelate rings. When data for an individual metal ion are necessary for the computations, the cobalt(III) ion will be considered.

M(en)a_2b_2

When a single ethylenediamine ring is present in a complex of this kind, it is of interest to determine the geometry of the stable conformations and how the other ligands in the complex influence the stereochemistry of the chelate ring.

Geometrical model. The complex is placed in a right-hand coordinate system with the metal atom at the origin and the two coordinate bonds lying in the *xy* plane (see Fig. 3-5). The conformation is usefully defined by the unique set of parameters α, β, z_1, and z_2, where α and β are the two ring angles $\angle$ N(1)MN(2) and $\angle$ MN(1)C(1), and z_1, z_2 are the z coordinates of the two ring carbon atoms C(1) and C(2). The other ring angles can remain unspecified but can be determined for any conformation. The position of any substituent to the ring is defined by the necessary bond lengths and bond angles, and the positions of *a* and *b* are defined by the Cartesian coordinates of the donor atoms. As stated above, because of the torsional arrangement of ligands, such as NH_3, about the coordinate bond in octahedral complexes, when these ligands are bound in positions *a* and *b*, they are free to orient themselves about the coordinate bond so that their van der Waals interactions with the rest of the system are minimized. By using this model variations in the stereochemistry are produced by varying bond lengths, bond angles, and z_1, z_2.*

Energy calculations. A range of conformations can be investigated for each set of ring parameters (α, β, and the five bond lengths) by choosing a value

* The ethylenediamine chelate ring can also be usefully specified by z_1, $\angle$ N(1)C(1)C(2), $\angle$ C(1)C(2)N(2), and ω, the dihedral angle between the N(1)C(1)C(2) and C(1)C(2)N(2) planes.

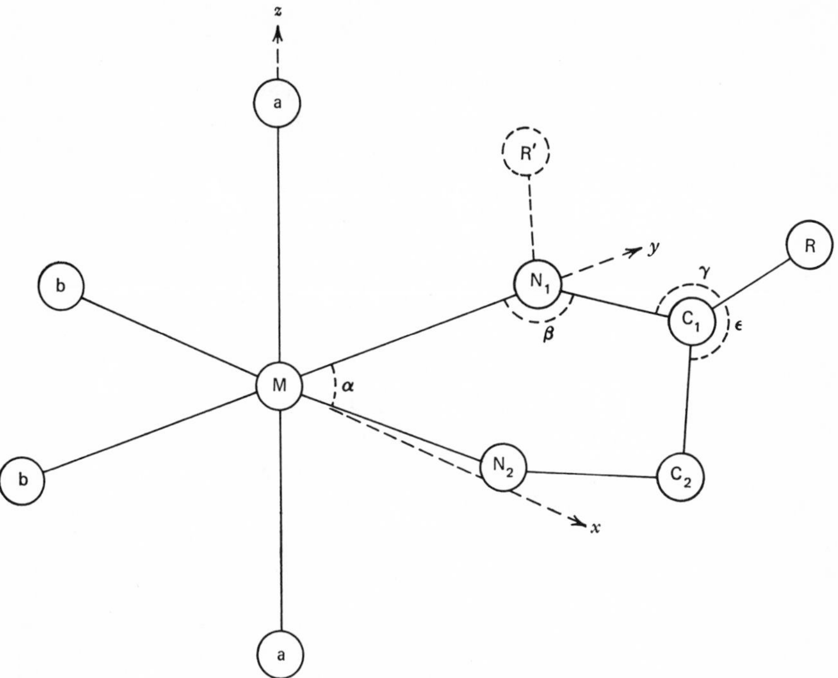

Fig. 3-5 Geometrical model for $M(en)a_2b_2$.

of z_1 and varying z_2 through its entire range of possible values and then repeating this procedure for all possible values of z_1. This can be repeated for other sets of ring parameters and for various positions of the ligands *a* and *b*. The variation of z_2 for a set of α, β, and z_1 corresponds to a variation in ω, the dihedral angle between the N(1)C(1)C(2) and N(2)C(2)C(1) planes. A computer program has been devised to calculate the relevant interatomic distances, torsional angles, and bond angles for each structure, and the associated energies [48].

First, let us consider a square planar complex where the ligands *a* are absent. As stated previously, the bond-length distortion is energetically too unfavorable to be of significance, and it has been found that with the van der Waals equations of Hill [65], Bartell [11], and Mason and Kreevoy [92] the nonbonded interactions are also insignificant. In this system, the total conformational energy comes from the two terms $\mathscr{E}_t$ and $\mathscr{E}_\theta$, which may be evaluated for each conformation defined by the parameters α, β, z_1, and z_2. For each set of α, β, and z_1 a set of graphs may be drawn showing the relationship among ω and $\mathscr{E}_t$, $\mathscr{E}_\theta$, and $(\mathscr{E}_t + \mathscr{E}_\theta)$. Gollogly and Hawkins [50] investigated these energies for $\alpha = 84°\text{–}90°$; $\beta = 104.5°\text{–}119.5°$; and $z_1 =$ 0–0.6 Å. Typical graphs are presented in Fig. 3-6.

For all these values of α, β, and z_1 the curves of $\mathscr{E}_t$ against ω are almost identical, suggesting that the total torsional energy of the ring depends not on the actual ring conformation but almost exclusively on the torsional arrangement about the C—C bond as defined by ω. As the significant torsional energies are associated with the N(1)C(1), C(1)C(2), and N(2)C(2) bonds, this means that if z_1 and z_2 are altered in such a way as to maintain a constant value of ω, any increase in $\mathscr{E}_t$(N(1)C(1)) is accompanied by a compensating decrease in $\mathscr{E}_t$(C(2)N(2)), and vice versa, giving rise to a constant value for $\mathscr{E}_t$. This has considerable significance, because it means that the torsional energies of symmetric and nonsymmetric skew conformations will be very similar provided that the angle ω is kept approximately constant.

$\mathscr{E}_\theta$ was found to be dependent on α, β, and ω and, to a first approximation, independent of z_1 and z_2. Therefore, the total ring-strain energy ($\mathscr{E}_t + \mathscr{E}_\theta$) depends on α, β, and ω and is largely independent of z_1, z_2.

For each set of α, β, and z_1 there are two stable enantiomeric conformations corresponding to the energy minima in the graph of ($\mathscr{E}_t + \mathscr{E}_\theta$) against positive and negative values of ω. It has been found that a number of these

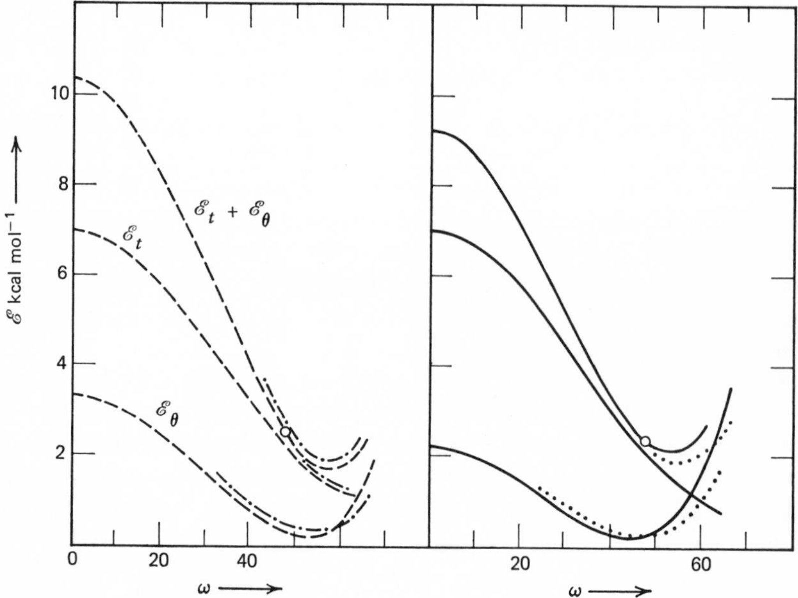

Fig. 3-6 Typical graphs illustrating variations of $\mathscr{E}_t$, $\mathscr{E}_\theta$, and ($\mathscr{E}_t + \mathscr{E}_\theta$) with ω: $\alpha = 86°$, $\beta = 109.5°$, $z_1 = -z_2$ (——); $\alpha = 89°$, $\beta = 109.5°$, $z_1 = 0.0$ (· · · ·); $\alpha = 90°$, $\beta = 104.5°$, $z_1 = -z_2$ (– – –); $\alpha = 90°$, $\beta = 104.5°$, $z_1 = 0.0$ (–·–·–). The small circle marks a Corey and Bailar conformation [50].

Table 3-5 Some minimum-energy cobalt(III) ethylenediamine conformations [50][a]

α	β	z_1, Å	z_2, Å	ω
90°	104.5°	0.1	−0.6	56°
90°	104.5°	0.1	−0.7	57.5°
90°	104.5°	0.2	−0.5	55°
88°	109.5°	0.1	−0.5	47°
88°	104.5°	0.2	−0.6	62°
88°	104.5°	0.4	−0.4	61°
86°	109.5	0.0	−0.7	55°
86°	109.5°	0.1	−0.6	53.5°
86°	104.5°	0.4	−0.4	60°

[a] Using a different geometrical model with finer parameter increments, Gollogly has found that the following geometries lie within 0.2 kcal mol^{-1} of the energy minimum [48]. The parameter values quoted for the various M—N bond lengths are expressed as a range of values found in conformations that corresponded to this energy. For M—N = 2.0 Å, α = 84.7 to 89.4°, β = 103.8 to 109.4°, z_1 = 0.1 to 0.65 Å, ω = 52.5 to 62.5°. For M—N = 2.1 Å, α = 83.1 to 87.4, β = 102.3 to 108.6°, z_1 = 0.15 to 0.60 Å, ω = 55.0 to 65.0°. For M—N = 2.2 Å, α = 80.0 to 83.0°, β = 104.6 to 108.8°, z_1 = 0.15 to 0.60 Å, ω = 57.5 to 67.5°. For M—N = 2.3 Å, α = 77.4 to 80.7°, β = 102.8 to 107.9°, z_1 = 0.20 to 0.60 Å, ω = 60.0 to 70.0°.

conformations have almost identical low energies. The data for some of these are given in Table 3-5. The most stable conformations of ethylenediamine include both symmetric and nonsymmetric skew conformations, and there is very little energy separating a whole range of conformations. The chelate is exceedingly flexible and not restricted to a pair of enantiomeric skew conformations with C_2 symmetry, as previously thought.

The amount of strain involved in forming an isolated ethylenediamine chelate ring is determined by the geometrical requirements of the NMN and NCCN fragments of the ring. If we assume that their optimum geometries are defined by the parameter values $\alpha = 90°$, $\angle$N(1)C(1)C(2) = 109.5°, $\angle$C(1)C(2)N(2) = 109.5°, and $\omega = 60°$, then the optimum NN distances for M—N = 2.0 Å are 2.83 and 2.79 Å respectively for the two fragments. Since the NN distances differ by only 0.04 Å, the ring can be formed without significant strain in either fragment. In this *unstrained* conformation, $\angle$MNC is 101.3° which is much smaller than the normal tetrahedral angle of 109.5°. In order to relieve the strain in the MNC angles, the ring tends to flatten causing a general slight decrease from the optimum values in the torsional and remaining bond angles. As the M—N bond length increases, the optimum NN distance associated with the NMN fragment increases markedly, and, before the ring can be formed at all, it is necessary to either decrease α or increase the NN distance in the NCCN fragment by increasing ω or the NCC angles. The last alternative is the least energetically favorable. The decrease in α, which also serves to relieve the increased strain in the MNC angles, and the increase in ω both are effective, and the minimum-energy conformations for M—N = 2.3 Å are found to have values of α and ω in the vicinity of 80° and 65°, respectively [48].

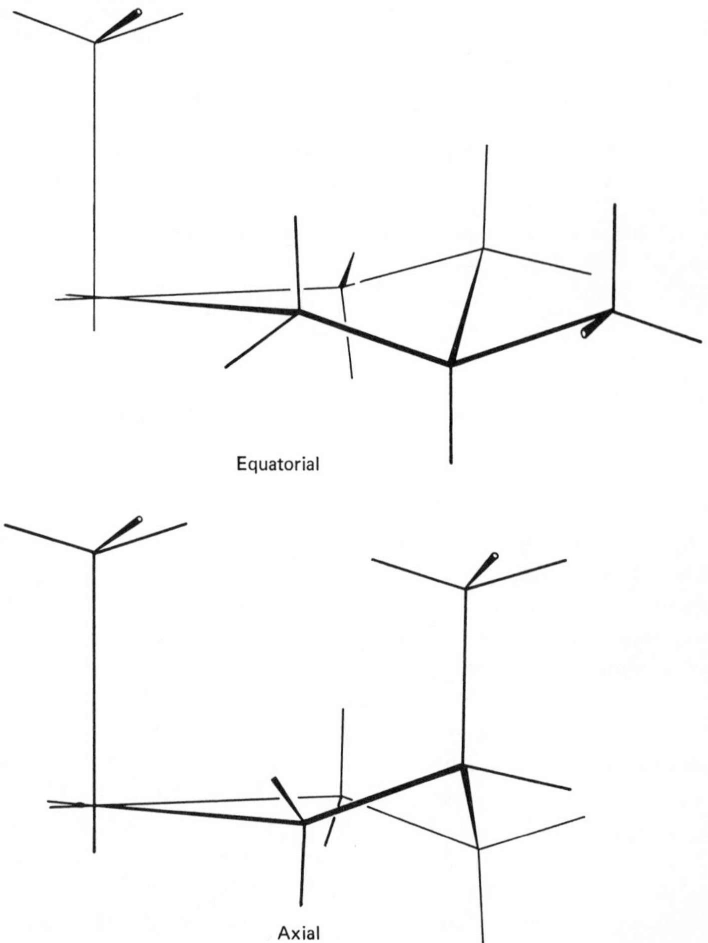

Fig. 3-7 Equatorial and axial orientations of the methyl group in R-propylenediamine complexes.

For the octahedral complex $M(en)a_2b_2$ or for the square planar complex $M(en)b_2$, in which solvent molecules are oriented along the z axis, the additional van der Waals interactions do not affect the previous conclusions except when the Mason and Kreevoy equation is used. For these high-energy equations the van der Waals energy slightly restricts the range of conformations.

M(R-pn)a_2b_2

The propylenediamine chelate ring has ring-strain characteristics that are identical to those of the ethylenediamine rings. The δ conformation of this chelate, however, has the methyl group in an axial orientation, for which the van der Waals term for the interaction with a is significant and is of great importance in limiting the range of geometries available to the chelate ring. When the methyl group is equatorial, its van der Waals interactions with a, b, and other atoms in the chelate ring are insignificant and do not restrict the range of conformations (see Fig. 3-7).

Geometrical model. The complex is situated in the Cartesian coordinates as described for the ethylenediamine complex and as illustrated in Fig. 3-5. The position of the methyl group is defined by the necessary bond lengths and bond angles. Because the staggered and eclipsed orientations of the methyl group with respect to the ring probably differ in torsional energy by about 3 kcal mol^{-1}, it has been assumed that the methyl group remains in a staggered orientation.

Energy calculations. The nonbonded interaction energies arise from the interaction of the methyl group with the ligands a and b and with other atoms in the chelate ring. Gollogly and Hawkins [50] calculated values of $\mathscr{E}_v$ using the van der Waals equations of Hill [65], Bartell [11], and Mason and Kreevoy [92] for the following sets of geometrical variables: α, β, z_1, z_2 (varied as in the ethylenediamine system); γ, ε = 109.5° and 114.5°; and $x_a y_a z_a$ = $(0,0,z_a)$, $(-0.1,-0.1,z_a)$, and $(-0.2,-0.2,z_a)$, where z_a corresponds to the Co-a bond length in Å for the particular ligand a (e.g., 2.0 for NH_3, 2.3 for Cl^-).

For each set of geometrical variables, the total conformational energy has been determined as the sum of the van der Waals energy, the appropriate ring-strain energy, and the additional angle-bending energies arising from distortions that occur outside the ring to relieve the van der Waals interactions. It has been shown that the interactions of the methyl group in both its axial and equatorial orientations with the ligand b and with other atoms in the chelate ring are of no significance to the calculations. In Fig. 3-8 graphs of the total conformational energy as a function of ω are drawn for a particular set of variables for the three kinds of van der Waals equations.

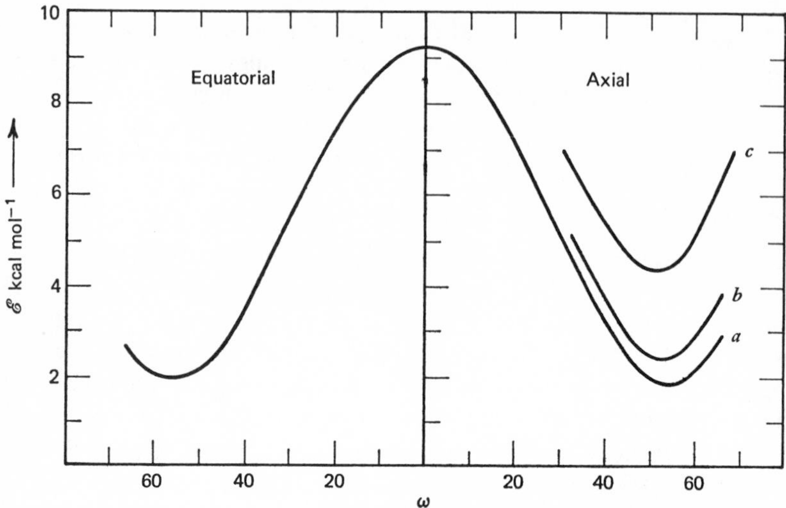

Fig. 3-8 Typical graphs of the variation of total conformational energy with ω for the complex $[Co(R\text{-}pn)(NH_3)_4]^{3+}$: (*a*) Hill, (*b*) Bartell, and (*c*) Mason and Kreevoy. [50]

This has been done for all combinations of the values above of the parameters, and the preferred conformations have again been determined by finding those whose energy minima are lowest in energy. The results are as follows.

When *a* is ammonia, the axial conformation is slightly higher in energy than the equatorial for Hill's (about 0.1 kcal mol^{-1}) and Bartell's (about 0.7 kcal mol^{-1}) equations. The minimum energy δ conformations include those with $\alpha = 90°$, $\beta = 104.5°$, γ and $\varepsilon = 109.5°$, $\omega = 57.5°$, $z_1 = 0.0$ Å, $z_2 = -0.7$ Å, and $x_a y_a z_a = 0, 0, 2.0$ (only for Hill's equation) or $-0.1, -0.1, 2.0$, and with $\alpha = 86°$, β, γ, and $\varepsilon = 109.5°$, $\omega = 55°$, $z_1 = 0.0$ Å, $z_2 = -0.7$ Å, and $x_a y_a z_a = 0, 0, 2.0$ (only for Hill's) or $-0.1, -0.1, 2.0$. With Mason and Kreevoy's equation, the energy difference is more marked (about 2.8 kcal mol^{-1}), and the axial conformations are more restricted. The minimum energy conformations included the following: $\alpha = 88°$, $\beta = 109.5°$, γ and $\varepsilon = 114.5°$, $\omega = 48°$, $z_1 = 0.0$ Å, $z_2 = -0.6$ Å, and $x_a y_a z_a = -0.1, -0.1, 2.0$ or $-0.2, -0.2, 2.0$; $\alpha = 86°$, $\beta = 109.5°$, γ and $\varepsilon = 114.5°$, $\omega = 55°$, $z_1 = 0.0$ Å, $z_2 = -0.7$ Å, and $x_a y_a z_a = 0, 0, 2.0$ or $-0.2, -0.2, 2.0$. It should perhaps be emphasized here that the λ conformation has the complete range of conformations available to it that are available to the ethylenediamine chelate ring.

When *a* is chloride, using Hill's equation the same conformations to the above corresponded to the minimum energy except that $x_a y_a z_a$ is distorted to $-0.1, -0.1, 2.3$ or $-0.2, -0.2, 2.3$, and the energy difference is about 0.3 kcal mol^{-1}.

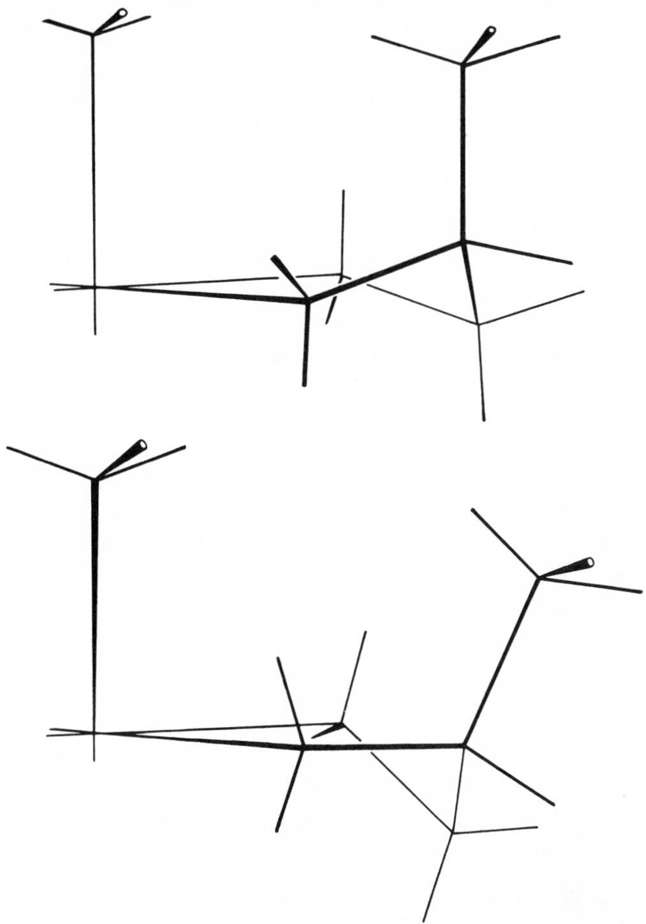

Fig. 3-9 Distortion of the conformation to alleviate nonbonded interactions of methyl group.

From the study of the trans diammine complex it has been concluded that (a) the actual rings for the conformations with the methyl group equatorial and axial are not enantiomeric, (b) the axial conformation is always nonsymmetrical, and (c) the difference in energy between the two kinds of conformations is somewhere between 0 and 3 kcal mol^{-1} depending on which nonbonded interaction equation is the most realistic. If the structure had not been allowed to distort to alleviate the van der Waals interactions between the axial methyl and the ligand *a*, the energy difference would have been much higher. Three means of decreasing the interactions were found to be energetically economical: distortion of the conformation, distortion of the position of the methyl group relative to the ring, and the distortion of the position of *a*. The first of these was found to be the most effective (see Fig. 3-9).

If the complex is square-planar with the ligand *a* missing, the energy difference is negligibly small. In solution, however, solvent molecules could orient themselves along the z axis and lead to a preference for the equatorial conformation.

The empirically derived van der Waals energy equation, Eq. (3-34), has also been applied to this system [52]. The results were similar to those derived using Bartell's equation. An enthalpy difference of 0.7 kcal mol^{-1} was found. An estimate of the entropy contribution to the free-energy difference between the axial and equatorial structures was also made for the complex, in which *a* is ammonia, by associating the variation of each of the geometrical parameters with a Hooke's-law vibration of the ring [52]. The energy levels and entropy of the vibration were calculated directly by using assumed values for the reduced mass of the atoms involved in the vibration. The most significant contribution to the entropy derives from the vertical oscillation of the C—C bond with the associated variation in z_1. The $-T\Delta S$ term was calculated to be approximately 0.3 kcal mol^{-1} at 300°K, making the total free-energy difference about 1.0 kcal mol^{-1}. An additional entropy contribution in favor of the equatorial structure could arise from the greater freedom of the ammonia molecules in position *a* to rotate about the coordinate bond for this orientation of the methyl group.

M(Meen)a_2b_2

When *N*-methylethylenediamine coordinates with the R configuration, the δ configuration has the methyl group in an equatorial conformation, whereas for the λ conformation the methyl group is axial (see Fig. 3-10).

Geometrical model. The model is similar to that previously described for propylenediamine (see Fig. 3-5). The methyl group is again assumed to remain in a staggered orientation with respect to the ring.

Energy calculations. The *N*-methylethylenediamine chelate system has the same ring-strain characteristics as the ethylenediamine ring. The interactions of the methyl group with *a* were minimized by the distortion of the angles γ' and ε', and of the positions of *a*, and by rotation of *a*, in the way previously described. The interactions of the methyl group with *b*, although more important than for propylenediamine, are not significant, especially since *b* can rotate away from high-interaction positions. For this system, additional positions of *a* were investigated ($x_a y_a z_a = 0, -0.1, z_a$, and $0, -0.2, z_a$). These were found to be more effective in reducing the total energy; both positions minimized the interaction energy for the ammonia and chloride

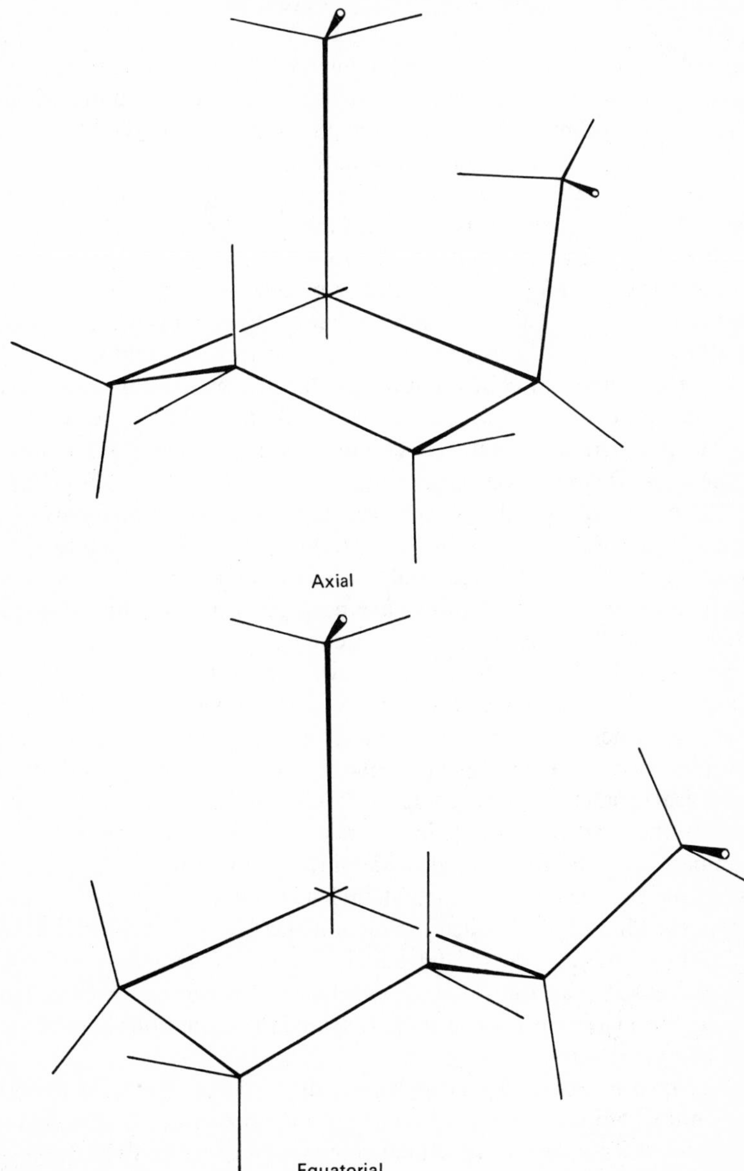

Fig. 3-10 Equatorial and axial orientation of the methyl group in R-*N*-methylethylenediamine complexes.

complexes for Hill's equation, whereas the latter, more distorted position was necessary when Bartell's expressions were used.

When Hill's and Bartell's expressions were used, a distortion of the angle γ' to 114.5° minimized the methyl-a interaction; it was not necessary to distort ε'. The position of C(1) is the most crucial in determining the energy of the conformation, because the interactions of the methyl group depend on z_1 rather than on the actual conformation of the ring. Therefore for each position of the methyl group (defined by z_1) the ring adjusts itself to one of the ethylenediamine conformations with minimized ring-strain energy.

With Hill's equation, axial conformations with z_1 between +0.4 and +0.2 Å and equatorial conformations with z_1 between −0.4 and 0.0 Å correspond to the lowest energy minima,* their actual energies lying within a range of 0.2 kcal mol^{-1}. Nevertheless, there would seem to be a slight energy preference for the equatorial conformation. If Bartell's expressions are used, the equatorial conformations with z_1 between 0.0 and −0.2 Å are more stable than the symmetrical equatorial conformations by about 0.5 kcal mol^{-1}, and the most favored axial conformations ($z_1 \sim 0.4$ Å) by about 0.8 kcal mol^{-1}. For the *trans*-dichloro complex the energy difference between the preferred equatorial (z_1 = −0.4 to −0.6 Å) and axial (z_1 = 0.0 to +0.6 Å) conformations is about 0.2 kcal mol^{-1} for Hill's equation.

No detailed energy calculations have been published for this system using the Mason and Kreevoy nonbonded interaction energies, because the interactions are so large that the complex would have to distort an unrealistic amount in order to minimize the large conformational energies.

It can be concluded from these results that the axial and equatorial conformations have available to them limited ranges of symmetrical and nonsymmetrical geometries that include conformations that are enantiomeric with respect to the rings themselves, in contrast to the propylenediamine system. In comparison with the propylenediamine system, the energy differences between the two classes of *N*-methylethylenediamine conformations depend more on the kind of nonbonded interaction energy expression used. It would appear, however, that for the tetrammine complex the energy difference is considerably less than the value of 4 kcal mol^{-1}, which had been calculated by Buckingham and his coworkers [19] using Hill's equations and Corey and Bailar conformations.

Again, in contrast to the propylenediamine system, in which distortions to the conformation were most effective in minimizing the nonbonded interactions, for the *N*-methyl substituted ring system distortions of the position of a and of the position of the methyl group with respect to the ring were

* For the R configuration with z_1 = 0.0 Å, the equatorial conformation has z_2 positive, whereas the axial has z_2 negative.

the most energetically economical way of alleviating van der Waals interactions.

M(R-chxn)a_2b_2

The optical isomers of *trans*-1,2-diaminocyclohexane have been used as conformational standards, because they are stereospecific with respect to the chelate ring when they coordinate as bidentates. For the R absolute configuration it is easily seen from molecular models that the ligand can coordinate as a bidentate only when the conformation is λ. When the cyclohexane is in its other chair conformation, the two NH_2 groups are trans axial and are too far apart to chelate. Although no detailed calculations have been published for this system, preliminary investigations have shown that, although the chirality is fixed, a range of geometries is still available to the ring. This range corresponds to the MNN plane flapping with respect to the remainder of the diaminocyclohexane structure.

trans-M(en)$_2$a$_2$

Corey and Bailar [26] studied the conformational energy differences between the dissymmetric *trans*-M(en)$_2$a$_2$, in which the conformations are either $\lambda\lambda$ or $\delta\delta$, and the meso form, in which the configuration is $\delta\lambda$. They found that the most significant interactions were those between the hydrogen atoms of the opposed NH_2 groups. In the meso form, these hydrogens are directly opposed, whereas, in the chiral form, they are more staggered. Using the Mason and Kreevoy interaction equation, they calculated that the chiral compound was more stable by about 1 kcal mol^{-1}. If the lower energy expressions are used, however, the enthalpy difference is insignificant, and the chelate rings have available to them the same range of geometries that have been found for the isolated chelate ring [51]. Although there is no enthalpy preference for one isomer, the meso form is preferred statistically over the chiral isomers by a factor of two, giving rise to an entropy and free energy difference of 0.4 kcal mol^{-1} at 298°K.

trans-M(R-pn)$_2$a$_2$

The three possible configurations, $\lambda\lambda$, $\lambda\delta$, and $\delta\delta$, have differences in free energy arising from the preference for the methyl group to be equatorial, and from the statistical preference for the meso isomer. The relative free energies of the above three configurations would be approximately 0, 0.6, and 2.0 kcal mol^{-1}, respectively.

cis-M(en)$_2$a$_2$ and M(en)$_3$

For reasons that will become obvious later, these two families of complexes have been considered together. Corey and Bailar calculated that for the D distribution of chelate rings the stability of the possible configurations for tris(ethylenediamine) complexes decreased in the order

$$(\delta\delta\delta) > (\delta\delta\lambda) > (\delta\lambda\lambda) > (\lambda\lambda\lambda)$$

with the D($\delta\delta\delta$) and D($\lambda\lambda\lambda$) configurations differing in energy by 1.8 kcal mol^{-1} [26]. The energy difference between D and L-$[Co(R\text{-}pn)_3]^{3+}$ has been determined experimentally to be 1.6 kcal mol^{-1} [35]. Because the methyl groups are expected to be equatorial and, in this orientation, do not interact with other atoms for both configurations, this energy difference has been equated to that between D($\lambda\lambda\lambda$) and L($\lambda\lambda\lambda$) {= D($\delta\delta\delta$)} $[Co(en)_3]^{3+}$. Because the calculated energy difference has been found to be very sensitive to the kind of van der Waals equation, the experimentally determined value has been used to test the kinds of equations commonly used.

Geometrical model. One chelate ring is placed in the xy plane of a Cartesian coordinate system so that the coordinate bonds make equal angles with the x and y axes (see Fig. 3-11). Once the coordinates of the atoms in the chelate ring have been determined in the normal way, the ring may then be moved into other positions in the complex by suitable coordinate transformations. Besides the intrinsic ring-strain energy of each chelate, the conformational energy of the various configurations derives from the interactions between three pairs of rings: (1,2), (2,3), and (3,1). Because of the symmetry of the D($\delta\delta\delta$) and D($\lambda\lambda\lambda$) configurations, their interaction systems

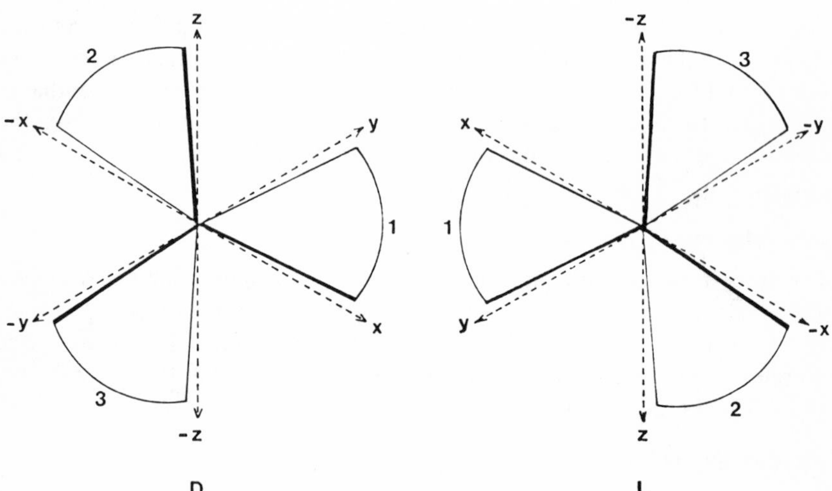

Fig. 3-11 Geometrical model for Men_3.

are related in the following way:

$$\text{D}(\delta\delta\delta) \equiv \text{L}(\lambda\lambda\lambda) \equiv 3\text{D}(\delta\delta) \qquad (3\text{-}46)$$

$$\text{D}(\lambda\lambda\lambda) \equiv \text{L}(\delta\delta\delta) \equiv 3\text{L}(\delta\delta) \qquad (3\text{-}47)$$

Thus, it is only necessary to consider the interactions between one pair of rings—1 and 2, for example—whose conformations are identical. For the configurations where the conformations have mixed chiralities, the interacting pairs of rings are not all equivalent, and the conformations are not necessarily related by any simple symmetry relationships. In order to simplify the calculations, it has been assumed that the rings of opposite chirality are strictly enantiomeric.* Thus

$$\text{D}(\delta\delta\lambda) \equiv \text{L}(\lambda\lambda\delta) \equiv \text{D}(\delta\delta) + 2\text{D}(\delta\lambda) \qquad (3\text{-}48)$$

$$\text{D}(\delta\lambda\lambda) \equiv \text{L}(\lambda\delta\delta) \equiv \text{L}(\delta\delta) + 2\text{L}(\delta\lambda) \qquad (3\text{-}49)$$

$$\text{D}(\delta\lambda) \equiv \text{L}(\delta\lambda) \qquad (3\text{-}50)$$

The interaction systems are shown in Fig. 3-12, in which the major nonbonded interactions are indicated.

Energy calculations. For this study chelate rings were chosen with $\alpha = 86°$ and $\beta = 109.5°$. Small distortions of these angles to relieve unfavorable interactions have been found not to affect the general conclusions. The rings were varied symmetrically ($z_1 = -z_2$) by varying z_1 between 0.00 and +0.50 Å with increments of 0.05 Å, and nonsymmetrically with z_1 held in turn at +0.20 and 0.0 Å, and z_2 varied from 0.0 to −0.80 Å with increments of 0.10 Å.

The variation of the van der Waals energy for D(δδ) and L(δδ) with the change in conformation is shown in Fig. 3-13. When the chelate rings are flat, the interactions between the rings are of course identical for the two chiral distributions. As the puckering increases, the interactions increase for both configurations but at different rates. When ω is small, the van der Waals interactions favor the L complex, but as ω increases further, the repulsive interactions in the L begin to increase more rapidly than in the D complex, until finally, the D becomes the more stable. The major contributions to the energy difference arise from the following interactions: for D(δδ), C(1)$\cdots$$H_{ax}$—N(1)′, N(2)—$H_{ax}$$\cdots$C(2)′; for L(δδ), C(1)$\cdots$$H_{eq}$—N(2)′, N(1)—$H_{eq}$$\cdots$C(2)′. Certain H$\cdots$H interactions make relatively large contributions to the total interaction energy but do not contribute significantly to the energy difference.

* A more detailed analysis has recently been completed in which this restriction was not applied [52].

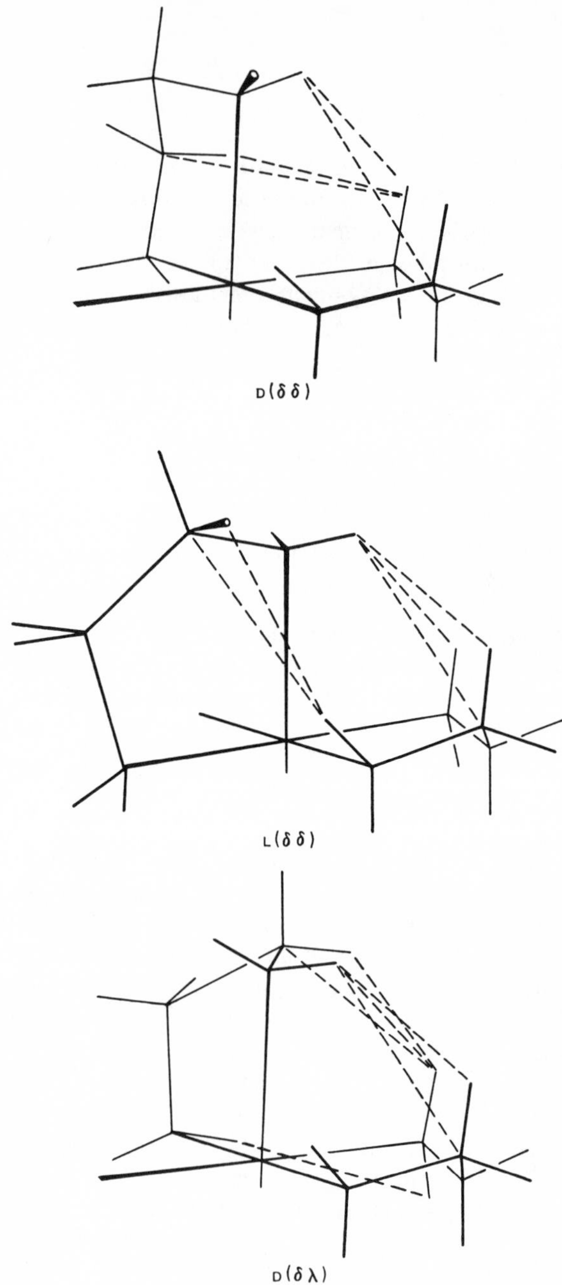

Fig. 3-12 Interaction system for *cis*-bis(ethylenediamine) complexes.

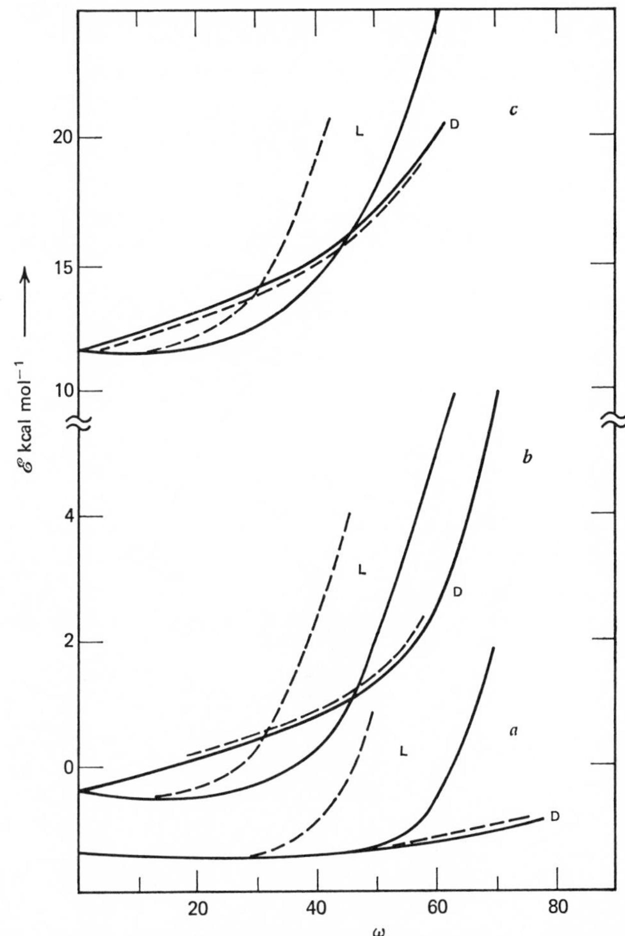

Fig. 3-13 Variation of van der Waals energy with change in chelate ring conformation for D($\delta\delta$) and L($\delta\delta$) using (*a*) Hill's, (*b*) Bartell's, and (*c*) Mason and Kreevoy's interaction equations (——) $z_1 = -z_2$; (– – –) $z_1 = 0.2$ Å [51].

The variation of the total energy (van der Waals, torsional, and angle-bending) with the conformation is presented in Fig. 3-14. The energy differences between the minimum energy structures for the L($\delta\delta$) {= D($\lambda\lambda$)} and D($\delta\delta$) configurations were found to be (a) for Hill's equation 0 kcal mol^{-1}, (b) for Bartell's, +0.27 kcal mol^{-1}, and (c) for Mason and Kreevoy's, −0.8 kcal mol^{-1}. To determine the relative energies of the D($\delta\delta\delta$) and D($\lambda\lambda\lambda$) configurations for the tris complex, the total energy curves for the

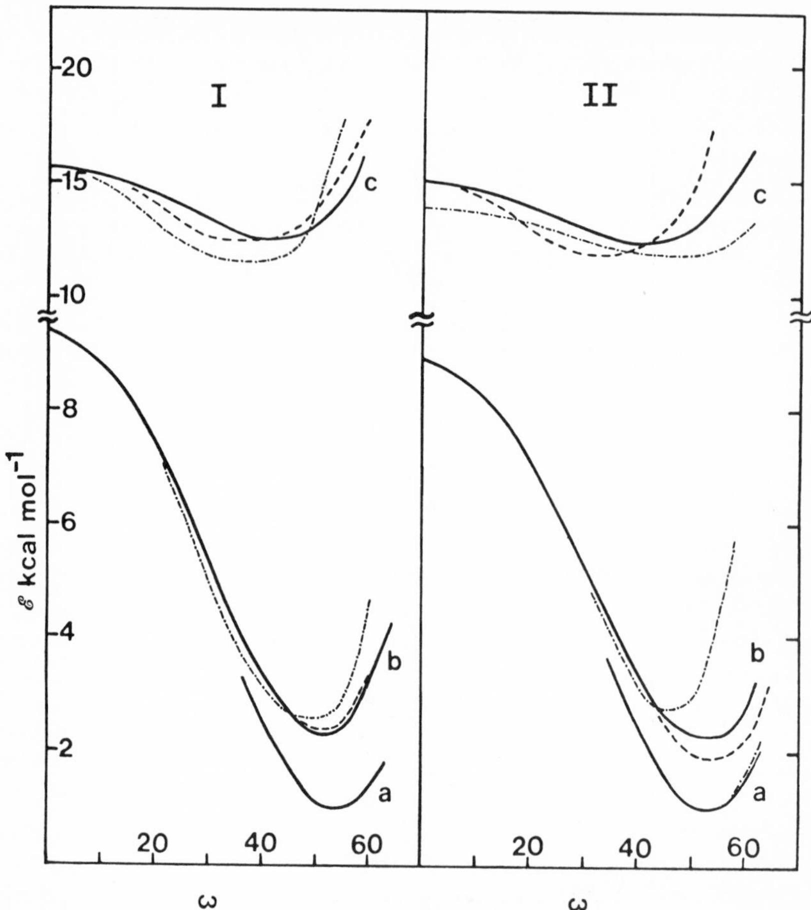

Fig. 3-14 Variation of total energy with the change in the chelate ring conformation for the D(δδ) (——), D(δλ) (– – –), and L(δδ) (–·–·–) interaction systems using (*a*) Hill's, (*b*) Bartell's, and (*c*) Mason and Kreevoy's interaction equations. I: symmetric distortion with $z_1 = -z_2$; II: nonsymmetric distortion with $z_1 = 0.2$ Å [51].

appropriate bis interaction systems were summed according to Eqs. (3-46) and (3-47) and each of the minimum energies determined: (a) for Hill's equation, the relative enthalpies are D(δδδ) = 0, D(λλλ) = 0 kcal mol^{-1}, (b) for Bartell's, D(δδδ) = 0, D(λλλ) = 0.8 kcal mol^{-1}, and (c) for Mason and Kreevoy's, D(δδδ) = 0, D(λλλ) = −2.4 kcal mol^{-1}.

Experimentally, this energy difference is known to be about 1.6 kcal mol^{-1} in favor of the D(δδδ) configuration [35]. The Hill and the Mason and Kreevoy equations are unable to account for this. Corey and Bailar had previously estimated a value for this energy difference of 1.8 kcal mol^{-1} [26], in good agreement with the observed value, using the Mason and Kreevoy nonbonded interaction equation. However, they had not allowed the chelate rings to distort to reduce the extremely high interaction energies. Gollogly and Hawkins found that, with the Mason and Kreevoy expression, the chelate ring had to flatten to, say, $z_1 = -z_2 = 0.25$ Å with $\omega = 38°$ to minimize the total energy [51]. This resulted in the D($\lambda\lambda\lambda$) becoming the more preferred.

The relative free energies for the four configurations of D-$[Coen_3]^{3+}$ have been calculated from experimental data at 293°K to be approximately D($\delta\delta\delta$) = 0, D($\delta\delta\lambda$) = 0, D($\delta\lambda\lambda$) = 0.5, and D($\lambda\lambda\lambda$) = 1.6 kcal mol^{-1}—see Section 3-10. When comparing these values with those from the a priori calculations, an entropy contribution to the free energy from the 1:3:3:1 statistical weighting of the four configurations must be taken into account. The calculated enthalpies and other entropy contributions must give rise to relative energies of the following order: 0, 0.65, 1.15, and 1.6 kcal mol^{-1}.

The Bartell equation yielded an energy difference of the correct order of magnitude for the D($\delta\delta\delta$) and D($\lambda\lambda\lambda$) configurations, but when applied to the *mixed* conformation species, it incorrectly predicted that the D($\delta\delta\lambda$) configuration is markedly more stable than the D($\delta\delta\delta$). An empirical equation (3-34) has been developed to account for the observed energy differences, and, except where expressly stated, it has been used for the following calculations [52].*

The four configurations of D-$[Coen_3]^{3+}$ differ in the range of conformations that may occur without significant change in energy. The D($\delta\delta\delta$) configuration has a large variety of conformational geometries that correspond to the lowest energy. These include both symmetric and nonsymmetric conformations. The D($\delta\delta\lambda$) configuration has a slight preference for a nonsymmetric conformation (e.g., $z_1 = 0.2$ Å, $z_2 = -0.4$ Å), in contrast to the D($\delta\lambda\lambda$) and D($\lambda\lambda\lambda$), which have a marked preference for the symmetric conformations. The high degree of flexibility of the chelate rings in the D($\delta\delta\delta$) configuration compared with the other three configurations tends to increase the free-energy difference between the configurations due to an entropy term.†

* The derivation of the empirical equation (3-34) was based on a more sophisticated geometrical model than the above. The calculated relative energies, which include an entropy contribution from the vibrational freedom of the chelate rings, but exclude the statistical contribution, are as follows: D($\delta\delta\delta$) = 0.0, D($\delta\delta\lambda$) = 0.8, D($\delta\lambda\lambda$) = 1.4, and D($\lambda\lambda\lambda$) = 1.5 kcal mol^{-1}. A derivation of Eq. (3-34) using the geometrical model above has been published [51]. However, it was based on the incorrect assumption that there is no statistical preference for the mixed conformations.

† The D($\delta\delta\lambda$) configuration possesses a C_2 axis, which as well as relating the geometries of the δ rings, should also preserve the intrinsic two-fold symmetry of the λ ring. Since the λ ring should remain symmetrical, the δ rings will adopt equivalent unsymmetrical

The relative enthalpies of the three configurations of D-*cis*-[Coen$_2$X$_2$] have been estimated to be (a) with Bartell's equation, D($\delta\delta$) = 0, D($\delta\lambda$) = -0.35, and D($\lambda\lambda$) = 0.27 kcal mol^{-1}, [51] and (b) with the empirical equation, D($\delta\delta$) = 0, D($\delta\lambda$) = 0.3, and D($\lambda\lambda$) = 0.4 kcal mol^{-1} [52]. For the mixed complex the major contributions to the van der Waals interaction energy come from the H···H interactions, C(1)—H_{ax}···H_{eq}—N(1)′, N(2)—H_{ax}···H_{eq}—N(1)′, and N(2)—H_{ax}···H_{ax}—C(1)′, and the C···H interactions, C(1)···H_{eq}—N(1)′, and N(2)—H_{ax}···C(1)′. The free energy differences between the three configurations also include entropy contributions from the freedom of vibration and from the statistical preference—0.4 kcal mol^{-1}—for the mixed conformations.

The geometrical model adopted for these calculations probably overestimates the interactions, because, for complexes where X$_2$ refers to two unidentates or to a planar or near planar chelate ring, the two ethylenediamine chelate rings should be able to distort more than for the tris complex to alleviate any interaction between them. In fact, it is conceivable that the order of preference above is in error because of this facility for distortion. It is apparent, however, that the energy differences between the three configurations are quite small, which is consistent with the small energy differences that have been found experimentally between D- and L-[Co(R-pn)$_2$ox]$^+$ [34].

It must be emphasized that the conclusions above regarding the preferred geometries of the ethylenediamine conformations have been arrived at by using a M—N distance of 2.0 Å, corresponding to the value for M ≡ Co. These results do not necessarily apply in every detail for other metal ions. For example, with longer M—N bond lengths, the degree of puckering is significantly increased, but the non-bonded interactions decrease because the interacting atoms on the different chelate rings are moved away from each other. The calculated relative free energies of the configurations of [Men$_3$] with M—N = 2.3 Å are D($\delta\delta\delta$) = 0, D($\delta\delta\lambda$) = -0.55, D($\delta\lambda\lambda$) = -0.45, and D($\lambda\lambda\lambda$) = 0.2 kcal mol^{-1} [52]. These energies are largely determined by the statistical preference for the mixed conformations.

M(R-pn)$_3$

For M(en)$_3$, the configurations D($\delta\delta\delta$) and L($\lambda\lambda\lambda$) are enantiomeric and, as such, have identical energies. Because of the preference, however, of R-propylenediamine chelates for the λ conformation, in which the methyl group is equatorial, these two configurations are no longer of equal energy.

---

conformations to minimize the interactions between the δ and λ rings. The same arguments apply to the D($\delta\lambda\lambda$) configuration, but the unsymmetrical distortions of the λ rings are limited by the inflexibility of the D($\lambda\lambda$) system.

The L configuration lies at lower energy by an amount equal to three times the energy difference between the axial and equatorial orientations in the tris complex. This energy difference for M(R-pn)$(NH_3)_4$ is about 1.0 kcal mol^{-1}, and it would have been higher except for the facility of the ammonia groups to rotate about their M—N bonds to relieve unfavorable interactions with the methyl group. A rotation of the —NH_2 group is not possible to the same extent in the tris complexes, and so the energy difference between D(δδδ) and L(λλλ) is likely to be slightly in excess of 3.0 kcal mol^{-1}. As stated above, the energy difference between D(λλλ) and L(λλλ) is equal to that between the same configurations for M$(en)_3$ (1.6 kcal mol^{-1} experimentally determined). Thus R-propylenediamine is relatively stereospecific in its tris complexes, because the L(λλλ) configuration is about 1 kcal mol^{-1} more stable than the next most stable configuration.

M(R-mepenten)

The stereospecificity of the sexadentate ligand R-*N*,*N*,*N'*,*N'*-tetrakis-(2'-aminoethyl)-1,2-diaminopropane (R-mepenten) has been determined by a conformational analysis of M(R-mepenten) [49]. Because of the complexity of the system, the analysis was not so detailed as for the simple bidentates. The intention, however, was to determine which configuration is preferred, and the approximate energy difference between the two configurations, which are shown in Fig. 3-15. The central chelate ring has a δ conformation with the methyl group axial in the D configuration and a λ conformation with the methyl group equatorial in the L complex. It is sterically impossible

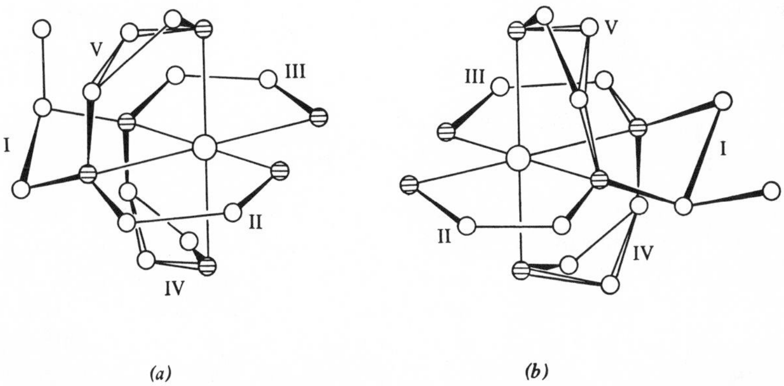

Fig. 3-15 Possible conformational arrangements for the standard structures of (*a*) D- and (*b*) L-[Co(R-mepenten)]$^{3+}$. In (*a*) the δ conformations of rings IV and V are completely shaded and the λ conformations, only partly shaded. For (*b*) this situation is reversed [49].

for the conformation of the central chelate ring to invert for either configuration. At the time of the study Hill's equations were commonly used and were applied to this problem. Because there was a desire, however, to underestimate rather than overestimate the energy difference, Hill's equation will provide a lower limit for the energy difference.

Geometrical model. Dreiding stereomodels were used to represent the standard structures.* It should be noted that care must be taken in setting up these models because of the *rod-and-cylinder* method of linking the atoms. The rods are more flexible than the cylinders, and, as a result, the strain involved in the formation of some rings can cause an irregular distortion of a few of the bond angles. For the two configurations of M(R-mepenten), this distortion of the model is noticeable only in the three rod kinds of *coordinate bonds* of the metal [cobalt(III)] ion. To remove this possible source of error the models were set up in such a way that in the region of interaction under investigation the bond-angle distortions for the alternative conformations were minimal and identical. In the closely related complex $[Co(EDTA)]^-$, X-ray analysis has shown that the *equatorial* ∠OCoO is 104° [133] and the corresponding NCoN angle in $Co(penten)^{3+}$ is similarly distorted [99]. When the more flexible rod kind of arms are used to represent these coordinate bonds in the Drieding models of $[Co(penten)]^{3+}$ and $[Co(EDTA)]^-$, the angles are found to be almost identical with those determined by X ray.

Energy calculations. There are three kinds of chelate rings:

1. Ring I, which has only one possible chirality
2. Rings II and III, which each have a single conformational type, which nevertheless permits a small range of conformations with similar bond angles and torsional strain
3. Rings IV and V, which may each exist in either of two distinct conformations that experience similar structural strain

Because the alternative conformations have similar ring-strain energies, the preferred conformations depend on the van der Waals interactions, which center mainly on the interactions of the methyl group with ring V (see Fig. 3-16), and these interactions determine the energy difference between the two configurations. The van der Waals interactions enforce a δ conformation on ring V for the D and a λ conformation for the L. These interactions could

* Standard structures are defined as molecular structures with minimized energies in which normal bond lengths are preserved and in which bond-angle distortion due to van der Waals interactions are negligible.

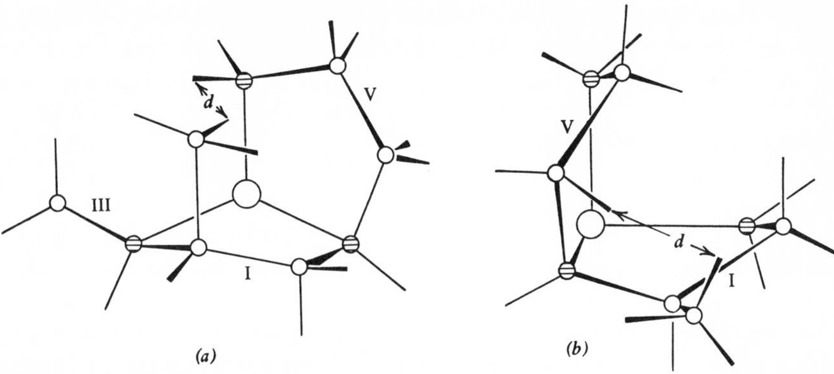

Fig. 3-16 Regions of interaction of the (*a*) axial and (*b*) equatorial methyl groups. Ring V is shown in the δ conformation [49].

be reduced by bond-angle distortions in the various rings, but such distortions would require larger energies than necessary for simple bidentate systems, because each angle distortion requires readjustment of other bond angles in regions that are already under considerable strain because of the steric limitations of the ligand. Therefore, distortions of the standard-structure bond angles are thought to be very small.

Table 3-6 Interactions of the methyl group in each of the standard structures of D- and L-[Co(R-mepenten)]$^{3+}$ [49]
In ring I the methyl group is bonded to the carbon atom C(I*a*). In other rings the carbon atoms are numbered *a*, *b* away from ring I. The nitrogen atoms in the four primary amine groups are identified by the rings to which they belong—N(V), for example. Substituent hydrogens are identified by the atom to which they are bound.

D isomer	r	$\mathscr{E}_v$	L isomer	r	$\mathscr{E}_v$
H(Me)—H(NV)	1.6	3.0	H(Me)—H(NV)	1.7	1.5
—H(CVa)	1.65	2.1	C(Me)—H(CV*a*)	2.15	1.4
—H(CIII*a*)	1.7	1.6	—C(V*a*)	2.7	0.9
—N(V)	2.1	0.7			
—C(V*a*)	2.1	2.0			
—C(III*a*)	2.15	1.5			
C(Me)—H(NV)	2.2	1.1			
—H(CIII*a*)	2.2	1.1			
—H(CV*a*)	2.3	0.6			
—C(III*a*)	2.75	0.6			
—C(V*a*)	2.9	0.2			
Total	···	14.5	··· ···	···	3.8

The important interactions are listed in Table 3-6. The energy difference between the two configurations was found to be about 10 kcal mol^{-1}. As stated above, this is probably the lower limit to the energy difference, but at this energy, the percentage of the D isomer in equilibrium with the L would be less than 0.01 percent.

SIX-MEMBERED DIAMINE RINGS

In the conformational analysis of organic molecules the system to which most attention has been directed is the six-membered carbocyclic ring. In contrast, the analogous diamine chelate ring, formed by 1,3-diaminopropane, has attracted little attention since the early comments of Corey and Bailar [26]. Recently, however, the detailed methods of analysis developed for the ethylenediamine complex systems have been applied to the mono, bis, and tris(1,3-diaminopropane) complexes [52].

The conformations available to the six-membered diamine chelate ring may be classified in a manner analogous to those of the cyclohexane ring. There are two basic kinds of conformations: (1) A chair form (see Fig. 1-7), in which the metal atom and the central carbon atom lie on different sides of a plane containing the other four atoms of the ring. In the absence of interactions external to the ring, this conformation is symmetrical. (2) A flexible boat that corresponds to a wide range of conformations that contains two special symmetrical forms: (*a*) the symmetrical boat (see Fig. 1-8), in which the metal and central carbon atoms lie on the same side of a plane containing the remaining four atoms, and (*b*) the symmetrical skew boat (see Fig. 1-5), in which the central carbon, the metal, and the two nitrogen atoms are coplanar, with the remaining carbon atoms lying, respectively, above and below this plane.

$M(tn)a_2b_2$

Geometrical model. As before, the chelate ring is placed in a right-handed Cartesian coordinate system with the coordinate bonds in the xy plane equidistant from the x and y axes (see Fig. 3-17). Bond lengths are as specified previously and are assumed to remain constant. For each conformational kind, a set of internal coordinates has been chosen to specify and to vary the conformation.

Symmetric chair. Conformations of this kind may be unambiguously defined by the set of six ring bond angles. From these and the set of bond lengths, the coordinates of the ring atoms can be generated by simple coordinate geometry. The conformation may be varied by varying the set of angles. Expansion of the angles results in a flattening, whereas contraction tends to increase the puckering of the ring.

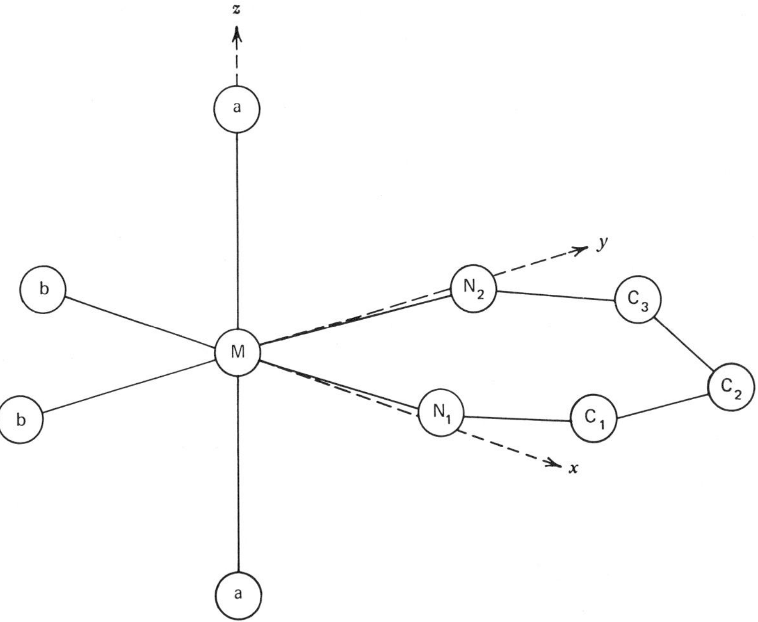

Fig. 3-17 Geometrical model for $M(tn)a_2b_2$.

Symmetric boat. The rings in the symmetrical-boat conformations have also been defined by the set of ring bond angles. The ring is constructed with the central carbon atom C(2) on the same side of the N(1)C(1)C(3)N(2) plane as the metal atom, instead of on the opposite side, as in the symmetrical-chair form.

Flexible boat. The symmetric skew boat and the asymmetric boat (see Fig. 1-6) come under this general classification. For them it is necessary to obtain a set of internal coordinates with which to define the basic conformational kind, and also allow the conformation to be varied to minimize unfavorable energies. The analogous method to that used for the five-membered diamine rings would be to specify three ring angles as well as the z coordinates of carbon atoms 1 and 3. Such a model presents some difficulties, however, in that it requires two parameters (z_1 and z_3) to specify the conformation of the ring and, in specifying only three ring angles, leaves three angles uncontrolled.

A more convenient method is to use z_2 in conjunction with five ring angles to define the ring. With these variables, it is possible to locate a preferred kind of conformation by varying z_2 with an approximate set of bond angles, and then the preferred conformation can be refined by varying the five ring angles. As found for the other conformational types, angular increase tends to flatten the ring, in contrast to angular decrease, which makes the ring more puckered. With this model, the symmetrical-skew-boat conformation occurs when $z_2 = 0$, $\angle$MN(1)C(1) = $\angle$MN(2)C(3), and $\angle$N(1)C(1)C(2) = $\angle$N(2)C(3)C(2).

The positions of *a* and *b* are determined and varied in the manner outlined for the ethylenediamine systems. The *a* ligands are permitted to bend back away from the *z* axis in order to relieve unfavorable interactions where necessary and are also allowed to rotate freely to minimize the interactions.

Energy calculations. When the ligands *a* are absent (e.g., a square planar complex), interactions external to the ring do not influence the results; the chelate ring can be considered as if it were isolated. The conformational energies of the symmetrical-chair and symmetrical-boat forms were investigated for this situation by independently varying the four different ring angles by increments of 2.5°. The geometries that were found to correspond to the minimum energies of 0.6 $\pm$ 0.1 kcal mol^{-1} for the symmetrical-chair form and 4.5 $\pm$ 0.1 kcal mol^{-1} for the symmetrical-boat form are given in Table 3-7. In the chair form, this energy derives from small torsional and angle-bending contributions. In the boat there is large torsional strain due to eclipsing of the substituents to the C—N bonds. This can only be reduced by angular distortions and results in a relatively large value for the minimum energy of the boat conformation. In both chair and boat forms, however, nonbonded interactions do not contribute significantly to the total energies.

For the flexible-boat conformation z_2 was varied from 0.0 to 2.0 Å with increments of 0.1 Å, maintaining the same chirality (say, δ) throughout. Typical graphs of the effect of this variation on the various energy terms are shown in Fig. 3-18. The variation of $\mathscr{E}_\theta$ reflects changes in the unspecified angle $\angle$C(1)C(2)C(3) and depends on the choice of the other five angles. If these angles are given their unstrained values (e.g., 90°, 109.5°), $\mathscr{E}_\theta$ increases slowly as z_2 is varied from 0, becoming significant only at large values of z_2. The torsional-energy term remains relatively constant for a range of small values of z_2 (0.0 Å to 0.2 Å) and then increases rapidly until the symmetrical-boat form is reached. The nonbonded interaction term $\mathscr{E}_v$ arises mainly from the interactions C(1)—$H_{ax}\cdots H_{ax}$—N(2) and C(3)—$H_{ax}\cdots H_{ax}$—N(1). These are relatively small at $z_2 = 0$ but increase with z_2 until they reach a maximum at $z_2 = 1.4$ Å, and then, as z_2 increases further, the interacting atoms move away from one another until, in the final symmetrical-boat conformation, $\mathscr{E}_v$ is insignificant again.

From these results it is apparent that the order of preference of the different conformational types is symmetrical chair, symmetrical skew boat, other flexible boats, and, finally, the symmetrical boat. This order depends mainly

Conformation	∠NMN	∠MNC	∠NCC	∠CCC	x_a	y_a	z_a	$\mathscr{E}$[a]
M(tn)								
Chair	90°	112°	112°	112°	...	...	...	0.6
Skew boat[b]	90°	109.5°	109.5°	110.9°	...	...	...	2.3
Sym. boat	90°	109.5°	109.5°	109.5°	...	...	...	4.5
$M(tn)a_2b_2$								
Chair ($a \equiv NH_3$)	95°	119.5°	112°	112°	−0.2	−0.2	2.0	2.6
Skew boat[b] ($a \equiv NH_3$)	90°	114.5°	109.5°	117.2°	−0.2	0.0	2.0	5.0
Chair ($a \equiv OH_2$)	95°	117°	112°	112°	−0.1	−0.1	2.0	2.0
Skew boat[b] ($a \equiv OH_2$)	90°	114.5°	109.5°	117.2°	−0.2	0.0	2.0	4.3
Chair ($a \equiv Cl^-$)	95°	119.5°	112°	112°	−0.2	−0.2	2.3	2.6
Skew boat[b] ($a \equiv Cl^-$)	90°	114.5°	109.5°	117.2°	−0.2	0.0	2.3	5.3
trans-$M(tn)_2a_2$								
Chair—as for mono								
Skew boat[b] ($a \equiv NH_3$)	90°	119.5	112°	119.5°	−0.3	+0.3	2.0	8.5
Skew boat[b] ($a \equiv OH_2$)	90°	114.5	109.5°	117.2°	−0.1	+0.1	2.0	5.0
Skew boat[b] ($a \equiv Cl^-$)	90°	114.5	109.5°	117.2°	−0.1	+0.1	2.3	6.2
$M(tn)_3$								
Chair	95°	119.5°	112°	109.5°	...	...	...	6.6
Skew boat[b]	90°	114.5°	114.5°	111°	...	...	...	7.1

[a] In kcal mol^{-1} per ring.
[b] With $z_2 = 0.0$.

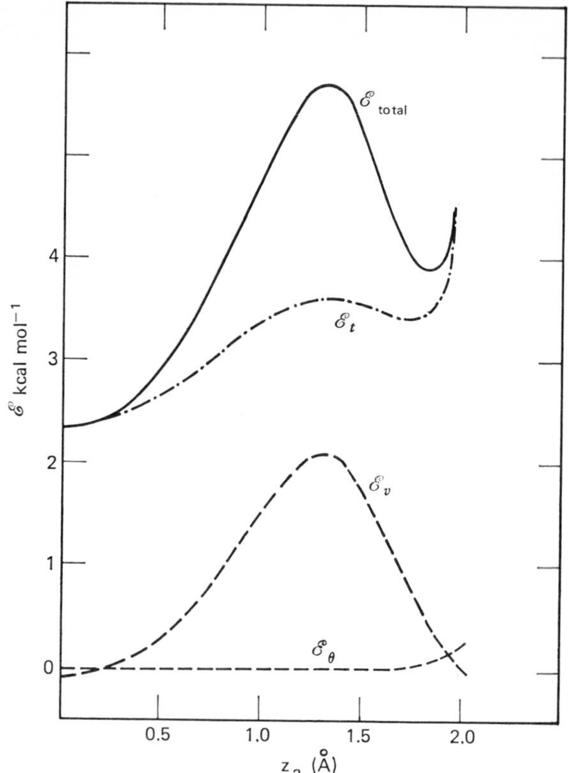

Fig. 3-18 Variation of $\mathscr{E}_\theta$, $\mathscr{E}_t$, and $\mathscr{E}_v$ with change in z_2 for $\angle$NMN = 90°, $\angle$MNC = 109.5°, and $\angle$NCC = 109.5° for flexible-boat conformations.

on the difference in the conformations' torsional characteristics. There is a small range of conformations that correspond to the energy minimum for the chair. This flexibility arises because the small torsional and angle-bending terms are not greatly affected by small changes (~2.5°) in the ring angles. The symmetrical-skew-boat conformation is also the center of a range of conformations with similar energies, because $\mathscr{E}_t$ and $\mathscr{E}_\theta$ are fairly insensitive to z_2 in the region close to $z_2 = 0$. The difference in flexibility between the isolated symmetrical chair and skew boat conformations leads to an energy difference from entropy at 300°K of about 0.3 kcal mol^{-1} favoring the symmetrical skew boat [52].*

* When ligands are present in the apical positions, interactions limit the flexibility of these conformations and the entropy difference is reduced.

The conformational preferences just described correspond to the chelate ring in a square-planar complex or in some other situation where external interactions are unimportant. When the ligands a occupy their octahedral positions, their nonbonded interactions with the atoms in the chelate seriously influence the conformations.

For the chair conformation strong interactions occur between the ligand a and the ring atoms C(1) and C(3) and their axial substituents. Because the ring and the interacting atoms are symmetrically placed with respect to the ligand a, the ring remains symmetrical, even as it distorts in order to relieve these interactions. Similarly, any distortions of the position of a is confined to a plane that bisects the xz and yz planes, thus making x_a and y_a always equal.

For these calculations the ring was varied in the same way as for the isolated ring, and the position of a was adjusted by changing x_a and y_a by increments of 0.1 Å. The calculations were performed for a series of ligands in the apical a positions and in the positions b—NH_3, H_2O, NO_2^-, Cl^-, for example. As found for the ethylenediamine systems, the interactions with b are insignificant. The preferred geometries are given in Table 3-7. The corresponding minimum energies were found to be 2.6 kcal mol^{-1} for $a \equiv NH_3$, 2.0 kcal mol^{-1} for $a \equiv OH_2$, and 2.6 kcal mol^{-1} for $a \equiv Cl^-$.

The symmetric-boat conformation is considerably destabilized relative to the chair conformation by the interactions between the ligand a and C(2) and its substituents. Therefore the symmetric-boat form was not considered further.

In the flexible-boat conformations the major external interactions are those involving ring atoms C(1) and C(3) and their substituents. When z_2 increases in the δ conformation, the interactions between one ligand a and the C(1) system increase, and those between the other ligand a and the C(3) system decrease. However, the former increase more rapidly than the latter decrease. Therefore in these complexes there is also a preference for symmetric-skew-boat conformations over other kinds of flexible-boat conformations. The twofold symmetry axis of the complex is retained while the ring distorts in order to relieve its interactions with the two ligands a.

The geometry of the minimum-energy symmetrical-skew-boat conformations where $a \equiv NH_3$, OH_2, and Cl^- is given in Table 3-7. The conformational energies of these structures are for $a = NH_3$, $\mathscr{E} = 5.0$ kcal mol^{-1}; for $a \equiv OH_2$, $\mathscr{E} = 4.3$ kcal mol^{-1}; and for $a \equiv Cl^-$, $\mathscr{E} = 5.3$ kcal mol^{-1}.

In conclusion, for these systems there is a marked preference for symmetrical-chair conformations. Interactions with the apical ligands a are relieved for these conformations by a general flattening of the ring, achieved mainly by distortion of $\angle$MNC and $\angle$N(1)MN(2), and by distortion of the position of a.

trans-$M(tn)_2a_2$

Most of the conclusions for the mono(1,3-diaminopropane) system are valid for the *trans*-bis(1,3-diaminopropane) system. There are two possible forms, however, for the chair conformation of the latter (see Fig. 3-19). In the structure where the two sets of carbons are on the same side of the NNMNN plane, the ligand *a* is unable to relieve its interactions with one ring without increasing its interactions with the other ring. The preferred trans, therefore, has the carbon atoms on opposite sides of the NNMNN plane. In this form, the interaction system of each ring is equivalent to that for the corresponding mono complex. This structure has been found for *trans*-$Nitn_2(NO_3)_2 \cdot 2H_2O$ [105] and *trans*-$Cutn_2(NO_3)_2$ [106]. As predicted, the angles $\angle N(1)MN(2)$ and $\angle MNC$ show distortions, and the NO_3^- groups are distorted away from their regular positions.

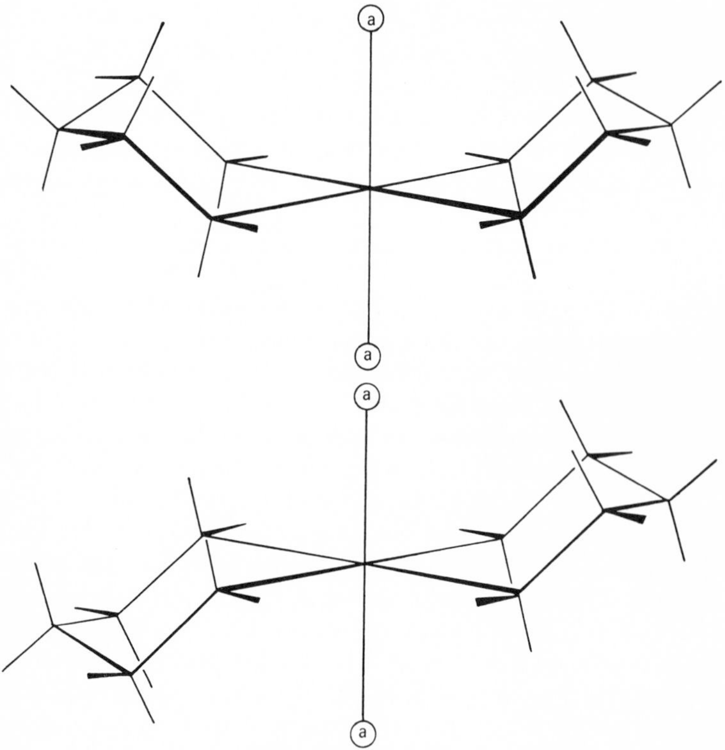

Fig. 3-19 Configurations of *trans*-$[M(tn)_2a_2]$.

$M(tn)_3$

Geometrical model. The geometrical model used for the tris complexes of 1,3-diaminopropane was similar to that already described for $M(en)_3$, the coordinates of one ring being determined in the *xy* position and transformed about the threefold axis into the other appropriate positions in the complex.

Energy calculations. For symmetric-chair conformations the main interactions are those of the ring atoms C(1) and C(3) and their substituents with the coordinated amine groups of an adjacent ring (see Fig. 3-20). This interaction system is directly comparable with that encountered in the mono complex with $a \equiv NH_3$. In the tris complex, however, the orientation and position of the amine group is not independent of the rest of the molecule, being related to the amine group of ring 1 by the C_3 axis. Nevertheless, the substituents of the interacting amine group are approximately symmetrical with respect to the ring with which it interacts, and, consequently, this ring has been assumed to remain symmetrical while under distortion. It should be noted that, because of the basic symmetry of the symmetrical-chair conformations, the D and L configurations of this complex are energetically equivalent.

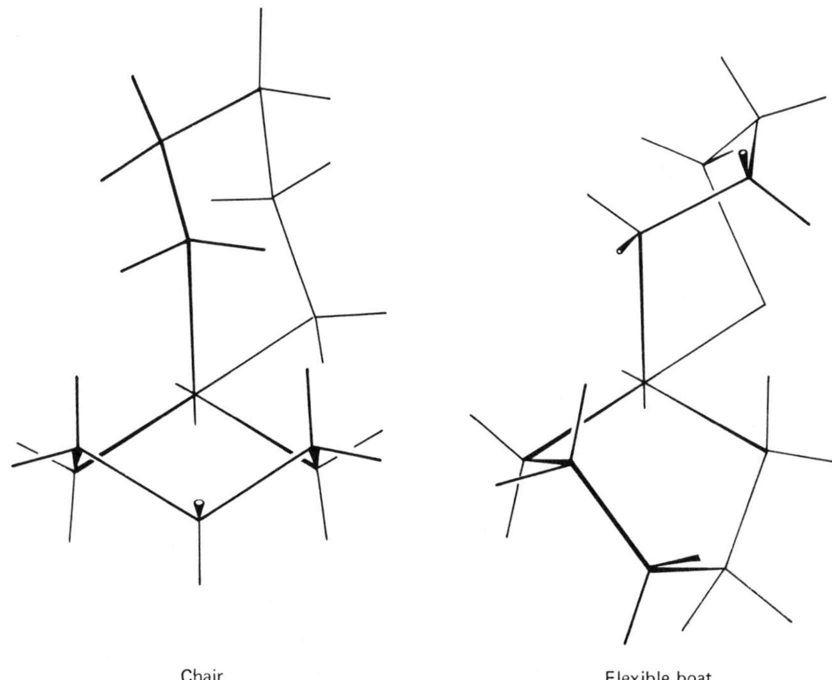

Fig. 3-20 Interaction system for *cis*-bis(1,3-diaminopropane) complexes.

The preferred geometry of these rings, found by varying the ring bond angles with increments of 2.5°, is given in Table 3-7. This corresponded to a minimum energy of 19.8 kcal mol^{-1}.

In complexes in which the rings have flexible-boat conformations and the chiralities of these are the same, the interaction system affecting each ring is comparable with that of the mono complex, and there is a preference within the range of flexible boats for conformations of the symmetric-skew type. This ring is therefore symmetrical about a twofold axis in the xy plane bisecting the $x0y$ angle. Because the two remaining chelate rings are also distributed symmetrically about this axis, each ring preserves its twofold symmetry as it undergoes distortion.

The ring geometry that was found to correspond to a minimum energy in the tris complexes is described in Table 3-7. In such complexes, each ring has a conformational energy of 7.1 kcal mol^{-1} (total energy = 21.3 kcal mol^{-1}).

In the tris complexes there is inevitably greater strain than in the mono and *trans*-bis complexes, because the interacting amine group is not permitted simply to move away from interacting atoms. The interactions are therefore relieved mainly by angular distortion centered on ∠MNC and ∠N(1)MN(2). In the minimum-energy conformation, the distortion is centered mainly on the MNC angles (119.5°). This conformation was found to correspond with a rather sharply defined energy minimum. At smaller values of ∠MNC the van der Waals interactions were larger, and at larger values of ∠MNC, $\mathscr{E}_\theta$ and $\mathscr{E}_t$ became too large. A second localized energy minimum was found to occur at higher energy with ∠MNC = 117.0° and ∠N(1)MN(2) = 100°.

FIVE-MEMBERED AMINO ACID RINGS

From X-ray structural studies of a number of amino acid complexes [45], it has been found that the amino acid chelate ring exists in either of two kinds of conformation (see Fig. 3-21):

1. A planar or slightly puckered conformation
2. An asymmetric-envelope conformation in which the NCCOO atoms are not far removed from coplanarity

The torsional arrangement of atoms in amino acid chelate rings is different from that in the diamines. It is important to note that when the C—O(—M) bond is eclipsed with the C—N bond, C═O is staggered with respect to R and H, and when C—O(—M) is staggered with respect to C—N and C—H the C═O bond is eclipsed with C—R (see Fig. 3-22). In diamine rings ω has values about 50°; for the amino acids, the X-ray structural studies have shown that the corresponding dihedral angle is between 0 and 30°

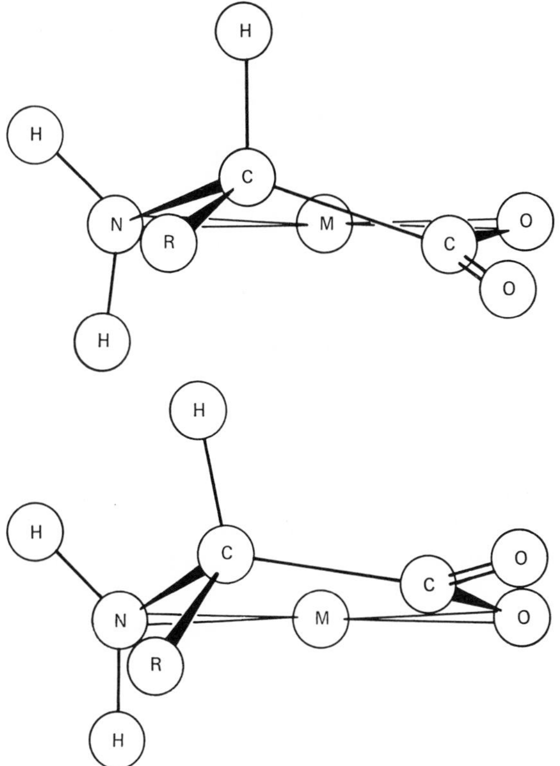

Fig. 3-21 Puckered and envelope configurations of amino acid chelate rings.

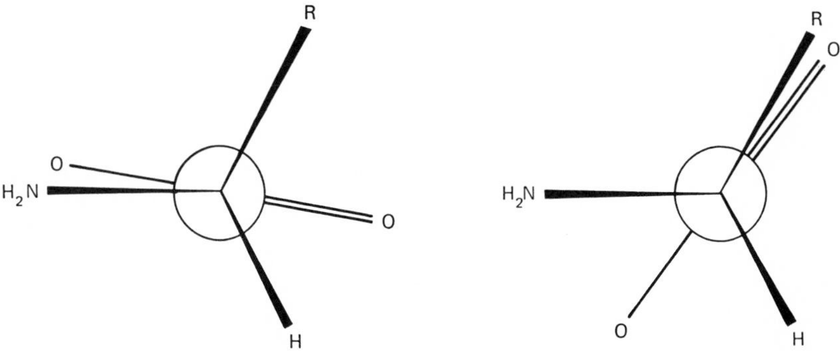

Fig. 3-22 Torsional arrangement about C—C$^\alpha$ bond in amino acid chelate rings.

[45]. This range of values is also found for the free amino acids in their zwitterionic or protonated forms (see Fig. 4-9) and for a number of carboxylic acids [71].

Thus it would seem that the torsional interactions in carboxylates are such that the C—O bond prefers to be eclipsed with a substituent on the α-carbon. The barrier to free rotation about the C—C$^{\alpha}$ bond has been found experimentally to be about 1 kcal mol^{-1} [1], and therefore for the amino acid chelates the torsional energy about the C—C$^{\alpha}$ bond has been determined from Eq. (3-51) with $V_0 = 1$ kcal mol^{-1} [52]:

$$\mathscr{E}_t = \tfrac{1}{2}V_0\{1 + \cos 3(60 - \omega)\} \tag{3-51}$$

It must be remembered, however, that when the C—O and C—N bonds are perfectly eclipsed and the chelate ring is flat, the torsional arrangement about the C—N bond is in its most unfavorable position with the substituents eclipsed (V_0(C—N) = 2 kcal mol^{-1}).

Geometrical model. For a complex of the type M(am)a_2b_2 the amino acid chelate ring has been oriented in a right-hand coordinate system, as shown in Fig. 3-23. The conformations have been specified by a constant set of

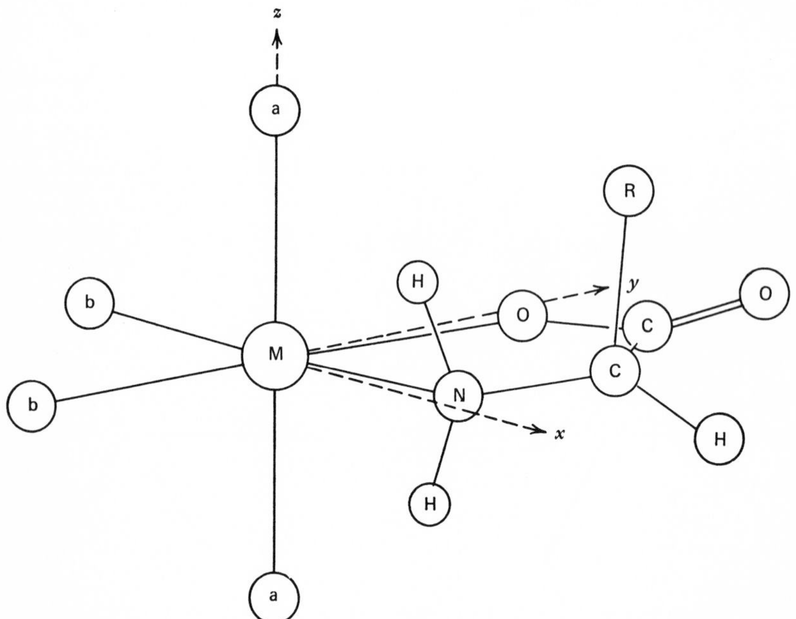

Fig. 3-23 Geometrical model for M(am)a_2b_2.

bond lengths and the internal coordinates $\angle$NMO ($= \alpha$), $\angle$MNC(1) ($= \beta$), z_1, and z_2. The positions of substituents to the ring have been defined by the appropriate bond lengths and bond angles, and the positions of groups a and b have been defined by the coordinates of the donor atoms.

Energy calculations. The torsional and angle-bending energies have been determined in the normal way for an extensive range of conformations. The angles α and β were varied from their unstrained values (90° and 109.5°) by increments of 2°, and z_1 and z_2 were varied independently in the range of 0.0 to ± 0.8 Å. The coordinated group a was permitted to bend away from nonbonded interactions with atoms on the chelate ring. The coordinates x_a and y_a were varied independently with increments of 0.1 Å.

Graphs of the variation of $\mathscr{E}_t + \mathscr{E}_\theta$ with ω for some typical values of α and β are shown in Fig. 3-24, and similar graphs for the total energy ($\mathscr{E}_v + \mathscr{E}_t + \mathscr{E}_\theta$) for $a \equiv NH_3$ and R $\equiv$ $-CH_2X$ are shown in Fig. 3-25. Each curve corresponds to a particular value of C(1) and a range of values of C(2). The full line in Figs. 3-24 and 3-25 represents the envelope of the energy minima for all the various curves. Geometric parameters for some of the conformations with minimum ring-strain energy are given in Table 3-8.

From Fig. 3-25 it is clear that the amino acid chelate ring is limited to a flattened or envelope conformation with ω in the range of 0 to 30°. Within the range $\omega = 0 \pm 30°$ there is no energy barrier to conformational change

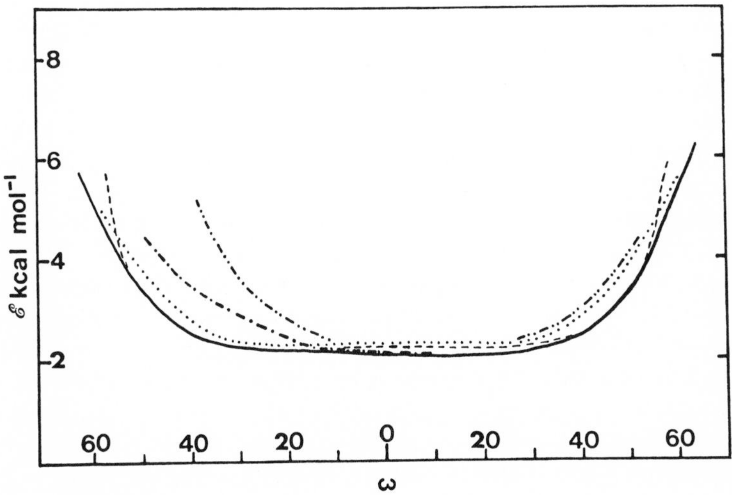

Fig. 3-24 Variation of $\mathscr{E}_t + \mathscr{E}_\theta$ with ω for $\alpha = 86°$ and $\beta = 109.5°$. (——) overall minimum energy for each value of ω; (– – –) $z_1 = -z_2$; (· · ·) $z_1 = 0.0$; (–·–·–) $z_1 = +0.2$ Å; and (–· ·–· ·) $z_1 = +0.4$ Å [52].

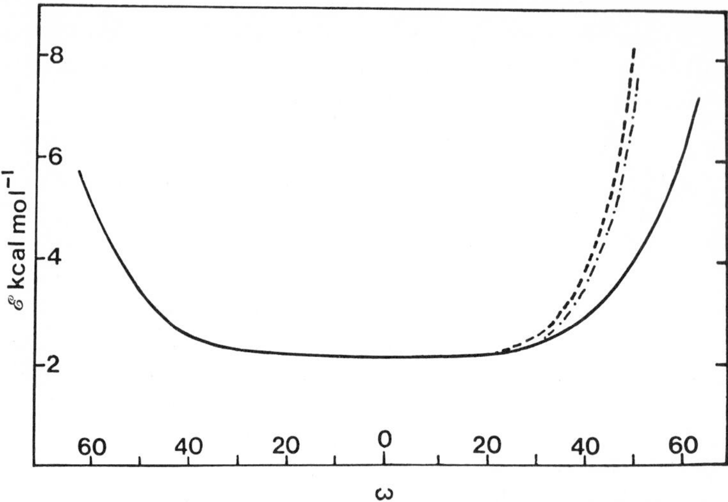

Fig. 3-25 Variation of the total energy with ω for $\alpha = 86°$, $\beta = 109.5°$: (——) overall minimum energy for each value of ω; (– – –) $z_1 = -z_2$; (· · ·) $z_1 = 0.0$, completely obscured by (——) curve; (–·–) $z_1 = 0.2$Å. ω values on the left of the diagram correspond to conformations in which the substituent is equatorial and on the right, axial [52].

because the energy remains at 2.0 ± 0.2 kcal mol^{-1}. The contributions of $\mathscr{E}_t$ and $\mathscr{E}_\theta$ to this energy depend on the kind of conformation as well as the degree of puckering of the ligand, as indicated by ω. In flattened conformations the substituents on N and C(1) are eclipsed giving rise to a large value of $\mathscr{E}_t$, and $\mathscr{E}_\theta$ is small because of the negligible angular strain in the ring. The eclipsing about the C—N bond can be relieved, while maintaining the favored torsional arrangement about the C—C$^\alpha$ bond, by the formation of an envelope conformation. The relief of torsional strain in this way, however, is accompanied by an increase in the angle-strain energy.

The nonbonded interactions between the substituents on the chelate ring and the group *a* are extremely small, even when the substituent of the type —CH_2X is axial. The only effect of the nonbonded interactions with the axial substituent is to limit the range of ω values for certain kinds of conformations. According to these calculations [52], there is no significant energy difference between the minimum energy axial and equatorial conformations, nor is there an energy barrier to ring inversion. Consistent with this conclusion, a recent X-ray study of *cis*-bis(L-isoleucinato)copper(II) has shown one α-substituent in the axial orientation [134].*

* Note added in proof: This has also been found for *cis*-bis(D-alaninato)copper(II) [R. D. Gillard, R. Mason, N. C. Payne, and G. B. Robertson, *J. Chem. Soc. A.*, 1864 (1969)].

Table 3-8 Some minimum-energy amino acid chelate ring conformations

α	β	z_1, Å	z_2, Å	ω
88°	107.5°	0.0	0.0	0°
88°	107.5°	+0.2	0.0	16.2°
88°	107.5°	+0.4	0.0	32.8°
88°	107.5°	+0.1	−0.1	17.4°
88	107.5°	+0.2	+0.2	2.6°
88°	107.5°	+0.4	+0.2	13.4°
88°	107.5°	0.0	−0.2	18.7°
86°	109.5°	+0.2	0.0	15.7°
86°	109.5°	+0.4	+0.2	2.5°
86°	109.5°	0.0	−0.2	18.1°
86°	109.5°	+0.1	−0.1	16.0°
86°	109.5°	+0.2	+0.2	2.5°
86°	107.5°	+0.4	+0.4	4.8°
86°	107.5°	+0.4	+0.2	13.8°
86°	107.5°	+0.4	+0.0	31.6°
84°	109.5°	+0.4	+0.4	4.5°

CORRELATION OF PREDICTED STRUCTURES WITH REAL STRUCTURES

The preferred conformations for a particular system predicted on the basis of the a priori calculations are unreal in the sense that external influences, such as solvation effects, or interactions peculiar to the crystalline state have not been taken into account. For most of the systems studied a range of structures has been found to correspond to the lowest energy, and another extensive range differing slightly from the former in geometry has been found to have energies that are only a little above the calculated minimum. In a crystal the system adopts the geometry that minimizes the sum of the *internal* and *external* interactions. Because such a large range of low-energy conformations is available to the molecule, it is highly probable that one of these enables the combined internal and external interactions to be minimized. Thus, it could be assumed that the conformations determined by X-ray diffraction studies should be among those predicted by the a priori calculations. However, certain interactions in the solid state—hydrogen bonding, for example—could well favor a high-energy conformation.

Compare the data in Table 3-5 with those in Table 4-4. The geometries of the ethylenediamine chelate rings found in various crystals correspond closely to those predicted. The six-membered diamine chelates provide another rather dramatic example of the good agreement between the predicted

Table 3-9 Ring angles in six-membered diamine chelate rings[a]

Angle	$Cutn_2(NO_3)_2$[b]	$Nitn_2(NO_3)_2 \cdot 2H_2O$[c]	Predicted bis	*l*-$Cotn_3Br_3 \cdot H_2O$[d]	Predicted tris
N(1)MN(2)	95	93	95	94	95
MN(1)C(1)	119	121	119.5	117	119.5
MN(2)C(3)	120	121	119.5	118	119.5
N(1)C(1)C(2)	110	110	112	114	112
N(2)C(3)C(2)	112	110	112	111	112
C(1)C(2)C(3)	112	115	112	114	109.5

[a] Averaged values in degrees.
[b] [106].
[c] [105].
[d] [103].

Table 3-10 Ring parameters for five-membered amino acid chelate rings from X-ray analysis

Complex	α	β	z_1, Å	z_2, Å	ω	Ref.
Ni(α-NH$_2$*i*but)$_2\cdot$4H$_2$O	83.0°	109.3°	−0.20	−0.33	+14.1°	102
	84.9°	109.1°	+0.04	−0.27	+28.1°	
Cu(gly)$_2\cdot$H$_2$O	85.4°	109.5°	+0.12	+0.06	+4.5°	46
	85.0°	109.3°	+0.08	−0.02	+8.3°	
Cu(L-(Et)$_2$ala)$_2$	84.1°	103.6°	+0.62	+0.22	+30.1°	101
Cu(L-isoleu)$_2$	84.2°	111.9°	+0.14	+0.04	+6.6°	134
	84.5°	111.2°	−0.08	+0.06	−10.5°	
Cu(L-ser)$_2$	82.3°	110.2°	+0.52	+0.36	+8.0°	62
	83.7°	109.6°	+0.52	+0.31	+13.4°	
Ni(L-ser)$_2\cdot$2H$_2$O	81.0°	108.1°	+0.55	+0.23	+23.5°	63

and found structures. The results are compared in Table 3-9. Various parameters for amino acid chelate rings determined by X-ray analysis are summarized in Table 3-10. These agree with the results from the a priori calculations (see above).

Although detailed information of the geometries of chelate rings in solution is presently lacking, it seems highly likely that at least one, and probably many more, of the computed minimum-energy conformations will still correspond to the lowest energy structures in solution.

The outstanding agreement between the predicted minimum-energy conformations and those found in the crystalline state shows that the main factors governing the conformational energies are understood. It is perhaps too much to hope for good agreement between the computed and experimentally found energy differences at this stage in the subject's development. It is hoped, however, that, as more experimental evidence becomes available, these calculations can be further refined.

EXPERIMENTAL CONFORMATIONAL ANALYSIS

Although various techniques have been used to obtain qualitative information concerning conformational preferences in metal complexes, little quantitative data have been determined. For some systems, methods are available that should enable conformational energy differences to be determined. These methods are reviewed here. The qualitative studies are discussed in the chapters dealing with the individual techniques.

Isolation Procedures

Consider the two diastereoisomers A and B. If they are brought to equilibrium and the two components separated without disturbing the equilibrium, the equilibrium constant can be determined from the ratio of the concentrations of the two species, and, consequently, the free-energy difference between the two can be computed according to Eq. (3-52):

$$\Delta G = -RT \ln K \tag{3-52}$$

This method has been applied to a number of cobalt(III) systems by Dwyer and his coworkers using activated charcoal for the equilibration.

Following the earlier work by Dwyer, Garvan, and Shulman [33], the equilibrium

$$\text{D-}[Co(\text{R-pn})_3]^{3+} \rightleftharpoons \text{L-}[Co(\text{R-pn})_3]^{3+}$$

has been studied by mixing cobalt(II) chloride (0.01 mole), R-propylenediamine (0.03 mole), and hydrochloric acid (0.01 mole) with activated charcoal in water and aerating the mixture for 3 hours [35]. Following the removal of the charcoal by filtration, the solution was chromatographed on paper with 1-butanol–H_2O–10*M* HCl (60:30:10) as eluent. The two bands that separated were collected and the relative concentrations determined spectrophotometrically. The equilibrium constant

$$K = \frac{[\text{L-}Co(\text{R-pn})_3{}^{3+}]}{[\text{D-}Co(\text{R-pn})_3{}^{3+}]} \tag{3-53}$$

was found to be equal to 14.6. This corresponds to an energy difference of -1.6 kcal mol^{-1}. Because the R-propylenediamine is assumed to exist almost exclusively in the λ conformation with the methyl group equatorial, this is equivalent to the energy difference between the L($\lambda\lambda\lambda$) and D($\lambda\lambda\lambda$) configurations. A similar study with the oxalato(R-propylenediamine)cobalt(III) system found {D(λ) − L(λ)} = 0.29 kcal mol^{-1} and {D($\lambda\lambda$) − L($\lambda\lambda$)} = 0.20 kcal mol^{-1} [34].

Dwyer and his group have also studied the relative energies of the four configurations D($\delta\delta\delta$) {= L($\lambda\lambda\lambda$)}, D($\delta\delta\lambda$) {= L($\lambda\lambda\delta$)}, D($\delta\lambda\lambda$) {= L($\lambda\delta\delta$)}, and D($\lambda\lambda\lambda$) {= L($\delta\delta\delta$)} for $[Co\text{-}en_3]^{3+}$ by equilibrating a mixture of cobalt, ethylenediamine, and R-propylenediamine, separating the various fractions chromatographically, and assuming that the λ conformation of the R-propylenediamine chelate ring with the methyl group equatorial has the same conformational energy as the λ conformation of the ethylenediamine chelate ring when placed in the same environment [35]. With an ethylenediamine–R-propylenediamine ratio of 1:2, the following percentage composition was obtained after correction for an incorrect identification of some of the fractions [88]:

	D	L
$[Co(\text{R-pn})_3]^{3+}$	*a* (2.1)	30.5
$[Co(\text{R-pn})_2en]^{3+}$	7.7	35.7
$[Co(\text{R-pn})en_2]^{3+}$	7.2	14.7 − *a* (12.6)
$[Co\text{-}en_3]^{3+}$	*b*	4.2 − *b*

The components D-$[Co(R\text{-}pn)_3]^{3+}$ and L-$[Co(R\text{-}pn)en_2]^{3+}$, and D- and L-$[Co\text{-}en_3]^{3+}$ were not separated. However, using the value $K = 14.6$ for Eq. (3-53) the concentrations of D-$[Co(R\text{-}pn)_3]^{3+}$ and L-$[Co(R\text{-}pn)en_2]^{3+}$ were estimated to be 2.1 and 12.6 respectively.

Dwyer and his coworkers assumed for the mixed complexes that, for the D configuration, the ethylenediamine chelate ring adopted the δ conformation and that, for the L, it adopted the λ conformation, whereas the R-propylenediamine chelate retained its λ conformation, irrespectively of the configuration. Based on this assumption, and with the corrected data, the following relative energies are obtained for the configurations above (the original values published are in parentheses): 0, 0.65 (0.45), 0.8 (1.2), and 1.6 kcal mol^{-1}, the last figure being taken directly from the study of D- and L-$[Co(R\text{-}pn)_3]^{3+}$ mentioned above.

Their basic assumption, however, that the conformation of the ethylenediamine is completely fixed by the configuration presupposes that there are big energy differences among the various configurations. In fact, the ethylenediamine can adopt both conformations, the relative concentrations of each being determined by the energy differences between the four kinds of configurations.

The relative energies can be determined from the ratios of the concentrations of the D and L isomers of $[Co(R\text{-}pn)en_2]^{3+}(D_1,L_1)$ and $[Co(R\text{-}pn)_2en]^{3+}$ (D_2,L_2) in the following way:

$$\frac{[L_1]}{[D_1]} = \frac{[L(\lambda\lambda\lambda)_1] + [L(\lambda\lambda\delta)_1] + [L(\lambda\delta\delta)_1]}{[D(\lambda\lambda\lambda)_1] + [D(\delta\lambda\lambda)_1] + [D(\delta\delta\lambda)_1]} \tag{3-54}$$

$$\frac{[L_2]}{[D_2]} = \frac{[L(\lambda\lambda\lambda)_2] + [L(\lambda\lambda\delta)_2]}{[D(\lambda\lambda\lambda)_2] + [D(\delta\lambda\lambda)_2]} \tag{3-55}$$

Because the R-propylenediamine ring is assumed to remain in the λ conformation, where the interactions of the equatorial methyl group are relatively small, the interaction system in each of the configurations in Eqs. (3-54) and (3-55) is equivalent to the interaction system in the corresponding configuration of $[Coen_3]^{3+}$. Consequently, the relative concentrations of the above configurations can be written in terms of the relative concentrations of the configurations of $[Coen_3]^{3+}$. Here, the ratios of the configurations of $[Coen_3]^{3+}$, which are independent of the statistical weighting for the mixed conformations, have been designated $1:a:b:c$ for the D($\delta\delta\delta$), D($\delta\delta\lambda$), D($\delta\lambda\lambda$), and D($\lambda\lambda\lambda$) configurations. One must take into account the fact, however, that there is a 2:1 statistical preference for configurations in which the two ethylenediamine rings of $[Co(R\text{-}pn)en_2]^{3+}$ have different chiralities. Equations (3-54) and (3-55) can then be rewritten as

$$\frac{[L_1]}{[D_1]} = \frac{1 + 2a + b}{c + 2b + a} \tag{3-56}$$

$$\frac{[\text{L}_2]}{[\text{D}_2]} = \frac{1 + a}{c + b} \tag{3-57}$$

Since c—$[\text{L}(\delta\delta\delta)]/[\text{L}(\lambda\lambda\lambda)]$—has been determined to be 1/14.6, and the relative concentrations of the complexes are known, a and b can be determined by solving Eqs. (3-56) and (3-57). Using the relative concentrations given above, the values obtained are

$$a = 1.53 \qquad b = 0.48$$

These values give rise to the following relative free energies:

$$\begin{aligned} \text{D}(\delta\delta\delta) &= \text{L}(\lambda\lambda\lambda) = 0 \\ \text{D}(\delta\delta\lambda) &= \text{L}(\lambda\lambda\delta) = -0.9 \\ \text{D}(\delta\lambda\lambda) &= \text{L}(\lambda\delta\delta) = -0.2 \end{aligned}$$

and from the previous study:

$$\text{D}(\lambda\lambda\lambda) = \text{L}(\delta\delta\delta) = 1.6 \text{ kcal mol}^{-1}$$

These values include the entropy contribution from the statistical preference for the mixed conformations.

For this system the calculated energies are very sensitive to small changes in the relative concentrations of the separated fractions and small experimental errors become magnified in terms of the low concentrations of some of the fractions. When this is taken into account along with the difficulty in identifying the isolated fractions, and the fact that the results above indicate the D($\delta\delta\lambda$) configuration of $[\text{Coen}_3]^{3+}$ is significantly more stable than D($\delta\delta\delta$) contrary to all other evidence, these results should be disregarded and the system perhaps reinvestigated.

Dwyer and his coworkers have redetermined these energy differences by a similar study of the tris complexes formed from a 2:3:3 mixture of cobalt, R-, and S-propylenediamine under equilibrium conditions [36]. Because the R- and S-propylenediamine will adopt λ and δ conformations respectively, with the methyl group equatorial, the interaction system can be equated with that for $[\text{Coen}_3]^{3+}$. Fractions were isolated containing the following racemates—the relative concentrations are also listed*:

$$\begin{aligned}
&\text{D-}[\text{Co(S-pn)}_3]^{3+} &&+ \text{L}[\text{Co(R-pn)}_3]^{3+} &&= \text{D}\ (\delta\delta\delta) + \text{L}(\lambda\lambda\lambda) = 40.5 \\
&\text{D-}[\text{Co(S-pn)}_2(\text{R-pn})]^{3+} &&+ \text{L-}[\text{Co(R-pn)}_2(\text{S-pn})]^{3+} &&= \text{D}\ (\delta\delta\lambda) + \text{L}(\lambda\lambda\delta) = 40.3 \\
&\text{D-}[\text{Co(S-pn)(R-pn)}_2]^{3+} &&+ \text{L-}[\text{Co(R-pn)(S-pn)}_2]^{3+} &&= \text{D}\ (\delta\lambda\lambda) + \text{L}(\lambda\delta\delta) = 16.5 \\
&\text{D-}[\text{Co(R-pn)}_3]^{3+} &&+ \text{L-}[\text{Co(S-pn)}_3]^{3+} &&= \text{D}\ (\lambda\lambda\lambda) + \text{L}(\delta\delta\delta) = 2.7
\end{aligned}$$

* The third and fourth fractions were not separated; the relative concentrations were estimated by assuming the concentration of the fourth fraction was 1/15 that of the first.

These concentrations are influenced by a 1:3:3:1 statistical ratio related to the exchange of R and S ligands, and the energy differences calculated from the relative concentrations are equal to the free energy differences between the various configurations of $[Coen_3]^{3+}$:

$$\begin{aligned} \text{D}(\delta\delta\delta) &= \text{L}(\lambda\lambda\lambda) = 0.0 \\ \text{D}(\delta\delta\lambda) &= \text{L}(\lambda\lambda\delta) = 0.0 \\ \text{D}(\delta\lambda\lambda) &= \text{L}(\lambda\delta\delta) = 0.5 \\ \text{D}(\lambda\lambda\lambda) &= \text{L}(\delta\delta\delta) = 1.6 \text{ kcal mol}^{-1} \end{aligned}$$

When the cobalt:R-propylenediamine:S-propylenediamine ratio was changed to 1:2:1, the relative concentrations of the fractions changed to 48.5, 36.0, 12.3 and 3.2. In this system, there is an additional statistical ratio 9:6:6:9 due to the 2:1 ratio of the ligands. When this is allowed for, the relative free energies are

$$\begin{aligned} \text{D}(\delta\delta\delta) &= \text{L}(\lambda\lambda\lambda) = 0.0 \\ \text{D}(\delta\delta\lambda) &= \text{L}(\lambda\lambda\delta) = -0.05 \\ \text{D}(\delta\lambda\lambda) &= \text{L}(\lambda\delta\delta) = 0.55 \\ \text{D}(\lambda\lambda\lambda) &= \text{L}(\delta\delta\delta) = 1.6 \quad \text{kcal mol}^{-1} \end{aligned}$$

The good agreement between the two sets of data lends some confidence to the analysis.

This experimental method of conformational analysis could be advantageously applied to other systems.

Nuclear Magnetic Resonance

Nmr can be used in at least two ways to determine conformational energy differences.

1. For two diastereoisomers at equilibrium, the ratio of the concentrations of the two species and, thus, the equilibrium constant and the free-energy difference can be determined directly from the ratio of the areas of corresponding nmr signals for the two compounds. This method requires that the two signals are separated from each other and, preferably, free of interference from other resonances. Examples of the method's application are discussed in Chap. 6 (e.g., the D-$[Coen_2(\text{L-val})]^{2+}$–D-$[Coen_2(\text{D-val})]^{2+}$ system). The band-area analysis technique is suitable for determining the energy difference between axial and equatorial orientations of substituents in chelate rings. Consider the propylenediamine chelate ring. The methyl group proton resonance for the equatorial orientation is expected to lie at a different field from that for the axial. Results with a related ligand suggest that the axial is at the higher field [138]. If the rate of ring inversion is slow compared with that of the nuclear transition being examined, the methyl

resonances for the two conformers appear as two signals, the relative peak areas of which give the ratio of the conformers. If the separation of the axial and equatorial resonances is small and if the rate of ring inversion is fast at room temperature, the two signals are merged into a weighted-average band. The equatorial-axial interconversion can be slowed down by lowering the temperature, and when the rate is less than the separation between the equatorial and axial resonances the two signals again separate. Thus the relative populations of the equatorial and axial conformations can be determined at the temperature in question from the peak areas. Unfortunately, the rate of chelate-ring inversion is fast, and a sufficiently low temperature has not yet been achieved to observe the spectra of the two conformers.

2. The equilibrium constant and thus the free-energy difference for an equatorial-axial equilibrium can be determined directly from the position of the average signal δ at room temperature according to Eq. (3-58):

$$K = \frac{(\delta_a - \delta)}{(\delta - \delta_e)} \tag{3-58}$$

This requires that the chemical shifts are known for the particular nucleus in the axial δ_a and equatorial δ_e conformations. This would again necessitate low temperature measurements for the chelate-ring systems.* Coupling

* Note added in proof: Reilley and his co-workers have recently reported conformational energy differences for a number of nickel(II) diamine and amino acid complexes determined by studying the effective chemical shift difference between axial and equatorial proton resonances (F. F.-L. Ho and C. N. Reilley, *Anal. Chem.*, **41**: 1835 (1969); F. F.-L. Ho and C. N. Reilley, *Anal.Chem.*, **42**: 600 (1970); F. F.-L. Ho, L. E. Erickson, S. R. Watkins, and C. N. Reilley, *Inorg. Chem.*, **9**: 1139 (1970)). If the chelate ring is inverting rapidly between energetically-nonequivalent conformations of opposite chirality, the effective chemical shift difference is a function of the populations of the conformations. The equilibrium constant and, hence the free energy difference, can be determined from the expression

$$K = \frac{1 - x}{1 + x}$$

where $x = \delta_{ae}{}^{eff}/\delta_{ae}{}^{int}$; $\delta_{ae}{}^{eff}$ and $\delta_{ae}{}^{int}$ are the effective and intrinsic chemical shift differences between the axial and equatorial protons—the intrinsic chemical shift difference is the difference for a fixed conformation. $\delta_{ae}{}^{int}$ is of the order of 100–200 ppm for the systems studied because the contact interaction in these paramagnetic nickel(II) complexes is large and is strongly dependent on the dihedral angle between the Ni—N—C and N—C—H planes. The enthalpy and entropy differences were also estimated by studying the temperature dependence of the free energy. Unfortunately, $\delta_{ae}{}^{int}$ is not known for the systems studied, and had to be determined indirectly. The method used for this was rather insensitive, and therefore accurate energetic data must await the more direct determination of $\delta_{ae}{}^{int}$. The published energy differences (in kcal mol^{-1}), however, between axial and equatorial conformations at 305°K for $[Ni(OH_2)_4L]^{n+}$

constants can also be used in a similar way for, say, a proton or fluorine atom that changes its orientation from axial to equatorial on ring inversion (see, for example, Ref. 128).

Infrared Spectroscopy

If two conformations that are in equilibrium in solution have a unique IR band characteristic of the conformation, the equilibrium constant can be determined from the measured extinctions A_1 and A_2 by

$$K = \frac{A_1 \varepsilon_2}{A_2 \varepsilon_1} \tag{3-59}$$

where ε_1 and ε_2 are the (integrated) molar extinctions. Unfortunately, the values for ε_1 and ε_2 are not usually known. Assumptions, such as $\varepsilon_1 = \varepsilon_2$ are unsatisfactory because there is little supporting evidence. The difference in enthalpy, however, can be determined by carrying out the measurements at two temperatures and using the equation [95]

$$\Delta H = \frac{RT_1T_2}{T_2 - T_1}\left[\ln\left(\frac{A_1}{A_2}\right)_{T_2} - \ln\left(\frac{A_1}{A_2}\right)_{T_1}\right] \tag{3-60}$$

assuming that ε_1 and ε_2 do not vary with temperature. Because ΔH is usually small and the band intensity ratio changes rather slowly with temperature, this method requires large temperature intervals and accurate measurements of band intensity. The application of the IR method to metal complexes is further limited by the complexes' low solubility in suitable solvents. Nevertheless, it could prove of value for some systems.

Thermodynamic Studies

Consider the equilibrium

$$M(OH_2)_6{}^{2+} + L \overset{K_1}{\rightleftharpoons} M(OH_2)_4L^{2+} + 2H_2O$$

where L represents the closely related set of ligands *dl*-2,3-diaminobutane, meso-2,3-diaminobutane, and 2-methyl-1,2-diaminopropane (i-bn). These ligands have very similar acid dissociation constants [13] (see Table 3-11) and, therefore, might be expected to form complexes with similar stabilities. Any differences in the measured constants should mirror the differences in

---

are: R-pn, $\Delta G = 1.1$; N,N′-$(CH_3)_2$en, $\Delta G = 0.86$, $\Delta H = 0.71$, $-T\Delta S = 0.15$; sar, $\Delta G = 0.5$, $\Delta H = 0.4$, $-T\Delta S = 0.1$. Relative energies of the various configurations of $Ni(en)_3{}^{2+}$ were also published: D(δδδ), 0; D(δδλ) 0.3; D(δλλ), 0.6; and D(λλλ), 0.9 kcal mol^{-1}. However, the derivation of these values was based on the unfounded assumption that the energies increase with identical increments.

Table 3-11 Stability constants for copper(II) and nickel(II) complexes with some C-substituted ethylenediamine ligands[a]

			Cu^{2+}		Ni^{2+}		
	pK_a[b]		log K_1	log β_2	log K_1	log β_2	log β_3
d-bn	6.91	10.00	11.39	21.21	7.71	14.19	18.50
meso-bn	6.92	9.97	10.72	20.06	7.04	12.74	15.63
i-bn	6.79	10.00	10.53	19.58	6.77	12.17	14.42

[a] At 25° from [14].
[b] At 25° from [13].

the conformational energies of the chelate rings. Racemic 2,3-diaminobutane is able to form a chelate with both methyl groups equatorial, but, for both the other ligands, one methyl group must be axial. If the thermodynamic properties of the ligands in solution (e.g., free energy of hydration) are identical, the ratio of the K_1 values for the racemic ligand and each of the other two ligands should be approximately equal to the equilibrium constant for an axial methyl $\rightleftharpoons$ equatorial methyl system, from which the free-energy difference can be calculated. Using the data given in Table 3-11 for the formation of the 1:1 complexes at 25°, a value of about 1 kcal was estimated for this energy for both copper(II) and nickel(II) complexes. In a similar way, the conformational-energy difference between $[Ni(meso\text{-}bn)_3]^{2+}$ and $[Ni(dl\text{-}bn)_3]^{2+}$ was calculated to be about 4 kcal mol^{-1}, and between $[Ni(i\text{-}bn)_3]^{2+}$ and $[Ni(dl\text{-}bn)_3]^{2+}$ about 5.7 kcal mol^{-1}.

Similar studies would be of value, but exceptional care must be taken in determining the constants. Values of ΔH and ΔS for the equilibria of interest could also be obtained by measuring the stability constants at a number of temperatures, but it must be emphasized that this requires work of high precision. Alternatively, ΔH values can be determined directly from calorimetric studies.

A measure of the conformational enthalpy difference between two conformers has been obtained for some organic systems from the difference in the heat of combustion of the two structures determined calorimetrically [39]. This method could usefully be applied to some chelate-ring systems.

It is important for the experimental conformational analysis studies to be designed to determine not only the free-energy difference between conformations but also the difference in enthalpy and entropy and the influence of the solvent. Although little has been accomplished to date in this field for chelate-ring systems, it is anticipated that rapid progress will be made in the near future.

GENERAL READING

1. Eliel, E. L., N. L. Allinger, S. J. Angyal, and G. A. Morrison, "Conformational Analysis," Wiley-Interscience, New York, 1965.
2. Newman, M. S., "Steric Effects in Organic Chemistry," Wiley, New York, 1956.
3. Sargeson, A. M., Conformations of Coordinated Chelates, in R. L. Carlin (ed.), "Transition Metal Chemistry," vol. 3, pp. 303–343, Marcel Dekker, New York, 1966.

REFERENCES

4. Abe, A., R. L. Jernigan, and P. J. Flory, *J. Am. Chem. Soc.*, **88**:631 (1966).
5. Allinger, N. L., M. A. Miller, F. A. Van Catledge, and J. A. Hirsch, *J. Am. Chem. Soc.*, **89**:4345 (1967).
6. Allinger, N. L., and W. Szkrybalo, *J. Org. Chem.*, **27**:4601 (1962).
7. Amdur, I., and A. L. Harkness, *J. Chem. Phys.*, **22**:664 (1954).
8. Amdur, I., and E. A. Mason, *J. Chem. Phys.*, **23**:415 (1955).
9. Amdur, I., and E. A. Mason, *J. Chem. Phys.*, **25**:630 (1956).
10. Aston, J. G., S. Isserow, G. J. Szasz, and R. M. Kennedy, *J. Chem. Phys.*, **12**:336 (1944).
11. Bartell, L. S., *J. Chem. Phys.*, **32**:827 (1960).
12. Barton, D. H. R., *J. Chem. Soc.*, **1948**:340.
13. Basolo, F., R. K. Murmann, and Y. T. Chen., *J. Am. Chem. Soc.*, **75**:1478 (1953).
14. Basolo, F., Y. T. Chen, and R. K. Murmann, *J. Am. Chem. Soc.*, **76**:956 (1954).
15. Blade, E., and G. E. Kimball, *J. Chem. Phys.*, **18**:626, 630 (1950).
16. Boer, J. de, *Physica*, **9**:363 (1942).
17. Bondi, A., *J. Phys. Chem.*, **68**:441 (1964).
18. Born, M., and J. E. Mayer, *Z. Physik*, **75**:1 (1932).
19. Buckingham, D. A., L. G. Marzilli, and A. M. Sargeson, *J. Am. Chem. Soc.*, **89**:825 (1967).
20. Buckingham, D. A., L. G. Marzilli, and A. M. Sargeson, *J. Am. Chem. Soc.*, **89**:5133 (1967).
21. Buckingham, D. A., L. G. Marzilli, and A. M. Sargeson, *Inorg. Chem.*, **7**:915 (1968).
22. Buckingham, R. A., *Trans. Faraday Soc.*, **54**:453 (1958).
23. Clementi, D., and D. R. Davis, *J. Chem. Phys.*, **45**:2593 (1966).
24. Clinton, W. L., *J. Chem. Phys.*, **33**:632 (1960).
25. Condrate, R. A., and K. Nakamoto, *J. Chem. Phys.*, **42**:2590 (1965).
26. Corey, E. J., and J. C. Bailar, *J. Am. Chem. Soc.*, **81**:2620 (1959).
27. Coulson, C. A., "Valence," 2d ed., p. 367, Oxford University Press, London, 1961.
28. Debye, P., *Physik. Z.*, **21**:178 (1920).
29. Dijkstra, A., *Acta Cryst.*, **20**:588 (1966).
30. Dostrovsky, I., E. D. Hughes, and C. K. Ingold, *J. Chem. Soc.*, **1946**:173.
31. Dows, D. A., *J. Chem. Phys.*, **35**:282 (1961).
32. Drude, P. K. L., "Lehrbuch der Optik," Hirzel, Leipzig, 1900; Eng. trans., Dover, New York, 1959.
33. Dwyer, F. P., F. L. Garvan, and A. Shulman, *J. Am. Chem. Soc.*, **81**:290 (1959).

34. Dwyer, F. P., T. E. MacDermott, and A. M. Sargeson, *J. Am. Chem. Soc.*, **85**:661 (1963).
35. Dwyer, F. P., T. E. MacDermott, and A. M. Sargeson, *J. Am. Chem. Soc.*, **85**:2913 (1963).
36. Dwyer, F. P., A. M. Sargeson, and L. B. James, *J. Am. Chem. Soc.*, **86**:590 (1964).
37. Eisenschitz, R., and F. London, *Z. Physik*, **60**:491 (1930).
38. Eliel, E. L., *Angew. Chem. Internat. Edn.*, **4**:761 (1965).
39. Eliel, E. L., N. L. Allinger, S. J. Angyal, and G. A. Morrison, "Conformational Analysis," pp. 137–139, Wiley-Interscience, New York, 1965.
40. Epstein, P. S., *Phys. Rev.*, **28**:695 (1928).
41. Eyring, H., *J. Am. Chem. Soc.*, **54**:3191 (1932).
42. Eyring, H., D. M. Grant, and H. Hecht, *J. Chem. Educ.*, **39**:466 (1962).
43. Eyring, H., G. H. Stewart, and R. P. Smith, *Proc. Natl. Acad. Sci. U.S.*, **44**:259 (1958).
44. Fink, W. H., and L. C. Allen, *J. Chem. Phys.*, **46**:2261 (1967).
45. Freeman, H. C., *Adv. Protein Chem.*, **22**:257 (1967).
46. Freeman, H. C., M. R. Snow, I. Nitta, and K. Tomita, *Acta Cryst.*, **17**:1463 (1964).
47. French, F. A., and R. S. Rasmussen, *J. Chem. Phys.*, **14**:389 (1946).
48. Gollogly, J. R., Ph.D. thesis, Univ. of Queensland, 1970.
49. Gollogly, J. R., and C. J. Hawkins, *Aust. J. Chem.*, **20**:2395 (1967).
50. Gollogly, J. R., and C. J. Hawkins, *Inorg. Chem.*, **8**:1168 (1969).
51. Gollogly, J. R., and C. J. Hawkins, *Inorg. Chem.*, **9**:576 (1970).
52. Gollogly, J. R., and C. J. Hawkins, to be published.
53. Gorin, E., J. Walter, and H. Eyring, *J. Am. Chem. Soc.*, **61**:1876 (1939).
54. Halford, J. O., *J. Chem. Phys.*, **15**:645 (1947).
55. Halford, J. O., *J. Chem. Phys.*, **16**:410, 560 (1948).
56. Harris, G. M., and F. E. Harris, *J. Chem. Phys.*, **31**:1450 (1959).
57. Hassé, H. R., *Proc. Camb. Phil. Soc.*, **27**:66 (1931).
58. Hecht, H., D. M. Grant, and H. Eyring, *Mol. Phys.*, **3**:577 (1960).
59. Heitler, H., and F. London, *Z. Physik.*, **44**:455 (1927).
60. Hellmann, H., *Acta Physicochim. URSS*, **2**:273 (1935).
61. Hellmann, H., *Trans. Faraday Soc.*, **33**:40 (1937).
62. Helm, D. van der, and W. A. Franks, *Acta Cryst.*, **B25**:451 (1969).
63. Helm, D. van der, and M. B. Hossain, *Acta Cryst.*, **B25**:457 (1969).
64. Hendrickson, J. B., *J. Am. Chem. Soc.*, **83**:4537 (1961).
65. Hill, T. L., *J. Chem. Phys.*, **16**:399 (1948).
66. Hirschfelder, J. O., C. F. Curtiss, and R. B. Bird, "Molecular Theory of Gases and Liquids," Wiley, New York, 1954.
67. Hirschfelder, J. O., and J. W. Linnett, *J. Chem. Phys.*, **18**:130 (1950).
68. Hornig, J., and J. O. Hirschfelder, *J. Chem. Phys.*, **20**:812 (1952).
69. Howlett, K. E., *J. Chem. Soc.*, **1957**:4353.
70. Kammerling Onnes, H., *Proc. Sect. Sci. Amsterdam*, **4**:125 (1902).
71. Kanters, J. A., J. Kroon, A. F. Peerdeman, and J. C. Schoone, *Tetrahedron*, **23**:4027 (1967).
72. Karplus, M., *J. Chem. Phys.*, **33**:316 (1960).
73. Karplus, M., and R. G. Parr, *J. Chem. Phys.*, **38**:1547 (1963).
74. Keesom, W. H., *Physik. Z.*, **22**:129 (1921).
75. Kemp, J. D., and K. S. Pitzer, *J. Am. Chem. Soc.*, **59**:276 (1937).
76. Ketelaar, J., "Chemical Constitution," Elsevier, Amsterdam, 1953.

77. Kirkwood, J. G., *Physik. Z.*, **33**:57 (1932).
78. Kistakowsky, G. B., J. R. Lacher, and W. W. Ransom, *J. Chem. Phys.*, **6**:900 (1938).
79. Kitaigorodsky, A. I., *Tetrahedron*, **9**:183 (1960).
80. Lassettre, E. N., and L. B. Dean, *J. Chem. Phys.*, **16**:151, 553 (1948).
81. Lassettre, E. N., and L. B. Dean, *J. Chem. Phys.*, **17**:317 (1949).
82. Lennard-Jones, J. E., in Fowler, R. H., "Statistical Mechanics," pp. 292–337, Camb. Univ. Press, 1929.
83. Lennard-Jones, J. E., *Proc. Roy. Soc. London, Ser. A*, **129**:598 (1930).
84. Lin, C. C., and J. D. Swalen, *Rev. Mod. Phys.*, **33**:632 (1960).
85. London, F., *Z. Physik. Chem.*, **B11**:222 (1930).
86. London, F., *Trans. Faraday Soc.*, **33**:8 (1937).
87. McCullough, R. L., and P. E. McMahon, *Trans. Faraday Soc.*, **60**:2089 (1964).
88. MacDermott, T. E., *Chem. Commun.*, **1968**:223.
89. Mack, E., *J. Am. Chem. Soc.*, **54**:2141 (1932).
90. Magnasco, V., *Nuovo Cimento*, **24**:425 (1962).
91. Margenau, H., *Phys. Rev.*, **38**:747 (1931).
92. Mason, E. A., and M. M. Kreevoy, *J. Am. Chem. Soc.*, **77**:5808 (1955).
93. Mason, E. A., and W. E. Rice, *J. Chem. Phys.*, **22**:843 (1954).
94. Mathieu, J.-P., *Ann. Phys.*, **19**:335 (1944).
95. Morino, Y., S. Mizushima, K. Kuratani, and M. Katayama, *J. Chem. Phys.*, **18**:754 (1950).
96. Morse, P. M., *Phys. Rev.*, **34**:57 (1929).
97. Müller, A., *Proc. Roy. Soc. London, Ser. A*, **154**:624 (1936).
98. Mulliken, R. S., *J. Am. Chem. Soc.*, **77**:887 (1955).
99. Muto, A., F. Marumo, and Y. Saito, *Inorg. Nucl. Chem. Letters*, **5**:85 (1969).
100. Nakagawa, I., personal communication, 1969.
101. Nash, C. P., and W. P. Schaefer, *J. Am. Chem. Soc.*, **91**:1319 (1969).
102. Noguchi, T., *Bull. Chem. Soc. Japan*, **35**:99 (1962).
103. Nomura, T., F. Marumo, and Y. Saito, *Bull. Chem. Soc. Japan*, **42**:1016 (1969).
104. Oosterhoff, L. J., *Disc. Faraday Soc.*, **10**:79, 87 (1951).
105. Pajunen, A., *Suomen Kemi.*, **B41**:232 (1968).
106. Pajunen, A., *Suomen Kemi.*, **B42**:15 (1969).
107. Pauling, L., *Proc. Natl. Sci. U.S.*, **44**:211 (1958).
108. Pauling, L., "The Nature of the Chemical Bond," 3d ed., Cornell Univ. Press, Ithaca, N.Y., 1960.
109. Pauling, L., and J. Y. Beach, *Phys. Rev.*, **47**:686 (1935).
110. Pauncz, R., and D. Ginsburg, *Tetrahedron*, **9**:40 (1960).
111. Pederson, L., and K. Morokuma, *J. Chem. Phys.*, **46**:3941 (1967).
112. Penney, W. G., *Proc. Roy. Soc. London, Ser. A*, **144**:166 (1934).
113. Pitzer, K. S., *Disc. Faraday Soc.*, **10**:66 (1951).
114. Pitzer, K. S., *J. Am. Chem. Soc.*, **78**:4565 (1956).
115. Pitzer, K. S., and E. Catalano, *J. Am. Chem. Soc.*, **78**:4844 (1956).
116. Pitzer, R. M., *J. Chem. Phys.*, **47**:965 (1967).
117. Pitzer, R. M., and W. N. Lipscomb, *J. Chem. Phys.*, **39**:1995 (1963).
118. Pritchard, H. O., and F. H. Sumner, *J. Chem. Soc.*, **1955**:1041.
119. Rosenblatt, R., and A. Schleede, *Liebigs Ann.*, **505**:54 (1933).
120. Scott, R. A., and H. A. Scheraga, *J. Chem. Phys.*, **42**:2209 (1965).
121. Slater, J. C., *Phys. Rev.*, **32**:349 (1928).
122. Slater, J. C., and J. G. Kirkwood, *Phys. Rev.*, **37**:682 (1931).

123. Smith, R. P., T. Ree, J. Magee, and H. Eyring, *J. Am. Chem. Soc.*, **73**:2263 (1951).
124. Smith, W. V., and R. Howard, *Phys. Rev.*, **79**:132 (1950).
125. Sovers, O. J., and M. Karplus, *J. Chem. Phys.*, **44**:3033 (1966).
126. Sugiura, Y., *Z. Physik*, **45**:484 (1927).
127. Theilacker, W., *Z. Anorg. Allgem. Chem.*, **234**:161 (1937).
128. Thomas, W. A., *Ann. Rev. NMR Spectroscopy*, **1**:43 (1968).
129. Urey, H. C., and C. A. Bradley, *Phys. Rev.*, **38**:1969 (1931).
130. Volkenstein, M. A., "Configurational Statistics of Polymer Chains," Interscience, New York, 1963.
131. Waals, J. D. van der, doctoral diss., Leiden, 1873.
132. Wang, S. C., *Physik. Z.*, **28**:663 (1927).
133. Weakliem, H. A., and J. L. Hoard, *J. Am. Chem. Soc.*, **81**:549 (1959).
134. Weeks, C. M., A. Cooper, and D. A. Norton, *Acta Cryst.*, **B25**:443 (1969).
135. Wiberg, K. B., *J. Am. Chem. Soc.*, **87**:1070 (1965).
136. Wilson, E. B., *Proc. Natl. Acad. Sci.*, **43**:816 (1957).
137. Wilson, E. B., *Adv. Chem. Phys.*, **2**:367 (1959).
138. Woldbye, F., personal communication to A. M. Sargeson, as quoted in Ref. 3, 1964.
139. Wyatt, R. E., and R. G. Parr, *J. Chem. Phys.*, **41**:3262 (1964); **43**:S217 (1965); **44**:1529 (1966).

4

Absolute Configurations By X-Ray Analysis

NORMAL DIFFRACTION AND FRIEDEL'S LAW

The three-dimensional structure of compounds in crystals can be determined in accurate detail by diffraction methods. For normal diffraction with X rays, characteristic radiation is used whose frequency is distant from any absorption edge of the atoms in the crystal—$\nu \gg \nu_K$ for atoms of low atomic number, such as carbon, or $<\nu_K$ for atoms of high atomic number, for example. Choice of radiation in either range avoids or minimizes the generation of excessive fluorescent radiation by the crystal specimen.

The experimental data obtained from an X-ray structure study, after application of the various correction factors appropriate to the particular experiment, are in the form of reduced intensities $I(\mathbf{hkl})$ $(= F^2(\mathbf{hkl}))$ of waves diffracted from the array of crystal planes **(hkl)**.† For *normal diffraction* reflection from a plane **(hkl)** in a polar crystal and that for the counterreflection $\overline{\mathbf{hkl}}$ are exactly equivalent, hence indistinguishable; for example, consider the simple case of zincblende. As shown in Fig. 4-1*a*, the zinc and sulfur atoms lie in periodically alternating planes with nonequal spacing between adjacent layers (specifically 1:3 here). For the reflection **hkl**, the wave scattered by the plane of zinc atoms leads that scattered by the plane of sulfur atoms by a quarter of the periodicity, that is, $\pi/2$ or 90°, whereas for the counterreflection $\overline{\mathbf{hkl}}$, the phase change is $3\pi/2$ or 270° (= −90°). Figure 4-1*b* indicates how the zinc and sulfur contributions combine to give the total wave amplitude $|F(\mathbf{hkl})|$ and $|F(\overline{\mathbf{hkl}})|$, respectively. In this situation,

† For terminology on reflections, planes, and so on, see International Tables for Crystallography, vol. 1.

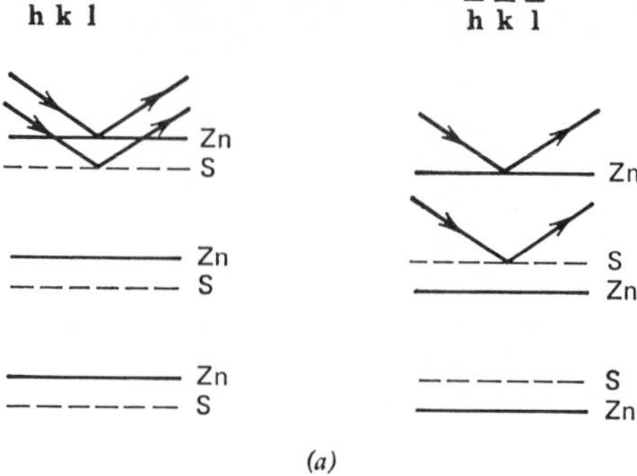

(a)

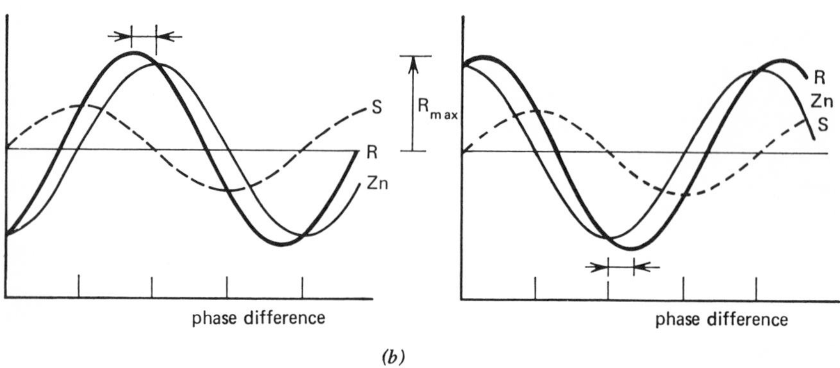

(b)

Fig. 4-1 (a) Spatial distribution of planes of zinc and sulfur atoms in zincblende showing **hkl** and $\overline{\mathbf{hkl}}$ reflections; (b) waves scattered by the zinc and sulfur atoms showing the magnitude and mutual phase relationship of the individual waves and the resultant R when Friedel's law is obeyed.

the amplitudes are equal and the front side of the ZnS crystal cannot be distinguished from the back side. In a simplified one-dimensional form this example is illustrative of the essential lack of capability of normal diffraction, in general, to distinguish between optical isomers.

For subsequent discussion it is advisable to outline the relationship between the intensity and the atomic structure of the crystal. The intensity corresponding to the wave diffracted by the plane (H) ($\equiv$ **hkl**) is related by Eq. (4-1) to its structure factor F_H:

$$I_H = kF_H \cdot F_H^* \tag{4-1}$$

where k is a proportionality constant applicable to all reflections in the structure and F^* is the complex conjugate of F.

This parameter can be expressed, to a good approximation, as a function of the atomic scattering factors f_j† of the individual atoms j and of their relative proportional position across the plane—Eq. (4-2):

$$F_H = \sum_j f_j(H)\cdot \exp\,[2\pi i\cdot H\cdot \mathbf{r}_j] \tag{4-2}$$

where $\mathbf{r}_j$ is the vector representing the position of the jth atom whose coordinates in the unit cell (of dimensions **a**, **b**, **c**) are $\mathbf{x}_j/\mathbf{a}$, $\mathbf{y}_j/\mathbf{b}$, $\mathbf{z}_j/\mathbf{c}$.

For normal diffraction atomic scattering factors are real, that is, $f = f^*$, and

$$\begin{aligned} F_H^* &= \sum_j f_j^*(H)\cdot \exp\,(-2\pi i\cdot H\cdot \mathbf{r}_j) \\ &= \sum_j f_j(H)\cdot \exp\,(2\pi i\cdot \overline{H}\cdot \mathbf{r}_j) = F_{\overline{H}} \end{aligned} \tag{4-3}$$

and

$$F_{\overline{H}}^* = F_H$$

Therefore $F_H\cdot F_H^* = F_{\overline{H}}\cdot F_{\overline{H}}^*$, and from Eq. (4-1), the measured intensities of the reflections **hkl** and $\overline{\textbf{hkl}}$ are exactly equal. This corresponds to *Friedel's law.*

ANOMALOUS DISPERSION AND ABSOLUTE CONFIGURATION

If the wavelength of the X rays used for the structure determination lies below that of the absorption edge of an atom in the structure, that is, $\lambda < \lambda_{abs}$ or $\nu > \nu_{abs}$, then the corresponding absorption level of that atom is excited. The scattering contribution due to this interaction f'' is shifted 90° in phase relative to the main scattering of the atom f', so that the atomic scattering factor for the atom j under such circumstances is given by

$$f_j = f_j' + if_j'' = (f_0)_j + \Delta f_j' + if_j'' \tag{4-4}$$

where $(f_0)_j$ is the scattering factor at infinite wavelength, $\Delta f_j'$ is the real part of the change of this factor at λ, and f_j'' is the imaginary part that obtains only on the short wavelength side of the absorption edge, $\lambda < \lambda_{abs}$ (see Ref. 1).

† At small diffracting angles the scattering factor for an atom is approximately proportional to its atomic number.

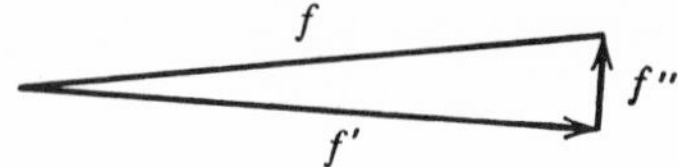

Fig. 4-2 Atomic scattering factor for anomalous scattering.

In the wavelength region, the atomic scattering factor is therefore complex (see Fig. 4-2), and, as a consequence, the equality derived in the previous section ($F_H \cdot F_H^* = F_{\bar{H}} \cdot F_{\bar{H}}^*$) does not hold for noncentric structures.

Thus in the case of ZnS referred to earlier, if the incident radiation has $\lambda < \lambda_{abs}$ (for K edge of Zn)—Au$L\alpha_1$, for example—the zinc K levels are excited. Not only does the normal scattering for Zn occur, but the anomalous scatter due to the K electrons produces a component leading by 90°. This introduces a phase change that, unlike the phase relationship determined by the relative atomic positions, is invariant with the direction of the incident radiation. Thus for ZnS Fig. 4-3*a* depicts in vectorial form the phase relationship for f'_{Zn} and f''_{Zn} and the resultant structure amplitudes F_H and $F_{\bar{H}}$

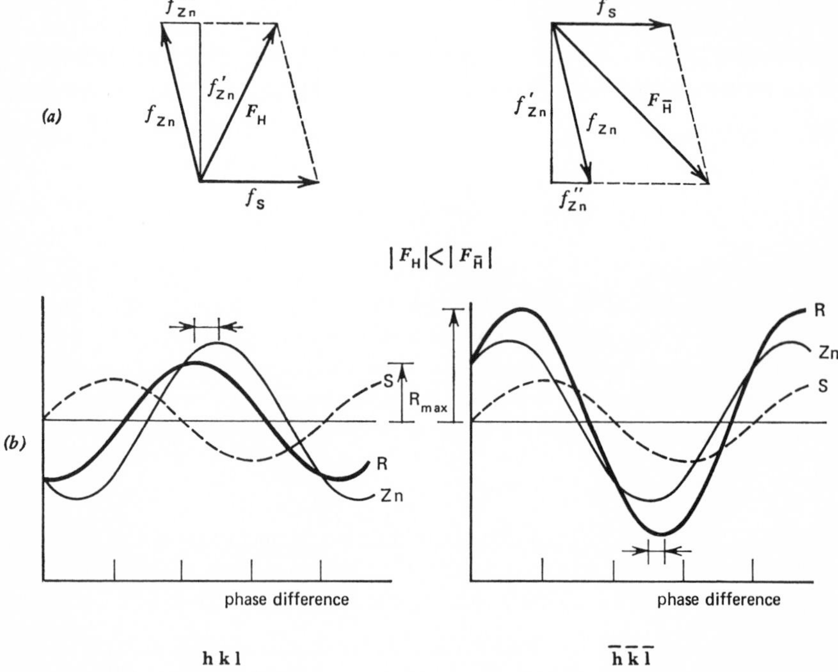

Fig. 4-3 Anomalous scattering from (**hkl**) and ($\overline{\mathbf{hkl}}$) planes of zincblende.

for the reflections detailed in Fig. 4-1. Figure 4-3*b* illustrates in an alternative manner the combination of scattered waves from S and Zn for the two reflections.

In structure factor formulation

$$f_{\mathrm{Zn}}(H) = f'_{\mathrm{Zn}} + if''_{\mathrm{Zn}} \tag{4-5}$$

$$F_H = (f'_{\mathrm{Zn}} + if''_{\mathrm{Zn}}) \exp (2\pi i \cdot H \cdot r_{\mathrm{Zn}}) + f_{\mathrm{S}} \cdot \exp (2\pi i \cdot H \cdot r_{\mathrm{S}}) \tag{4-6}$$

$$r_{\mathrm{S}} = 0 \qquad \text{and} \qquad r_{\mathrm{Zn}} = \tfrac{1}{4} \tag{4-7}$$

$$F_H = (f'_{\mathrm{Zn}} + if''_{\mathrm{Zn}})\left(i \sin \frac{H \cdot \pi}{2}\right) + f_{\mathrm{S}} \tag{4-8}$$

For

$$H = 2n + 1$$

that is,

$$H = 1$$

$$F_H = (f'_{\mathrm{Zn}} + if''_{\mathrm{Zn}}) \cdot i + f_{\mathrm{S}}$$

$$= if'_{\mathrm{Zn}} - f''_{\mathrm{Zn}} + f_{\mathrm{S}}$$

$$= (f_{\mathrm{S}} - f''_{\mathrm{Zn}}) + if'_{\mathrm{Zn}} \qquad \text{(see Fig. 4-3}a\text{)} \tag{4-9}$$

$$F^*_H = (f_{\mathrm{S}} - f''_{\mathrm{Zn}}) - if'_{\mathrm{Zn}} \tag{4-10}$$

$$F_{\bar{H}} = (f'_{\mathrm{Zn}} + if''_{\mathrm{Zn}}) \cdot -i + f_{\mathrm{S}}$$

$$= (f_{\mathrm{S}} + f''_{\mathrm{Zn}}) - if'_{\mathrm{Zn}} \qquad \text{(see Fig. 4-3}a\text{)} \tag{4-11}$$

$$F^*_{\bar{H}} = (f_{\mathrm{S}} + f''_{\mathrm{Zn}}) + if'_{\mathrm{Zn}} \tag{4-12}$$

$$F_H \cdot F^*_H = (f_{\mathrm{S}} - f''_{\mathrm{Zn}})^2 + (f'_{\mathrm{Zn}})^2 \tag{4-13}$$

$$F_{\bar{H}} \cdot F^*_{\bar{H}} = (f_{\mathrm{S}} + f''_{\mathrm{Zn}})^2 + (f'_{\mathrm{Zn}})^2 \tag{4-14}$$

that is,

$$F_H \cdot F^*_H \neq F_{\bar{H}} \cdot F^*_{\bar{H}} \dagger$$

Intensities of reflections $H = 1$ and $H = -1$ are thus no longer identical. Herein lies the possibility of establishing, by a direct experimental procedure,

† It should be noted that Fig. 4-3*a* refers to $H = 1$. For the second order $H = 2$, the values F_H and $F_{\bar{H}}$ are equivalent. The inequality condition is therefore not observed with all reflections. Its observation is conditional on certain relationships holding and in any specific structure it may be necessary to be selective to ensure clear-cut results.

the absolute back and front of a simple system such as zincblende or related compounds. In fact, Coster, Knol, and Prins provided a beautiful experimental demonstration of this effect as early as 1930 [19].

It is an interesting commentary on the development of the subject that the wide-ranging implications of this simple experiment were not grasped until about 1949, when Bijvoet with his coworkers [10,59] realized the possibilities of the technique and, with carefully selected radiation, established the first absolute configuration, that of a tartaric acid derivative.

BIJVOET'S METHOD: METHOD A

Bijvoet and his coworkers determined the first absolute configuration by X-ray analysis when they studied sodium rubidium *d*-tartrate using zirconium K_α radiation to excite the rubidium atom [59].

Their method requires a preliminary structure determination by normal X-ray analysis. A suitable target material is then chosen to produce X rays with a wavelength a little shorter than the absorption edge of a heavy atom in the crystal—M in Fig. 4-4, for example. When the crystal is irradiated with these X rays, the waves scattered by the M atoms are advanced relative to the waves from the other atoms. This leads to reinforcement for $\overline{\mathbf{hkl}}$ and interference for **hkl**, enabling the two enantiomers to be distinguished. In practice a number of sets of $|F_{\mathbf{hkl}}|^2$ and $|F_{\overline{\mathbf{hkl}}}|^2$ values for different planes are obtained and compared with the predicted behavior for the two chiralities. It should be noted that after the normal analysis has been completed, the determination of the absolute configuration is a relatively minor step. It is important in applying this method to ensure correct indexing in relation to a specified chiral set of axes. The appropriate procedure has been detailed by Peerdeman and Bijvoet [58].

As devised originally, the technique was likely to be somewhat restricted by the requirement of a radiation of wavelength just below the absorption edge of the heavy atom. It was pointed out by Peterson [64], however, that with measurement of intensities using quantum counters, it was possible to estimate detectable differences quite distant from the absorption edge—chlorine absorption edge ($\lambda = 4.397$ Å) with copper radiation ($\lambda = 1.54$ Å), for example. In Table 4-1 the scattering parameters are given for a list of atoms with the chromium, copper, and molybdenum K_α radiations. If a film method is used, it is preferable for the anomalous scatterer to have f'' in excess of $3K$ electrons [51], though detectable Bijvoet differences have been observed for values of f'' less than this value. For example, Fridrichsons and Mathieson [29] determined the absolute configuration of the fungal product gliotoxin using copper K_α radiation and the sulphur atoms with $f'' = 0.6$ as the anomalous scatterers. Very recently, oxygen has been used as the

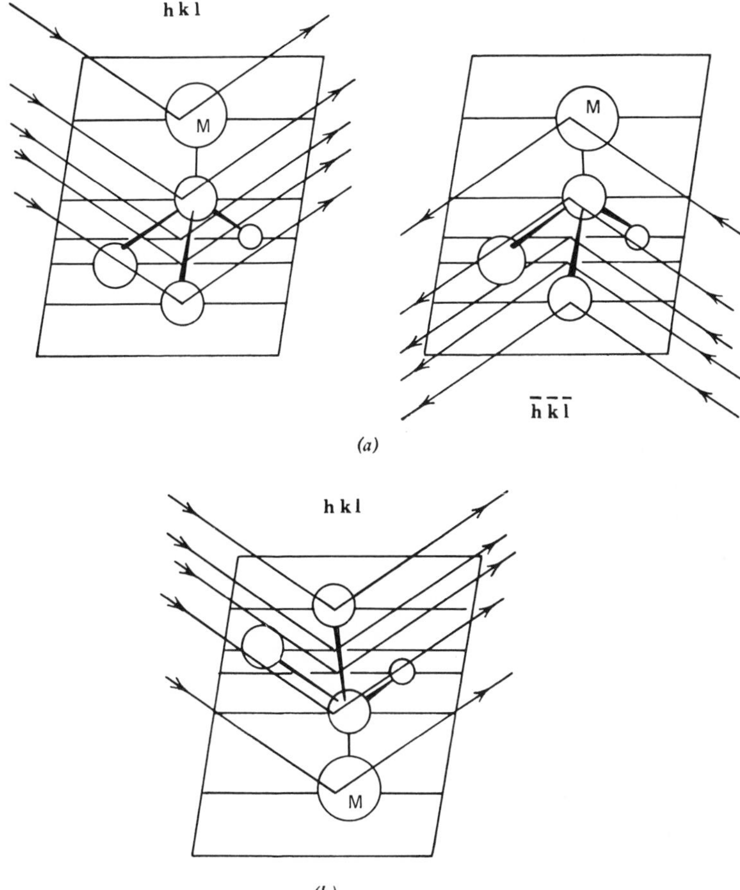

Fig. 4-4 Equivalence of $\overline{\mathbf{hkl}}$ reflection from (*a*) and **hkl** reflection from the enantiomer (*b*).

anomalous scatterer in a determination of the absolute configuration of *d*-tartaric acid [31]. The value of f'' for the copper K_α radiation is in the range of 0.03 to 0.1, but detectable Bijvoet differences were found using an automatic diffractometer. A list of targets for the first-row transition elements with f'' equal to or in excess of 3 is given in Table 4-2.

The Bijvoet method has been successfully applied to a number of metal complexes, following the pioneering work of Saito and his coworkers, who studied the absolute configuration of *d*-$[Coen_3]^{3+}$ in the double salt (*d*-$[Coen_3]Cl_3)_2 \cdot NaCl \cdot 6H_2O$. The results, first reported to the 3rd International

Table 4-1[a] Scattering parameters f' and f''[b]

Radiation			^{24}Cr K_α = 2.291 Å		^{29}Cu K_α = 1.542 Å		^{42}Mo K_α = 0.7107 Å	
Anomalous scatterer	Atomic number	K abs edge, Å[c]	$\Delta f'$	f''	$\Delta f'$	f''	$\Delta f'$	f''
Ti	22	2.497	−1.7	3.8	0.2	1.9	0.3	0.6
V	23	2.269	(−4.4)	0.6	0.1	2.3	0.3	0.7
Cr	24	2.070	−2.2	0.7	−0.1	2.6	0.3	0.8
Mn	25	1.896	−1.8	0.8	−0.5	3.0	0.4	0.9
Fe	26	1.743	−1.6	0.9	−1.1	3.4	0.4	1.0
Co	27	1.608	−1.4	1.0	−2.2	3.9	0.4	1.1
Ni	28	1.488	−1.2	1.2	(−3.1)	0.6	0.4	1.2
Cu	29	1.380	−1.1	1.3	−2.1	0.7	0.3	1.4
Zn	30	1.283	−1.0	1.5	−1.7	0.8	0.3	1.6
Zr	40	0.689	−0.7	4.6	−0.6	2.5	−2.8	0.8
Nb	41	0.653	−0.8	5.1	−0.6	2.8	−2.1	0.9
Mo	42	0.620	−0.9	5.6	−0.5	3.0	−1.7	0.9
Tc	43	0.589	−1.0	6.2	−0.5	3.3	−1.4	1.0
Ru	44	0.560	−1.2	6.7	−0.5	3.6	−1.2	1.1
Rh	45	0.533	−1.3	7.3	−0.5	4.0	−1.1	1.2
Pd	46	0.509	−1.6	7.9	−0.5	4.3	−1.0	1.3
Ag	47	0.486	−1.9	8.6	−0.5	4.7	−0.9	1.4
Cd	48	0.464	−2.2	9.2	−0.6	5.0	−0.8	1.6
Hf	72	0.190	−5	10	−6	5	−0.7	7.3
Ta	73	0.184	−5	11	−6	6	−0.8	7.6
W	74	0.178	−5	11	−6	6	−1.0	8.0
Re	75	0.173	−5	12	−5	6	−1.2	8.3
Os	76	0.168	−5	13	−5	7	−1.4	8.8
Ir	77	0.163	−5	14	−5	7	−1.7	9.2

Radiation			^{24}Cr $K_\alpha = 2.291$ Å		^{29}Cu $K_\alpha = 1.542$ Å		^{42}Mo $K_\alpha = 0.7017$ Å	
Anomalous scatterer	Atomic number	K abs edge, Å[c]	$\Delta f'$	f''	$\Delta f'$	f''	$\Delta f'$	f''
Pt	78	0.158	−5	15	−5	8	−1.9	9.6
Au	79	0.153	−5	15	−5	8	−2.2	10.1
Hg	80	0.149	−5	16	−5	9	−2.6	10.6
C	6	43	0.0	0.1				
N	7	30	0.0	0.1	0.0	0.0		
O	8	23	0.1	0.2	0.0	0.1		
F	9	. . .	0.1	0.2	0.0	0.1		
Na	11	9.512	0.2	0.4	0.1	0.2	0.0	0.1
Mg	12	7.951	0.2	0.5	0.1	0.3	0.0	0.1
P	15	5.787	0.3	1.0	0.2	0.5	0.1	0.2
S	16	5.018	0.3	1.2	0.3	0.6	0.1	0.2
Cl	17	4.397	0.3	1.5	0.3	0.7	0.1	0.2
K	19	3.436	0.0	2.2	0.3	1.4	0.2	0.3
Ca	20	3.070	−0.2	2.7	0.3	1.6	0.2	0.4
As	33	1.044	−0.7	2.2	−1.2	1.2	0.1	2.2
Br	35	0.920	−0.6	2.7	−0.9	1.5	−0.3	2.6
I	53	0.373	(−7.1)	13.6	−1.1	7.2	−0.5	2.4
Ba	56	0.331	−11	8	−2.1	8.9	−0.4	3.0
Pb	82	0.141	−6	18	−4	10	−3.8	11.7
Bi	83	0.137	−6	19	−4	10	−4.5	11.2

[a] Data taken from Ref. 5, vol. III.

Tables compiled by (b) Templeton, D. H., pp. 213–216 [for $(\sin\theta)/\lambda = 0$]; (c) Rieck, G. D., pp. 60–62.

Table 4-2 Suitable targets for exciting first-row transition elements

Anomalous scattering atom	Atomic number	K abs edge, Å[a]	Target materials	Atomic number	Emission series	Wavelength, Å			
						α_1	α_2	$\Delta f'$	f''
Ti	22	2.497	Cr	24	K	2.290	2.294	−1.9[c]	3.0[c]
V	23	2.269	Mn	25	K	2.102	2.106	−2.3[c]	3.3[c]
Cr	24	2.070	Fe	26	K	1.936	1.940	−2.5[c]	3.3[c]
Mn	25	1.896	Co	27	K	1.789	1.793	−3.8[c]	3.5[c]
			Cu	29	K	1.541	1.544	−0.5[b]	3.0[b]
Fe	26	1.743	Ni	28	K	1.658	1.662	−2.8[c]	3.5[c]
			Cu	29	K	1.541	1.544	−1.1[b]	3.4[b]
Co	27	1.608	Hf	72	L	1.570	1.580	−3.6[c]	3.5[c]
			Cu	29	K	1.541	1.544	−3.0[c]	3.4[c]
Ni	28	1.488	W	74	L	1.476	1.487	−5.0[c]	3.5[c]
			Zn	30	K	1.435	1.439	−3.2[c]	3.2[c]
Cu	29	1.380	Ir	77	L	1.351	1.363	−3.7[c]	3.3[c]
			Ga	31	K	1.340	1.344	−3.4[c]	3.4[c]

[a] Rieck, G. D., in Ref. 5, vol. III, pp. 60–65.
[b] Templeton, D. H., *ibid.*, pp. 213–216.
[c] Okaya, Y., and Pepinsky, R., in Ref. 51, p. 292.

Congress of IUCr in Paris July 1954 [71] and published in detail subsequently [72,48], showed this isomer to have structure **XXXIII**, the D configuration. The number of complexes studied in this way is now quite large and is growing at an increasing rate. The technique has also been extensively applied to organic compounds; a total of 94 absolute configurations have been published by the end of 1967 according to two recent compilations [6,7].

PEPINSKY AND OKAYA'S METHOD: METHOD B

Whereas the procedure just described necessitated the prior application of classical X-ray structure analysis, there are alternative methods that do not. With these, the absolute configuration can be determined directly from sets of $(|F_H|^2 - |F_{\bar{H}}|^2)$ values, and in so doing the complexities of the phase problem are largely bypassed.

In translating experimental data into structural parameters, use is made of the Patterson function, which for normal scatterers is centrosymmetric:

$$
\begin{aligned}
P(r) &= 2 \sum_H{}' |F_H|^2 \exp(-2\pi i \cdot H \cdot \mathbf{r}) \\
&= 2 \sum_H{}' |F_H|^2 \cos(2\pi H \cdot \mathbf{r})
\end{aligned}
\qquad (4\text{-}15)
$$

where $\sum'$ is the summation over half the reciprocal lattice. From the formula for the structure factor for anomalous scatterers

$$
F_H = \sum_j (f_j' + i f_j'') \exp(2\pi i \cdot H \cdot \mathbf{r}_j) \qquad (4\text{-}16)
$$

it is found that

$$
\begin{aligned}
|F_H|^2 &= \sum_j \sum_k (f_j' f_k' + f_j'' f_k'') \exp[2\pi i \cdot H \cdot (r_j - \mathbf{r}_k)] \\
&\quad - i \sum_j \sum_k (f_j' f_k'' - f_j'' f_k') \exp[2\pi i \cdot H \cdot (\mathbf{r}_j - \mathbf{r}_k)] \\
&= \sum_j \sum_k (f_j' f_k' + f_j'' f_k'') \cos[2\pi H \cdot (\mathbf{r}_j - \mathbf{r}_k)] \\
&\quad + \sum_j \sum_k (f_j' f_k'' - f_j'' f_k') \sin[2\pi H \cdot (\mathbf{r}_j - \mathbf{r}_k)]
\end{aligned}
\qquad (4\text{-}17)
$$

$$
|F_H|^2 + |F_{\bar{H}}|^2 = 2 \sum_j \sum_k (f_j' f_k' + f_j'' f_k'') \cos[2\pi H \cdot (\mathbf{r}_j - \mathbf{r}_k)] \qquad (4\text{-}18)
$$

$$
|F_H|^2 - |F_{\bar{H}}|^2 = 2 \sum_j \sum_k (f_j' f_k'' - f_j'' f_k') \sin[2\pi H \cdot (\mathbf{r}_j - \mathbf{r}_k)] \qquad (4\text{-}19)
$$

Substituting the expression for $|F_H|^2$—Eq. (4-17)—into the Patterson function—Eq. (4-15)—one finds that

$$
P(r)_{\text{an. scatt.}} = P_c(r) - i P_s(r) \qquad (4\text{-}20)
$$

where

$$P_c(r) = \sum_H |F_H|^2 \cos [2\pi H\cdot(\mathbf{r}_j - \mathbf{r}_k)]$$
$$= \sum_H{}' (|F_H|^2 + |F_{\bar{H}}|^2) \cos [2\pi H\cdot(\mathbf{r}_j - \mathbf{r}_k)] \qquad (4\text{-}21)$$

and

$$P_s(r) = \sum_H |F_H|^2 \sin [2\pi H\cdot(\mathbf{r}_j - \mathbf{r}_k)]$$
$$= \sum_H{}' (|F_H|^2 - |F_{\bar{H}}|^2) \sin [2\pi H\cdot(\mathbf{r}_j - \mathbf{r}_k)] \qquad (4\text{-}22)$$

In the absence of anomalous scattering, the Patterson function has peaks corresponding to the transform of $f_j\cdot f_k$ at $\mathbf{r} = \mathbf{r}_j - \mathbf{r}_k$. With anomalous scattering, $P_c(r)$ is a centrosymmetric function that has peaks of height related to the form factors $(f_j'f_k' + f_j''f_k'')$ at the points $\mathbf{r} = \pm(\mathbf{r}_j - \mathbf{r}_k)$ where $\mathbf{r}_j$ and $\mathbf{r}_k$ are the vectors locating the jth and kth atoms, and $P_s(r)$ is an antisymmetric function with peaks of height depending on $(f_j'f_k'' - f_j''f_k')$ at the points $\mathbf{r} = \mathbf{r}_j - \mathbf{r}_k$, and of heights corresponding to $-(f_j'f_k'' - f_j''f_k')$ at the points $\mathbf{r} = \mathbf{r}_k - \mathbf{r}_j$.

The $P_s(r)$ function is the interesting one for absolute configuration determination, because for a system having one anomalous scatterer k, a positive peak appears at $\mathbf{r} = \mathbf{r}_j - \mathbf{r}_k$ and a negative peak at $\mathbf{r} = \mathbf{r}_k - \mathbf{r}_j$, the two having the same magnitude (f_k'' and f_j' are positive and $f_j'' = 0$). If both atoms do not scatter anomalously, the peak will be nonexistent ($f_j'' = f_k'' = 0$). Thus the $P_s(r)$ function indicates the direction of vectors between unlike anomalous scatterers and between anomalous and normal scatterers. The positions of the nonexcited atoms relative to the anomalous scatterers are therefore known, and hence the absolute configuration is established.

Usually the irradiating X ray is chosen to excite only one element in the crystal. If there are N of these atoms k in the unit cell,

$$|F_H|^2 - |F_{\bar{H}}|^2 = 2Nf_k'' \sum_j \sum_k f_j' \sin [2\pi H\cdot(\mathbf{r}_j - \mathbf{r}_k)] \qquad (4\text{-}23)$$

The usefulness of this method depends on the magnitude of $(|F_H|^2 - |F_{\bar{H}}|^2)$, which itself depends on N, f_k'', f_j', and the distribution of the normal scatterers about the excited atoms, although the effect of $f_j'(= (f_0)_j + \Delta f_j')$ on $P_s(r)$ is generally negligible.

The absolute configurations of a number of complexes have been determined in this way: d-$[\mathrm{Coen_3}]^{3+}$ (in $(d\text{-}[\mathrm{Coen_3}]\mathrm{Cl_3})_2\cdot\mathrm{NaCl}\cdot 6\mathrm{H_2O}$ [73] and in $[\mathrm{Coen_3}]\mathrm{Br}\cdot d$-tartrate$\cdot 5\mathrm{H_2O}$ [61]), which was found to have the D configuration in agreement with the results from method A [48] and the aspartate complexes of cobalt(II) and zinc(II) [21].

For greater practical detail of the application of this technique, the reader is referred to Okaya and Pepinsky [51].

THE QUADRATIC EQUATION METHOD [50,62]: METHOD C

The structure factors for a crystal with anomalous scatterers present can be expressed as

$$\begin{aligned} F_H &= F_H^{\text{n}\cdot\text{s}} + F_H^{\text{a}\cdot\text{s}} \\ &= A_H^{\text{n}\cdot\text{s}} + iB_H^{\text{n}\cdot\text{s}} + A_H^{\text{a}\cdot\text{s}} + iB_H^{\text{a}\cdot\text{s}} \end{aligned} \quad (4\text{-}24)$$

For the normal scatterers,

$$A_H^{\text{n}\cdot\text{s}} = A_{\overline{H}}^{\text{n}\cdot\text{s}} \quad (4\text{-}25)$$

$$B_H^{\text{n}\cdot\text{s}} = -B_{\overline{H}}^{\text{n}\cdot\text{s}} \quad (4\text{-}26)$$

From these relationships it follows that

$$|F_H|^2 = (A_H^{\text{a}\cdot\text{s}} + A_H^{\text{n}\cdot\text{s}})^2 + (B_H^{\text{a}\cdot\text{s}} + B_H^{\text{n}\cdot\text{s}})^2 \quad (4\text{-}27)$$

$$|F_{\overline{H}}|^2 = (A_{\overline{H}}^{\text{a}\cdot\text{s}} + A_H^{\text{n}\cdot\text{s}})^2 + (B_{\overline{H}}^{\text{a}\cdot\text{s}} - B_H^{\text{n}\cdot\text{s}})^2 \quad (4\text{-}28)$$

If the position of the anomalous scatterers has been established by usual methods, $A_H^{\text{a}\cdot\text{s}}$, $B_H^{\text{a}\cdot\text{s}}$, $A_{\overline{H}}^{\text{a}\cdot\text{s}}$, and $B_{\overline{H}}^{\text{a}\cdot\text{s}}$ are known. Further, $|F_H|^2$ and $|F_{\overline{H}}|^2$ can be determined from the original density measurements. Thus Eqs. (4-27) and (4-28) are two simultaneous quadratic equations in $A_H^{\text{n}\cdot\text{s}}$ and $B_H^{\text{n}\cdot\text{s}}$. For a given plane ($H$) there are two solutions, $(A_H^{\text{n}\cdot\text{s}})_{\text{I}}$, $(B_H^{\text{n}\cdot\text{s}})_{\text{I}}$ and $(A_H^{\text{n}\cdot\text{s}})_{\text{II}}$, $(B_H^{\text{n}\cdot\text{s}})_{\text{II}}$. There are several procedures that theoretically allow a choice to be made. For example, (1) if a wavelength of the incident X ray is selected so that no atom is excited, $A_H^{\text{n}\cdot\text{s}}$ and $B_H^{\text{n}\cdot\text{s}}$ remain unaltered, but $A_H^{\text{a}\cdot\text{s}}$ and $B_H^{\text{a}\cdot\text{s}}$ are transformed to $A_H^{\text{h}\cdot\text{a}}$ and $B_H^{\text{h}\cdot\text{a}}$, where h·a represents a heavy atom without anomalous scattering. This gives rise to equations corresponding to (4-27) and (4-28) and enables a choice to be made between the two possible solutions. (2) If the anomalous scatterer is replaced isomorphously by an atom with a considerably different normal scattering power, another set of equations corresponding to (4-27) and (4-28) can be obtained, and this again enables a choice to be made. It is important, however, that the two structures are strictly isomorphous, so that the coordinates and hence the $F_H^{\text{n}\cdot\text{s}}$ contributions, remain unaltered.

The absolute configurations can be deduced from a knowledge of $A_H^{\text{a}\cdot\text{s}}$, $A_H^{\text{n}\cdot\text{s}}$, $B_H^{\text{a}\cdot\text{s}}$, and $B_H^{\text{n}\cdot\text{s}}$. The method, however, has a number of disadvantages compared with method B. Perhaps the most important is that F_H data must be on an absolute scale. This is not necessary for the P_s function. Further, the choice between the roots is sometimes very difficult; the positions of some atoms could lead to undesirable values for the parameters in Eqs. (4-27) and (4-28).

THE CRYSTAL-ENGINEERING METHOD [40,60]: METHOD D

The inclusion of a dissymmetric moiety of known absolute configuration in a crystal containing the molecule whose absolute configuration is required allows the absolute configuration to be determined by normal X-ray structural analysis. The procedure is, of course, not direct but involves comparison with the reference molecule.

The inclusion of the dissymmetric moiety is often very simple. For a charged complex species it may be only necessary to form a diastereomeric salt with a resolving agent of known absolute configuration; for example, the absolute configuration of *d*-$[Coen_3]^{3+}$ could be obtained by the normal analysis of the compound *d*-$[Coen_3]Br \cdot$ *d*-tartrate $\cdot 5H_2O$, in which the configuration of tris(ethylenediamine)cobalt(III) is related to the known configuration of *d*-tartrate.

This approach was used when Wunderlich [80] determined the absolute configuration of the alkaloid retusamine by studying the base's salt with α'-bromo-*d*-camphor-*trans*-π-sulfonic acid. The result was later confirmed by the normal Bijvoet method [81].

In some compounds the moiety of known configuration is an intrinsic part of the molecule whose absolute configuration is to be determined. Such was the case for *d*-*trans*-[Co(L-ala)$_3$] [22] and *d*-[Coen$_2$(L-glu)]ClO_4 [23], whose configurations were found by this method to be D.

The technique is very useful, because it does not require a heavy-atom anomalous scatterer, nor does it require the use of rare target materials. The information emerges from a normal analysis.

A number of compounds that have been studied by the above method are listed in Table 4-3.

CONFORMATIONS OF CHELATE RINGS

X-ray diffraction studies can provide geometrical details of chelate rings—bond lengths, bond angles, dihedral angles, for example—that are not obtainable by other means and, in addition, can yield other detailed information of the conformations of chelate rings.

Five-Membered Diamine Chelate Rings

Ethylenediamine and substituted ethylenediamine systems have been extensively studied. Scouloudi and Carlisle [74,75] were the first to verify the

Table 4-3 Some complexes studied by X-ray analysis

Complex ion (X)	Crystal	Space group	Abs. conf.	Method	Ref.
d-$[Coen_3]^{3+}$	$(XCl_3)_2 \cdot NaCl \cdot 6H_2O$	P3	D($\delta\delta\delta$)	A	71,72,48
			D($\delta\delta\delta$)	B	73
	XBr · *d*-tartrate · $5H_2O$	P1	D($\delta\delta\delta$)	B	61
	$XBr_3 \cdot H_2O$	$P4_32_12$	D($\delta\delta\delta$)	A	47
l-$[Co(R\text{-}pn)_3]^{3+}$	XBr_3	$P6_3$	L($\lambda\lambda\lambda$)	A	32
l-$[Cotn_3]^{3+}$	$XBr_3 \cdot H_2O$	$P2_1$	D	A	49
d-$[Copenten]^{3+}$	$X[Co(CN)_6] \cdot 2H_2O$	$P2_12_12_1$	D	A	45
l-$[Coen_2sar]^{2+}$	$XI_2 \cdot 2H_2O$	C2	L($\delta\lambda\lambda$)	A	11
d-$[Coen_2(L\text{-}glu)]^{+}$	$XClO_4$	$P2_12_12_1$	D($\delta\delta$)	D	23
d-$[Coen_2(CN)_2]^{+}$	$XCl \cdot H_2O$	$P2_1$	D($\lambda\lambda$)	A	42
d-*cis*-$[Co(R\text{-}pn)_2(NO_2)_2]^{+}$	XCl	$P2_12_12_1$	L($\lambda\lambda$)	A	8
trans-$[Co(R\text{-}pn)_2Cl_2]^{+}$	$XCl \cdot HCl \cdot 2H_2O$	C2	($\lambda\lambda$)	A	70
d_{436}-$[Co(NH_3)_4sar]^{2+}$	XNO_3	$P2_12_12_1$	S	A	38
d-*trans*-$[Co(L\text{-}ala)_3]$	X	$P2_12_12_1$	D	D	22
l-$[Fe(phen)_3]^{2+}$	$X(SbC_4H_2O_6)_2 \cdot 8H_2O$	$P3_22_1$	D	A	76

puckered arrangement for the chelate ring. They found an asymmetric-skew conformation for the ethylenediamine in $[Cuen_2][Hg(SCN)_4]$ with one carbon 0.35 Å above the CuNN plane and the other 0.55 Å below. The complex was centrosymmetric with the $\delta\lambda$ configuration, which was also found by Nakahara and coworkers [46] for *trans*-$[Coen_2Cl_2]Cl \cdot HCl \cdot 2H_2O$ and has subsequently been found for all *trans*-bis(ethylenediamine) complexes [9,14,15,25,37,44,53,54,55,64].† Geometric data for a selection of bis and tris complexes are given in Table 4-4. It is apparent from these results that asymmetric puckering is not unusual and that a great variety of conformations exist in the crystalline state. This would suggest that the energy differences between these various conformations is quite small, as proposed in Chap. 3.

The discovery of the $\delta\lambda$ configuration for the *trans*-bis(ethylenediamine) complexes was in conflict with the calculated energy preference of 1 kcal mol^{-1} for the $\delta\delta$ and $\lambda\lambda$ over the $\delta\lambda$ conformations as determined by Corey and Bailar [18] but is consistent with the negligible difference proposed by Gollogly and Hawkins [30]. The definite preference for the "meso" form in the crystalline state has been explained in terms of the requirements for close packing of rigid molecules [64], but no doubt it is also due in part to the statistical preference for this configuration.

trans-Dichlorobis(*dl*-propylenediamine)cobalt(III) chloride hydrochloride is also centrosymmetric [69], but *trans*-$[Co(l\text{-pn})_2Cl_2]^+$ has the $\lambda\lambda$ configuration with the two methyl groups trans to one another and equatorial [70]. The two methyl groups are also trans and equatorial in the complex *l-trans*-$[Co(Meen)_2Cl_2]ClO_4 \cdot \frac{1}{2}H_2O$ [68].

In the *cis*-bis(ethylenediamine) complexes with the remaining two octahedral positions occupied by unidentates, or planar, or nearly planar chelates, all possible configurations have been found; for example, *d*-$[Coen_2(CN)_2]^+$ has the D($\lambda\lambda$) configuration [42], *d*-$[Coen_2(\text{L-glu})]^+$ the D($\delta\delta$) [23], and *cis*-[Nien$_2$Cl(NCS)] the ($\delta\lambda$) [64]. This suggests that the energy difference between the configurations is very small, which is consistent with the experimental results of Dwyer and coworkers [24] and with the conformational analysis calculations [30].

Prior to 1968, all the X-ray structures of tris(ethylenediamine) complexes had either the D($\delta\delta\delta$) or L($\lambda\lambda\lambda$) configurations. These are of equal energy and have been estimated to be more stable than the D($\delta\lambda\lambda$) {= L($\lambda\delta\delta$)} and D($\lambda\lambda\lambda$) {= L($\delta\delta\delta$)} configurations [18,30], although they would seem to be of similar energy to the D($\delta\delta\lambda$) {= L($\lambda\lambda\delta$)} [30]. Recently, Ibers and his

† Note added in proof: In a recent communication the X-ray determined structure of *trans*-$[Coen_2Cl(NO)]ClO_4$ shown in a figure has the $\lambda\lambda$ configuration [D. A. Snyder and D. L. Weaver, *Chem. Commun.*, 1425 (1969)].

Table 4-4 Some X-ray structural data on ethylenediamine and propylenediamine chelate rings

Complex	∠N(1)MN(2)	∠MN(1)C(1)	∠MN(2)C(2)	ω	Z_1, Å	Z_2, Å	Ref.
trans-$[Coen_2Cl_2]NO_3$	85.6°	106.7°	112.0°	48.7°	±0.48	∓0.18	55
trans-$[Coen_2Br_2]Br \cdot HBr \cdot 2H_2O$	86.5°	109.2°	108.3°	45.8°	±0.35	∓0.27	54
d-$[Coen_3]Br_3 \cdot H_2O$	87.0°[a]	108.8°	104.5°	50.8°	−0.54	+0.14	47
trans-$[Co(\text{R-pn})_2Cl_2]Cl \cdot HCl \cdot 2H_2O$	88.8°	103.0°	101.5°	63.0°	+0.34	−0.42	70
	87.9°	107.5°	106.5°	46.0°	+0.37	−0.21	
l-$[Co(\text{R-pn})_3]Br_3$	86.5°	108.0°	110.0°	50.2°	+0.31	−0.39	32
trans-$[Cren_2Cl_2]Cl \cdot HCl \cdot 2H_2O$	85.2°	105.8°	108.3°	49.1°	+0.22	−0.40	53
$[Cuen_2](SCN)_2$	85.0°	108.6°	110.5°	50.3°	−0.53	+0.16	15
$[Cuen_2](NO_3)_2$	86.2°	108.5°	109.1°	43.6°	+0.39	−0.19	37
$[Cren_3][Ni(CN)_5] \cdot 1 \cdot 5H_2O$	82.6°[a]	109.5°	108.8°	52.6°	+0.27	−0.42	66
	82.1°[a]	108.8°	110.6°	50.6°	−0.40	+0.26	

[a] Average value.

coworkers have found the D($\delta\delta\lambda$) {or L($\lambda\lambda\delta$)} configuration and the other "less stable" configurations in tris(ethylenediamine)chromium(III) such as the pentacyanonickelate and the hexacyanocobaltate(III) salts [65,66,67]. They proposed that hydrogen bonding between the cyanide groups and the amine protons was responsible for the presence of the otherwise unfavorable configurations.

The crystal structure of the more stable (levo) isomer of $[Co(R\text{-}pn)_3]^{3+}$ has been determined for the bromide salt [32]. The expected configuration L($\lambda\lambda\lambda$) was found with the three methyl groups equatorial and cis to one another. The less favorable configuration has been studied for *l*-$[Co(S\text{-}pn)_3]$-$[Co(CN)_6]\cdot 3H_2O$ [33]. The three methyl groups were again cis, and equatorial, and the configuration was found to be L($\delta\delta\delta$).

Saito and his coworkers have determined the anisotropic thermal parameters of the atoms in the chelate rings in *d*-$[Coen_3]Cl_3\cdot H_2O$ [34]. The results for the ethylenediamine, which lies on a twofold axis in the crystal, are shown in Fig. 4-5. Whereas the vibrations of the cobalt and the nitrogens have relatively small amplitudes with the cobalt atoms being nearly isotropic, the two carbon atoms are found to oscillate perpendicularly to the C—C bond with a mean amplitude of about 0.3 Å. The other two chelate rings interact with the chloride ions, and the vibrations of their carbon atoms are smaller. This result lends strong support to the results of the conformational analyses described in Chap. 3. From the a priori calculations it was estimated that, for the ethylenediamine chelate ring, a whole range of conformations corresponded to the lowest energy, and these conformations could be interrelated by holding the cobalt and nitrogen atoms steady while flapping the two carbons up and down and maintaining a relatively constant dihedral angle ω. This equality in energy between various asymmetric and symmetric

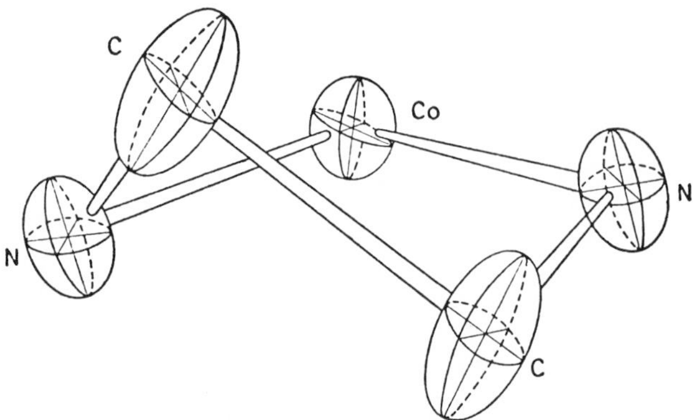

Fig. 4-5 View of the ellipsoids of thermal vibration [34].

skew conformations and the consequent freedom of the two carbon atoms to oscillate is not expected to exist to the same extent in the D(λλλ) or L(δδδ) configurations (see Chap. 3). It would therefore be interesting to compare the anisotropic thermal parameters for the ring atoms in, say, L- and D-$[Co(R\text{-}pn)_3]^{3+}$ for crystals in which the anions do not significantly interact with the chelate rings.

The structure of ethylenediamine as a bridging ligand has also been studied by X-ray diffraction. Brodensen found it to have a trans conformation in $Hgen(Cl)_2$ [13].

Six-Membered Diamine Chelate Rings

Although a greater variety of conformational types are available for six-membered diamine chelates than for the five-membered ring systems, the complexes studied so far by X-ray diffraction have had very similar conformations, with the rings adopting a slightly flattened-chair structure [43,49,56,57]. In *l*-$[Cotn_3]Br_3 \cdot H_2O$, the complex ion has an approximate threefold axis of rotation [49]. Its projection along this axis is shown in Fig. 4-6. The flattening of the rings was predicted from the conformational analysis described in Chap. 3. As a result of ring strain and nonbonded interactions the ring bond angles increase from their normal values: $\angle CoNC = 117.5°$, $\angle NCC = 112.3°$, $\angle CCC = 114.0°$, and $\angle NCoN = 94.3°$ (average values). Structural data for the three chelate rings are given in Table 4-5. Chair conformations were also found for the six-membered diamine ring systems in dichloro-1,4,8,11-tetraazacyclotetradecanenickel(II) [12]. The structure of this complex is discussed further under "Multidentate Ring Systems."

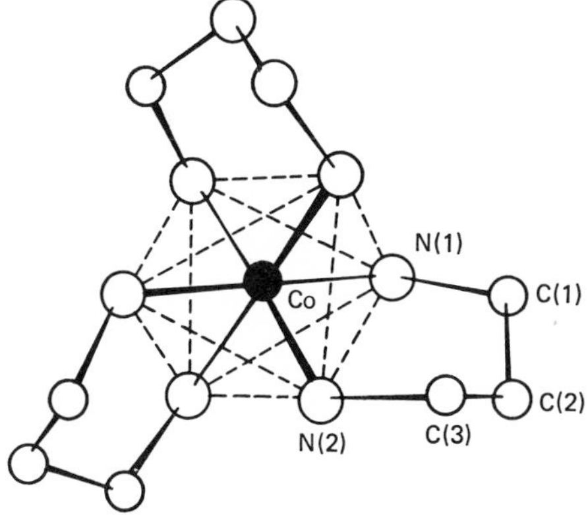

Fig. 4-6 Structure of *l*-$[Cotn_3]^{3+}$ viewed along the approximate threefold axis [49].

Table 4-5 Structural data for *l*-$[Cotn_3]Br_3 \cdot H_2O$ [49]

Bond length, Å	Ring 1	Ring 2	Ring 3
Co—N(1)	2.00	2.00	2.00
Co—N(2)	2.00	2.00	2.00
N(1)—C(1)	1.47	1.48	1.47
N(2)—C(3)	1.45	1.47	1.46
C(1)—C(2)	1.54	1.54	1.55
C(2)—C(3)	1.54	1.54	1.54
Bond angle, deg			
N(1)—Co—N(2)	96	92	95
Co—N(1)—C(1)	117	118	116
Co—N(2)—C(3)	118	119	117
N(1)—C(1)—C(2)	114	113	114
N(2)—C(3)—C(2)	110	111	112
C(1)—C(2)—C(3)	114	114	114

Five-Membered Amino Acid Chelate Rings

The crystal structures of complexes of α-amino acids and peptides have recently been very thoroughly reviewed by Freeman [27]. Much of the following discussion is based on material from this excellent source.

In the amino acid complexes the bond lengths and bond angles differ little from the values for the free amino acids. Average values are shown in Fig. 4-7. The distance between the two donor atoms seems to be independent of the metal ion. This means that, as the M—N and M—O bond lengths

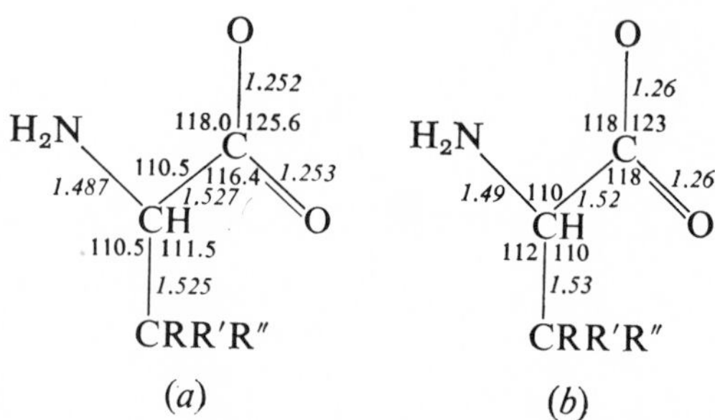

Fig. 4-7 Averaged structural data for (*a*) free amino acids and (*b*) chelated α-aminocarboxylates [27].

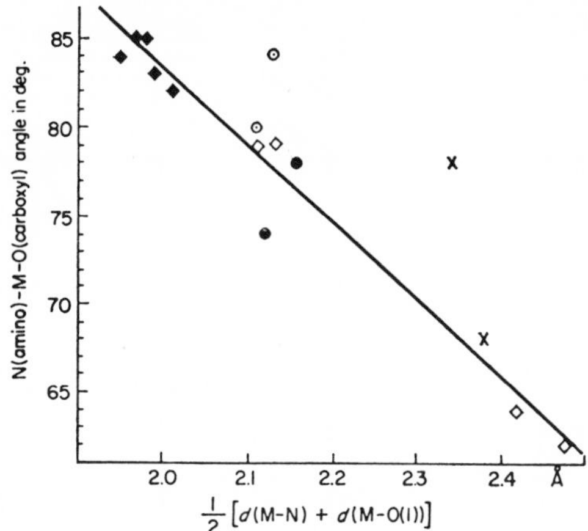

Fig. 4-8 Plot of N(1)–M–O(1) angles in α-amino acid chelate rings versus mean metal ligand distances. The metals in the chelates are distinguished as follows: ●, Co(II); ⊙, Ni(II); ♦, Cu; ◇, Zn; ×, Cd [27].

are increased, the angle ∠NMO is decreased (see Fig. 4-8). Copper(II) complexes, for which the average values of the metal to donor atom bond lengths are 1.99 Å for N and 1.96 Å for 0, have an average value of ∠NMO = 84°.

In the free amino acids the dihedral angle between the C—N and C—O bonds about the C—C bond is found to vary between 0 and about 30° (see Fig. 4-9), although in a few peptides, values of up to 67° have been found [39]. The same range of dihedral angles is found for the metal complexes. These are shown in Fig. 4-10. The positions of the metal ion relative

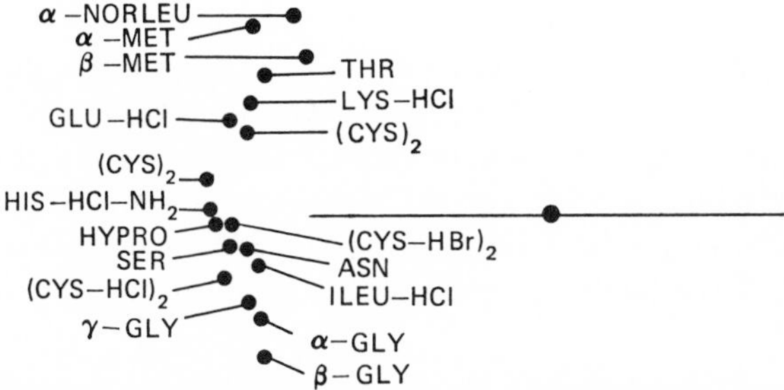

Fig. 4-9 The conformation about the C^{α}—C bond in amino acids. The view is along the C^{α}—C bond, which is indicated by the *large dot*. The horizontal line represents the carboxylate group and the *small dots* are the projected positions of the nitrogen atoms. The dihedral angle of interest is the angle between the horizontal line and a line joining the dots representing the nitrogen and carbon atoms [39].

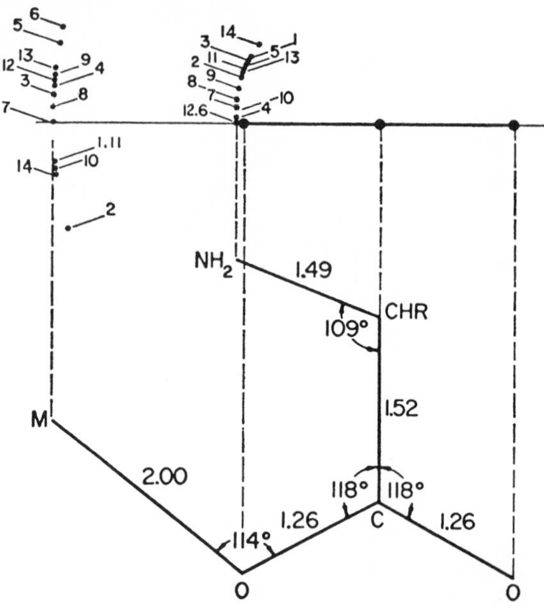

Fig. 4-10 Conformation about the C^{α}—C bond in some amino acid complexes. The positions of the N and M atoms are projected in a direction parallel to the C^{α}—C bond, average dimensions being assumed as shown in the lower half of the figure. Code for points: 1, his(1), in Co(L-his)$_2 \cdot H_2O$; 2, his(2) in Co(L-his)$_2 \cdot H_2O$; 3, Ni(gly)$_2 \cdot 2H_2O$; 4, ligand (1) in Ni(α-NH$_2$*i*but)$_2 \cdot 4H_2O$; 5, ligand (2) in Ni(α-NH$_2$*i*but)$_2 \cdot H_2O$; 6, Ni(DL-his)$_2 \cdot H_2O$; 7, gly(1) in Cu(gly)$_2 \cdot H_2O$; 8, gly(2) in Cu(gly)$_2 \cdot H_2O$; 9, Cu(pen)$_2$; 10, Cu(glu)$\cdot 2H_2O$; 11, Zn(asp)$\cdot 3H_2O$; 12, Zn(glu)$\cdot 2H_2O$; 13, Cd(gly)$_2 \cdot H_2O$; 14, Cd(L-his)$_2 \cdot 2H_2O$ [27].

to the CCOO planes are also given. This large variation is due to the low torsional barrier to rotation about the C—C^{α} bond (see Chap. 3).

All the *trans,trans*-bis(*dl*-α-aminocarboxylato) and -(glycinato) complexes that have been studied have centrosymmetric structures. As mentioned for the diamines, this is probably the result of the requirements for close packing.

A number of complexes of interest have been selected for more detailed discussion.

Bis(glycinato)copper(II) hydrate. Three crystalline bis(glycinato)copper(II) complexes have been isolated. Two of these, a dihydrate [77] and a monohydrate [82], are reported to have the trans structure. The third modification, which crystallizes as a deep-blue monohydrate, has been shown by X-ray diffraction to have the cis structure (see Fig. 4-11) [28]. The two chelates are

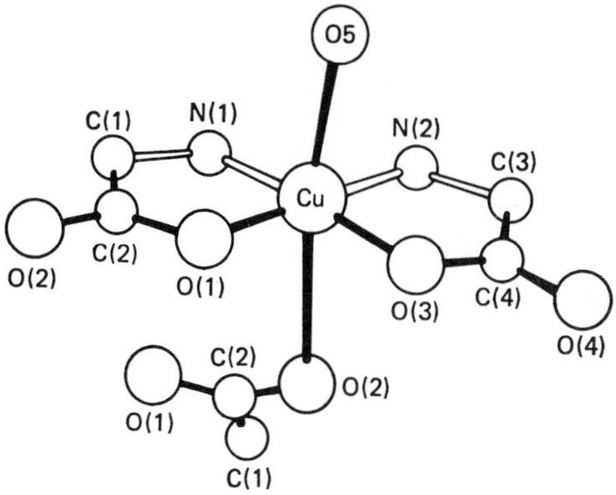

Fig. 4-11 Bis(glycinato)copper(II) hydrate. H_2O molecule represented by 0(5) (from [27]).

crystallographically independent: in one, the copper and nitrogen atoms are 0.006 and −0.103 Å from the CCOO plane, and in the other −0.126 and −0.162 Å. The remaining octahedral positions are occupied by a water molecule and a carboxyl oxygen atom from a neighboring molecule. Structural data for the chelate rings are given in Table 4-6.

Table 4-6 Structural data for chelate rings in $Cu(gly)_2H_2O$ [28]

Bond length	Ring 1	Ring 2
Cu—O	1.95	1.94
Cu—N	1.98	2.02
N—C(1)	1.47	1.48
C(1)—C(2)	1.49	1.54
C(2)—O(1)	1.27	1.29
C(2)—O(2)	1.22	1.24
Bond angle	**Ring 1**	**Ring 2**
O(1)—Cu—N	85.0	85.4
Cu—N—C(1)	109.3	109.6
Cu—O(1)—C(2)	115.3	115.8
N(1)—C(1)—C(2)	112.6	111.3
C(1)—C(2)—O(1)	117.4	117.5
C(1)—C(2)—O(2)	118.3	119.7
O(1)—C(2)—O(2)	124.3	122.8

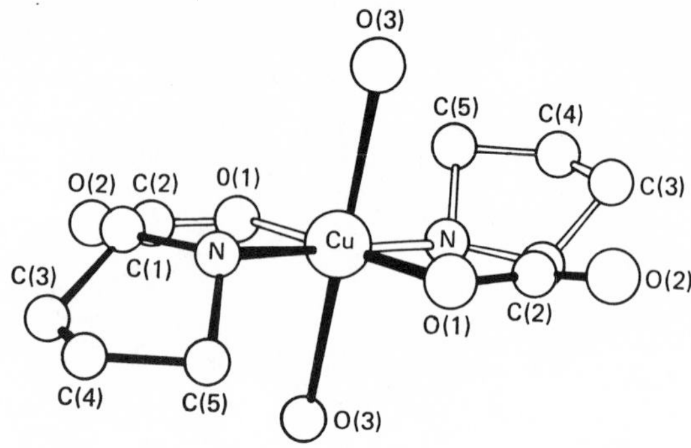

Fig. 4-12 Bis(DL-prolinato)copper(II) dihydrate. H_2O molecule is represented by O(3) (from [27]).

Bis(*DL*-prolinato)copper(II) dihydrate [41]. Proline and hydroxyproline are unique as naturally occurring amino acids in that their amino group is part of a pyrrolidine ring. The heterocyclic ring has an envelope conformation with C(4) (see Fig. 4-12) out of the plane of the other four atoms by about 0.5 Å [39]. This conformation is retained in bis(DL-prolinato)copper(II) dihydrate with C(4) out of the C(2)C(3)NC(5) plane by 0.60 Å and situated trans to the carboxyl group with respect to the plane of the ring. The heterocyclic ring has a restrictive influence on the possible conformations of the chelate ring. The structure of the complex, which is centrosymmetric, is shown in Fig. 4-12, and the structural details of the chelate rings are given in Table 4-7.

Table 4-7 Structural data for chelate rings in Cu(DL-pro)$_2$ $2H_2O$ [41]

Bond length, Å					
Cu—O	2.03	Cu—N	1.99	C(1)—O(1)	1.24
C(1)—O(2)	1.24	C(1)—C(2)	1.50	C(2)—C(3)	1.52
C(3)—C(4)	1.50	C(4)—C(5)	1.52	C(5)—N	1.53
C(2)—N	1.52				
Bond angle, deg					
N—Cu—O(1)	82	Cu—N—C(2)	112	Cu—N—C(5)	113
Cu—O(1)—C(1)	116	O(1)—C(1)—O(2)	122	C(2)—C(1)—O(1)	120
C(2)—C(1)—O(2)	118	C(1)—C(2)—C(3)	112	C(1)—C(2)—N	108
C(2)—C(3)—C(4)	97	C(3)—C(4)—C(5)	109	C(4)—C(5)—N	96
C(5)—N—C(2)	108	N—C(2)—C(3)	108		

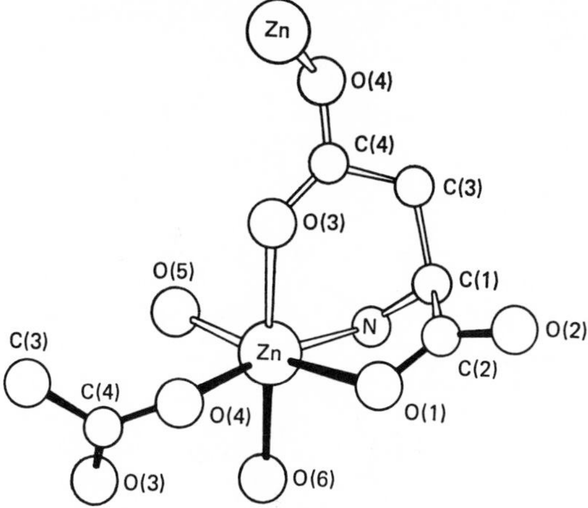

Fig. 4-13 L-Aspartatozinc(II) trihydrate. Coordinated H_2O molecules are represented by O(5) and O(6), free H_2O molecule is omitted (from [27]).

(L-Aspartato)diaquazinc(II) hydrate [20,21] The aspartate dianion NH_2—CH(COO^-)—CH_2—COO^- chelates as a tridentate about one of the octahedral faces in this zinc complex, which has the remaining coordination positions occupied by two water molecules and a carboxyl oxygen from a neighboring molecule (see Fig. 4-13). The chelation of the —CH_2COO^- side group enforces an asymmetric envelope conformation onto the five-membered chelate ring. Structural data for the chelate rings are given in Table 4-8.

Table 4-8 Structural data for chelate rings in Zn(L-asp)·$3H_2O$ [20]

Bond length, Å					
Zn—N(1)	2.06	Zn—O(1)	2.20	Zn—O(3)	2.15
N—C(1)	1.49	C(1)—C(2)	1.52	C(2)—O(1)	1.26
C(2)—O(2)	1.27	C(1)—C(3)	1.54	C(3)—C(4)	1.50
C(4)—O(3)	1.29	C(4)—O(4)	1.26		
Bond angle, deg					
N—Zn—O(1)	78.9	N—Zn—O(3)	90.3	O(1)—Zn—O(3)	84.6
Zn—N—C(1)	104.2	N—C(1)—C(2)	109.5	N—C(1)—C(3)	113.7
C(2)—C(1)—C(3)	110.4	C(1)—C(2)—O(1)	119.6	C(1)—C(2)—O(2)	117.4
O(1)—C(2)—O(2)	123.0	Zn—O(1)—C(2)	108.4	C(2)—C(3)—C(4)	116.7
Zn—O(3)—C(4)	127.6				

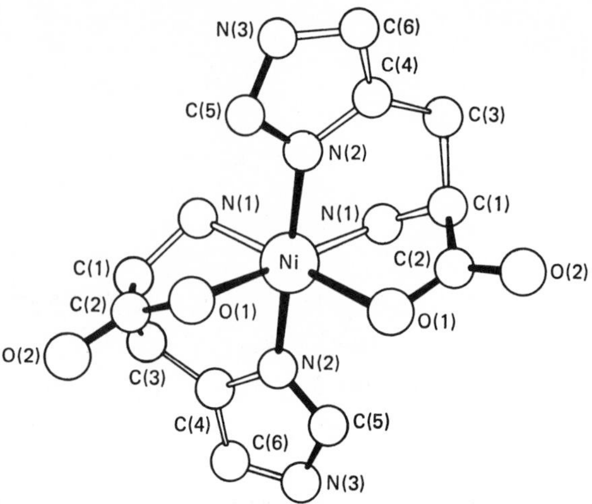

Fig. 4-14 Bis(L-histidinato)nickel(II) complex in bis(DL-histidinato)nickel(II) hydrate (from [27]).

Bis(*DL*-histidinato)nickel(II) hydrate [26]. The two chelates are situated cis to each other and are related by a crystallographic twofold axis that passes through the nickel atom, and, therefore, the two ligands are of the same chirality. The crystal contains equal numbers of [Ni(L-his)$_2$] and [Ni(D-his)$_2$] molecules. As found for the zinc aspartate, the steric requirements for chelation by way of the side group, in this instance, the imidazole nucleus, give rise to an asymmetric envelope conformation of the five-membered chelate ring. Structural data for the chelate rings are given in Fig. 4-14 and in Table 4-9.

Table 4-9 Structural data for the chelate rings in Ni(DL-his)$_2$·H_2O [26]

Bond length, Å					
Ni—N(1)	2.10	Ni—N(2)	2.09	Ni—O(1)	2.11
C(1)—C(2)	1.53	C(1)—C(3)	1.54	C(3)—C(4)	1.47
C(4)—N(2)	1.39	C(4)—C(6)	1.38	C(5)—N(2)	1.33
C(5)—N(3)	1.34	C(6)—N(3)	1.36	C(2)—O(1)	1.28
C(2)—O(2)	1.25	C(1)—N(1)	1.48		
Bond angle, deg					
N(1)—Ni—N(2)	87.6	N(1)—Ni—O(1)	79.7	N(2)—Ni—O(1)	87.8
Ni—N(1)—C(1)	103.6	Ni—N(2)—C(5)	126.9	Ni—O(1)—C(2)	110.9
O(1)—C(2)—O(2)	123.6	O(1)—C(2)—C(1)	117.2	O(2)—C(2)—C(1)	119.2
N(1)—C(1)—C(2)	112.3	C(2)—C(1)—C(3)	109.9	N(1)—C(1)—C(3)	110.3
C(1)—C(3)—C(4)	113.5	C(3)—C(4)—N(2)	123.6		

d_{436}-Sarcosinatotetramminecobalt(III) nitrate [38]. A preliminary report on the absolute configuration of d_{436}-sarcosinatotetramminecobalt(III) nitrate has been published. The coordinated asymmetric nitrogen has the S configuration, and the amino acid chelate ring, which is slightly puckered, has the λ conformation (see Fig. 4-15). The puckering is such that C(1) is on the opposite side, and C(2) on the same side, of the O(1)—Co—N(5) plane as the methyl group.

Six-Membered Amino Acid Chelate Rings

Only a limited number of complexes containing six-membered amino acid chelate rings have been studied. The aspartate complex discussed above has a six-membered as well as a five-membered ring. The structure of its six-membered ring does not differ very much from those in the bis(β-alaninato) complexes of cobalt(II) [36], nickel(II) [35], and copper(II) [78,27] and in bis(DL-β-aminobutyrato)copper(II) dihydrate [16]. In the latter complex the copper(II) ion occupies a center of symmetry, chelated by one D and one L ligand.

Structural details for these complexes are given in Table 4-10 and in Fig. 4-16. The ring bond angles are slightly larger than for the α-amino acid chelates, except for M—O(1)—C(3), which is significantly and consistently larger in the β-amino acids. The N—M—O(1) angle does not vary from metal to metal, being $91 \pm 1°$, in contrast to the α-amino acids, for which N—M—O(1) was found to decrease in order to alleviate strain on increasing the metal to ligand bond length (see Fig. 4-8).

The six-membered rings have very distorted boat conformations with the ends at C(1) and O(1). Dihedral angles around the ring are given in Table 4-11. The metal ion does not lie in the plane of the carboxyl group but, rather, out of the plane by between 0.27 and 0.72 Å.

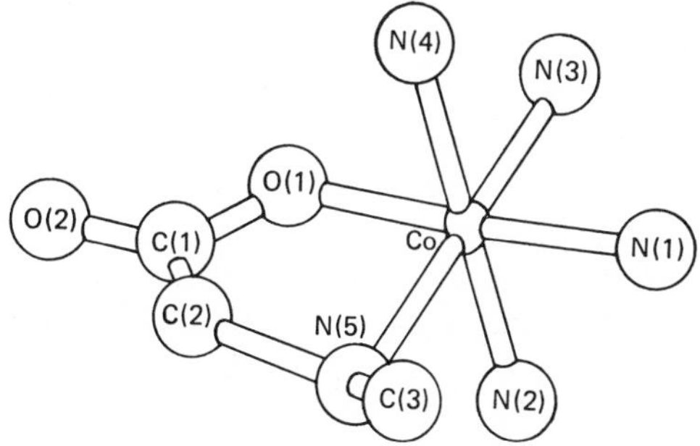

Fig. 4-15 d-$_{436}$-Sarcosinatotetramminecobalt(III) nitrate [38].

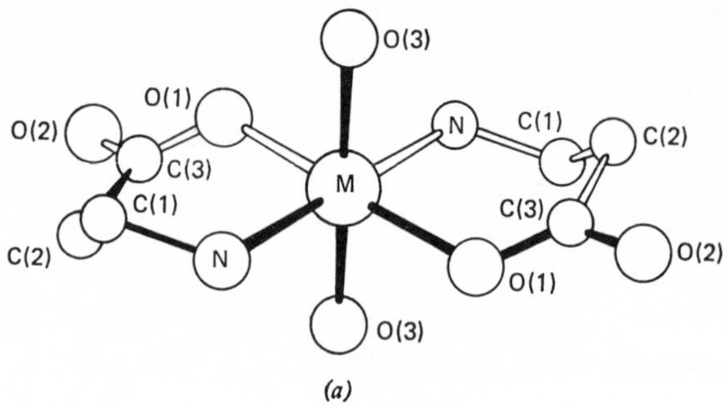

(a)

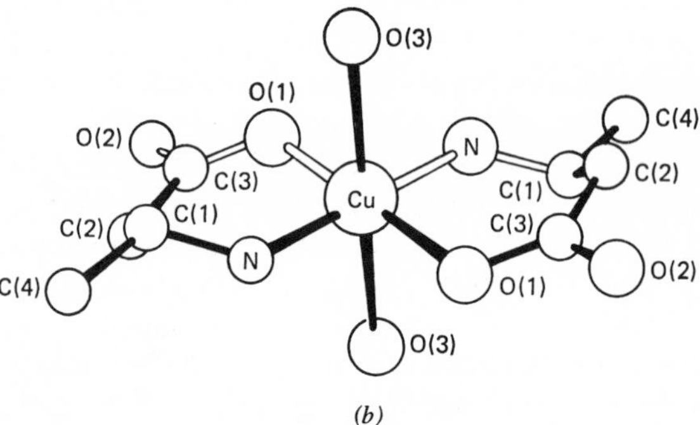

(b)

Fig. 4-16 Structures of complexes with six-membered amino acid chelate rings: (a) bis(β-alaninato)cobalt(II) [or nickel(II)] dihydrate; (b) bis(DL-β-aminobutyrato)copper-(II)dihydrate. H_2O molecule is represented by O(3) (from [27]).

Ring Systems in Multidentate Chelates

Interest in the configurations and conformations of multidentate chelates has recently increased rather sharply. This is probably a result of the realization that complexes containing multidentates have available to them a large variety of possible isomeric forms—[Co(trien)X_2] [17], for example. Because of the complexity of some of these systems, X-ray structural studies have had to play an extremely important part in the determination of the configurations. Unfortunately, at present, much of this recent work has appeared only in the form of preliminary communications, and the conformational details required for this review are not available.

Two related structures that are of considerable interest are those of $[CoEDTA]^-$ and $[Co(penten)]^{3+}$. Both ligands are known to be capable of

Table 4-10 Structural data for six-membered amino acid chelate rings [27]

Bond length, Å	Co(β-ala)$_2$	Ni(β-ala)$_2$	Cu(β-ala)$_2$	Cu(β-NH$_2$but)$_2$	Zn(asp)[a]
M—N	2.14	2.10	2.01	1.99	2.06
M—O(1)	2.13	2.14	2.04	2.00	2.15
N—C(1)	1.50	1.50	1.51	1.44	1.49
C(1)—C(2)	1.58	1.57	1.58	1.49	1.54
C(2)—C(3)	1.56	1.55	1.54	1.49	1.50
C(3)—O(1)	1.28	1.22	1.22	1.30	1.29
C(3)—O(2)	1.25	1.28	1.28	1.23	1.26
N···O(1)	3.02	3.03	2.90	2.87	2.99
Bond angle, deg					
N—M—O(1)	90	91	91	92	90
M—N—C(1)	115	115	115	117	104
M—O(1)—C(3)	125	123	140	126	126
N—C(1)—C(2)	111	110	115	114	114
C(1)—C(2)—C(3)	112	113	117	113	118
C(2)—C(3)—O(1)	119	126	114	121	122
C(2)—C(3)—O(2)	116	114	116	123	115
O(1)—C(3)—O(2)	124	123	129	117	123

[a] Atoms labeled as for β-alanine.

coordination as sexadentates. A projection of the electron density in $NH_4[CoEDTA]\cdot 2H_2O$ onto the (100) plane is given in Fig. 4-17 [79]. Features of the structure are clearly visible in this projection. The complete structure is shown in Fig. 4-18. The two apical chelate rings N(2)—O(6) and N(1)—O(8) are almost planar, in contrast to the two equatorial glycinate rings N(1)—O(5) and N(2)—O(7), which have envelope conformations

Table 4-11 Dihedral angles, in degrees,[a] about bonds in six-membered amino acid chelate rings [27]

A—B—C—D	Ni(β-ala)$_2$	Cu(β-ala)$_2$	Cu(β-NH$_2$but)$_2$	Zn(asp)·3H$_2$O
O(1)—M—N—C(1)	10	22	6	47
M—N—C(1)—C(2)	−56	−55	−50	−80
N—C(1)—C(2)—C(3)	74	62	67	60
C(1)—C(2)—C(3)—O(1)	−27	−39	−28	−6
C(2)—C(3)—O(1)—M	−29	−6	−23	−14
C(3)—O(1)—M—N	35	9	32	−7

[a] The dihedral angle is defined as positive if the rotation is clockwise when looking along BC.

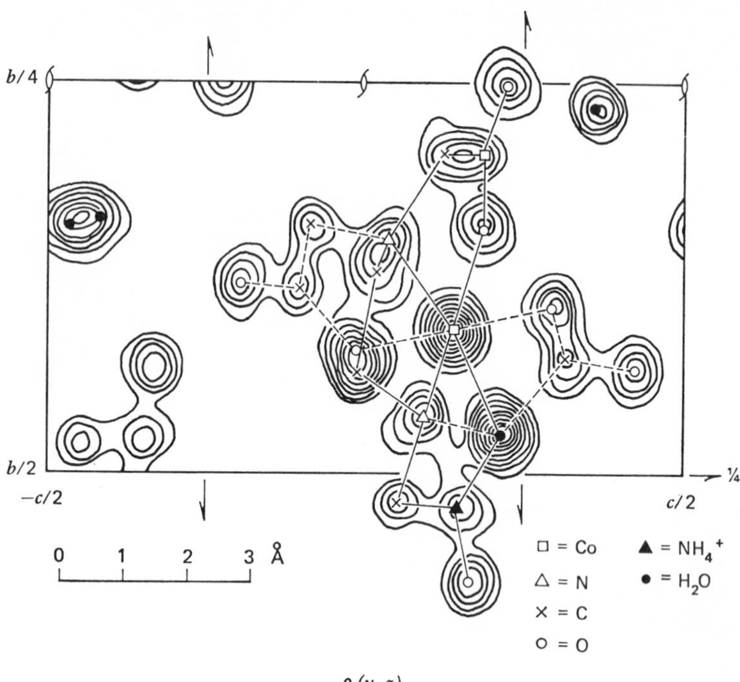

Fig. 4-17 Electron density in $NH_4[CoEDTA]\cdot 2H_2O$ projected onto (100). Contours at intervals of $2e/\text{Å}^2$, starting with $4e/\text{Å}^2$ contour, except around Co where the interval is $5e/\text{Å}^2$, starting with $5e/\text{Å}^2$ contour. The skeleton of one complex anion is indicated. [79].

and are positioned on opposite sides of the N—Co—N plane with the atoms O(5) and O(7) distorted out of the N—Co—N plane. The considerable strain in the equatorial glycinate rings has been greatly relieved by the opening up of ∠O(5)—Co—O(7) to 104°. The bond angles found for the glycinate rings (see Table 4-12) are not much different from those normally found in bidentate α-amino acid complexes (see Table 4-5).

The amine equivalent of EDTA, *N,N,N′,N′*-tetrakis(2′-aminoethyl)-1,2-diaminoethane (penten), has been studied in the complex *d*-[Co(penten)]-$[Co(CN)_6]\cdot 2H_2O$ [45]. A perspective drawing of the complex ion is presented in Fig. 4-19. In contrast to the EDTA complex, none of the chelate rings are nearly planar. Each ethylenediamine chelate ring is unsymmetrically puckered, and it would appear from the structure published that the equatorial terminal rings are of the asymmetric envelope type positioned on opposite sides of the N—Co—N plane. The dextro isomer has the D configuration with the central ethylenediamine ring having a δ conformation.

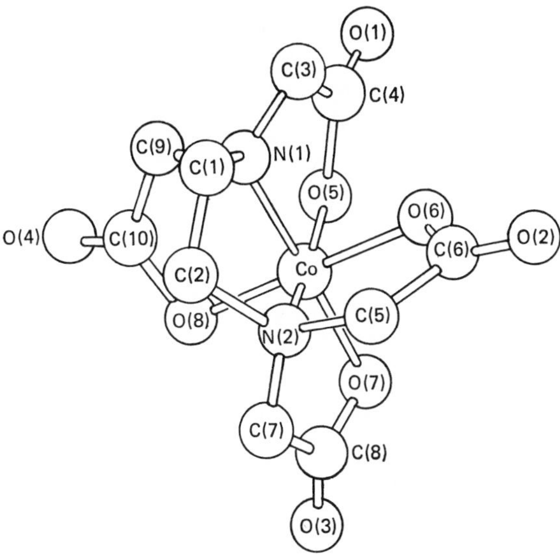

Fig. 4.18 Structure of $[CoEDTA]^-$ [79].

Table 4-12 Structural data[a] for chelate rings in $NH_4[CoEDTA]\cdot 2H_2O$ [79]

Bond length, Å					
Co—N	1.93	C—N	1.49	C—O	1.30
C—C	1.53	Co—O	1.90	C═O	1.22

Bond angle, deg					
Apical		Equatorial		Ethylenediamine	
CO—N—C	108.1	Co—N—C	105.9	Co—N—C	107.5
N—C—C	111.1	N—C—C	106.4	N—C—C	108.1
C—C—O	116.9	C—C—O	115.1	N—Co—N	89.7
Co—O—C	113.1	Co—O—C	112.9		
N—Co—O	88.5	N—Co—O	83.2		

[a] Averaged values.

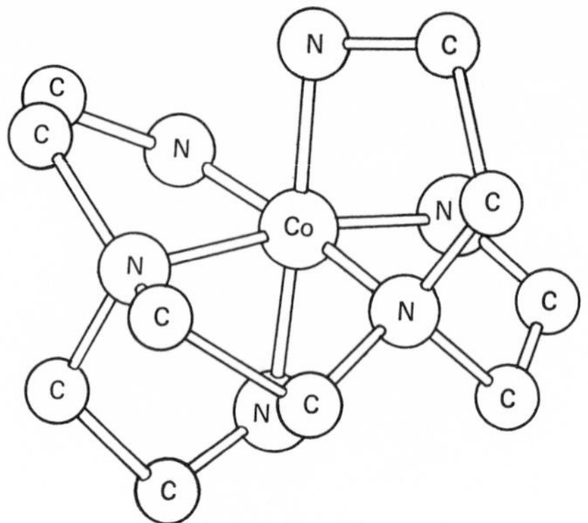

Fig. 4.19 Structure of $[Co(penten)]^{3+}$ [45].

The structure of the cyclic tetradentate amine ligand, 1,4,8,11-tetraazacyclotetradecane, has been determined for its *trans*-dichloronickel(II) complex [12]. This ligand forms two five-membered and two six-membered chelate rings, which have very similar conformations to their simple bidentate equivalents. Structural data for the rings are given in Fig. 4-20. The six-membered rings are in the chair conformation. The ring angles ∠N(1)—Ni—N(2), ∠Ni—N(1)—C(1), and ∠Ni—N(2)—C(3) are significantly greater for the six-membered rings than for the same angles in the five-membered rings, as is also found for the bidentate equivalents.

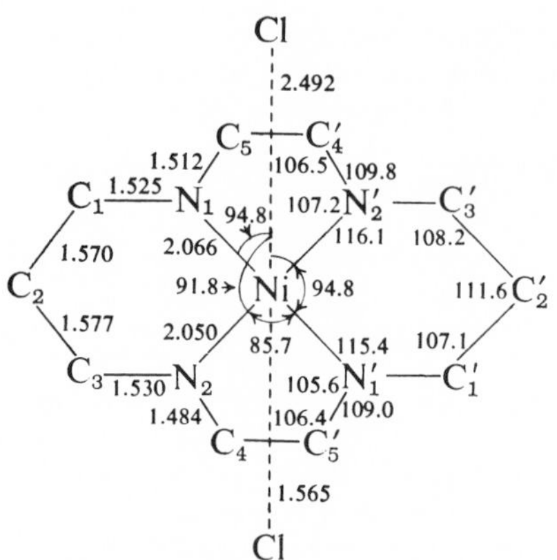

Fig. 4.20 Structural data for *trans*-dichloro-1,4,8,11-tetraazacyclotetradecanenickel(II) [12].

GENERAL READING

X-ray Crystallography

1. James, R. W., "The Optical Principles of the Diffraction of X-rays," G. Bell, London, 1954.
2. Stout, G. H., and L. H. Jensen, "X-ray Structure Determination," Macmillan, New York, 1968.

Anomalous scattering and absolute configuration

3. Bijvoet, J. M., *Endeavour*, **14**:71 (1955).
4. Ramaseshan, S., The Use of Anomalous Diffraction in Crystal Structure Analysis, in G. N. Ramachandran (ed.), "Advanced Methods of Crystallography," pp. 67–95, Academic, New York, 1964.

General Data

5. International Tables for X-ray Crystallography, vols. I, II, and III, Kynoch Press, Birmingham.

REFERENCES

6. Allen, F. H., S. Niedle, and D. Rogers, *Chem. Commun.*, **1968**:308.
7. Allen, F. H., and D. Rogers, *Chem. Commun.*, **1966**:838.
8. Barclay, G. A., E. Goldschmied, N. C. Stephenson, and A. M. Sargeson, *Chem. Commun.*, **1966**:540.
9. Becker, K. A., G. Grosse, and K. Plieth, *Z. Krist.*, **112**:375 (1959).
10. Bijvoet, J. M., *Proc. Roy. Soc. Amsterdam*, **52**:313 (1949).
11. Blount, J. F., H. C. Freeman, A. M. Sargeson, and K. R. Turnbull, *Chem. Commun.*, **1967**:324.
12. Bosnich, B., R. Mason, P. J. Pauling, G. B. Robertson, and M. L. Tobe, *Chem. Commun.*, **1965**:97.
13. Broderson, K., *Z. Anorg. Allgem. Chem.*, **298**:142 (1959).
14. Brown, B. W., and E. C. Lingafelter, *Acta Cryst.*, **16**:753 (1963).
15. Brown, B. W., and E. C. Lingafelter, *Acta Cryst.*, **17**:254 (1964).
16. Bryan, R. F., R. J. Poljak, and K. Tomita, *Acta Cryst.*, **14**:1125 (1961).
17. Buckingham, D. A., P. A. Marzilli, and A. M. Sargeson, *Inorg. Chem.*, **6**:1032 (1967).
18. Corey, E. J., and J. C. Bailar, *J. Am. Chem. Soc.*, **81**:2620 (1959).
19. Coster, D., K. S. Knol, and J. Prins, *Z. Physik.*, **63**:345 (1930).
20. Doyne, T., Ph.D. thesis, Pennsylvania State Univ., 1957.
21. Doyne, T., R. Pepinsky, and T. Watanabe, *Acta Cryst.*, **10**:438 (1957).
22. Drew, M. G. B., J. H. Dunlop, R. D. Gillard, and D. Rogers, *Chem. Commun.*, **1966**:42.
23. Dunlop, J. H., R. D. Gillard, N. C. Payne, and G. B. Robertson, *Chem. Commun.*, **1966**:874.
24. Dwyer, F. P., T. E. MacDermott, and A. M. Sargeson, *J. Am. Chem. Soc.*, **85**:661 (1963).
25. Foss, O., and K. Marøy, *Acta Cryst.*, **13**:201 (1959).
26. Fraser, K. A., and M. M. Harding, *J. Chem. Soc. A.*, **1967**:415.
27. Freeman, H. C., *Adv. Protein Chem.*, **22**:257 (1967).

28. Freeman, H. C., M. R. Snow, I. Nitta, and K. Tomita, *Acta Cryst.*, **17**:1463 (1964).
29. Fridrichsons, J., and A. McL. Mathieson, *Acta Cryst.*, **23**:439 (1967).
30. Gollogly, J. R., and C. J. Hawkins, *Inorg. Chem.*, **9**:576 (1970).
31. Hope, H., and U. de la Camp, *Nature*, **221**:54 (1969).
32. Iwasaki, H., and Y. Saito, *Bull. Chem. Soc. Japan*, **39**:92 (1966).
33. Iwasaki, H., K. Tano, and Y. Saito, to be published.
34. Iwata, M., K. Nakatsu, and Y. Saito, *Acta Cryst.*, **B25**:2562 (1969).
35. Jose, P., L. M. Pant, and A. B. Biswas, *Acta Cryst.*, **17**:24 (1964).
36. Jose, P., L. M. Pant, and A. B. Biswas, to be published. (Data taken from Ref. 27.)
37. Komiyama, Y., and E. C. Lingafelter, *Acta Cryst.*, **17**:1145 (1964).
38. Larsen, S., K. J. Watson, A. M. Sargeson, and K. R. Turnbull, *Chem. Commun.*, **1968**:847.
39. Marsh, R. E., and J. Donohue, *Adv. Protein Chem.*, **22**:235 (1967).
40. Mathieson, A. McL., *Acta Cryst.*, **9**:317 (1956).
41. Mathieson, A. McL., and H. K. Welsh, *Acta Cryst.*, **5**:599 (1952).
42. Matsumoto, K., Y. Kushi, S. Ooi, and H. Kuroya, *Bull. Chem. Soc. Japan*, **40**:2988 (1967).
43. Matsumoto, K., S. Ooi, and H. Kuroya, presented at the 17th Annual Meeting of the Chemical Society of Japan, 1964.
44. Mazzi, F., *Rend. Soc. Mineralog. Ital.*, **9**:148 (1953).
45. Muto, A., F. Marumo, and Y. Saito, *Inorg. Nucl. Chem. Letters*, **5**:85 (1969).
46. Nakahara, A., Y. Saito, and H. Kuroya, *Bull. Chem. Soc. Japan*, **25**:331 (1952).
47. Nakatsu, K., *Bull. Chem. Soc. Japan*, **35**:832 (1962).
48. Nakatsu, K., M. Shiro, Y. Saito, and H. Kuroya, *Bull. Chem. Soc. Japan*, **30**:158 (1957).
49. Nomura, T., F. Marumo, and Y. Saito, *Bull. Chem. Soc. Japan*, **42**:1016 (1969).
50. Okaya, Y., and R. Pepinsky, *Phys. Rev.*, **103**:1645 (1956).
51. Okaya, Y., and R. Pepinsky, in R. Pepinsky, J. M. Robertson, and J. C. Speakman (eds.), "Computing Methods and the Phase Problem in X-ray Crystal Analysis," pp. 273–299, Pergamon, New York, 1960.
52. Okaya, Y., Y. Saito, and R. Pepinsky, *Phys. Rev.*, **98**:1857 (1955).
53. Ooi, S., Y. Komiyama, and H. Kuroya, *Bull. Chem. Soc. Japan*, **33**:354 (1960).
54. Ooi, S., Y. Komiyama, Y. Saito, and H. Kuroya, *Bull. Chem. Soc. Japan*, **32**:263 (1959).
55. Ooi, S., and H. Kuroya, *Bull. Chem. Soc. Japan*, **36**:1083 (1963).
56. Pajunen, A., *Suomen Kemi.*, **B41**:233 (1968).
57. Pajunen, A., *Suomen Kemi.*, **B42**:15 (1969).
58. Peerdeman, A. F., and J. M. Bijvoet, *Acta Cryst.*, **9**:1012 (1956).
59. Peerdeman, A. F., A. J. van Bommel, and J. M. Bijvoet, *Proc. Roy. Soc. Amsterdam*, **54**:16 (1951).
60. Pepinsky, R., *Rec. Chem. Prog.*, **17**:163 (1956).
61. Pepinsky, R., and Y. Okaya, *Proc. Natl. Acad. Sci. U.S.*, **42**:286 (1956).
62. Pepinsky, R., and Y. Okaya, *Phys. Rev.*, **108**:1231 (1957).
63. Peterson, S. W., *Nature*, **176**:395 (1955).
64. Porai-Koshits, M. A., *Russ. J. Inorg. Chem.*, **13**:644 (1968).
65. Raymond, K. N., P. W. R. Corfield, and J. A. Ibers, *Inorg. Chem.*, **7**:842 (1968).
66. Raymond, K. N., P. W. R. Corfield, and J. A. Ibers, *Inorg. Chem.*, **7**:1362 (1968).
67. Raymond, K. N., and J. A. Ibers, *Inorg. Chem.*, **7**:2333 (1968).
68. Robinson, W. T., D. A. Buckingham, G. Chandler, L. G. Marzilli, and A. M. Sargeson, *Chem. Commun.*, **1969**:539.

69. Saito, Y., and H. Iwasaki, in S. Kirschner (ed.), "Advances in the Chemistry of Coordination Compounds," p. 562, Macmillan, New York, 1961.
70. Saito, Y., and H. Iwasaki, *Bull. Chem. Soc. Japan*, **35**:1131 (1962).
71. Saito, Y., K. Nakatsu, M. Shiro, and H. Kuroya, *Acta Cryst.*, **7**:636 (1954).
72. Saito, Y., K. Nakatsu, M. Shiro, and H. Kuroya, *Acta Cryst.*, **8**:729 (1955).
73. Saito, Y., Y. Okaya, and R. Pepinsky, *Phys. Rev.*, **100**:970 (1955).
74. Scouloudi, H., *Acta Cryst.*, **6**:651 (1953).
75. Scouloudi, H., and C. H. Carlisle, *Nature*, **166**:357 (1950).
76. Templeton, D. H., A. Zalkin, and T. Ueki, *Acta Cryst.*, **21**:A154 (1966) (suppl.).
77. Tomita, K., *Bull. Chem. Soc. Japan*, **34**:280 (1961).
78. Tomita, K., *Bull. Chem. Soc. Japan*, **34**:297 (1961).
79. Weakliem, H. A., and J. L. Hoard, *J. Am. Chem. Soc.*, **81**:549 (1959).
80. Wunderlich, J. A., *Chem. and Ind. London*, **1962**:2089.
81. Wunderlich, J. A., *Acta Cryst.*, **23**:846 (1967).
82. Yasui, T., and Y. Shimura, *Bull. Chem. Soc. Japan*, **39**:604 (1966).

5

Circular Dichroism

HISTORICAL

The early studies of the polarization of light soon led to the discovery in the nineteenth century of the twin phenomena of *circular birefringence* and *circular dichroism*. It was found that certain crystalline materials, such as quartz, and solutions containing particular compounds—oils of turpentine, laurel, and lemon, for example—had the ability to rotate the plane of polarization of plane-polarized light and that this ability depended on the energy of the light:

> "Luminous molecules of different kinds, which have passed through the plate rock crystal, have, by the action of this plate, turned their axis of polarization into different azimuths. . . . The violet molecules turn faster than the blue, the blue faster than the green, the green faster than the yellow, and so on to the red molecules which will be slowest of all" [13].

Fresnel realized that this was due to the fact that the plane-polarized light is composed of two circularly polarized rays of opposite sign, which travel with unequal velocities within these media—circular birefringence [60]. It was also observed that these rays were absorbed unequally while passing through such media—circular dichroism [37,75]. Crystals and solutions that exhibited these properties were classified as *optically active*.

In an attempt to explain the circular birefringence measured along the optic axis of quartz Fresnel suggested that the arrangement of the molecules in the crystal is helicoidal, so that the two circularly polarized rays were not presented with the same aspect while traversing the medium [61]. This proposal was shown to be correct, at least in principle, almost a century later, when the X-ray analysis of quartz showed that the silicon and oxygen atoms formed a giant helix [23,24,65].

Pasteur [157] studied the symmetry requirements for an object, such as a hemihedral crystal or a molecule, to have a nonsuperimposable mirror image. He proposed that if molecules are to be optically active in solution they must be dissymmetric. Since then the study of the stereochemical problem of the dissymmetry of molecules and the study of the spectroscopic problem of optical activity have proceeded almost completely interdependently.

INTERACTION OF LIGHT WITH AN OPTICALLY ACTIVE MEDIUM

As the two manifestations of optical activity, circular birefringence and circular dichroism, arise from the interaction of light with matter, we need to consider how this interaction occurs.

Plane-polarized light has an electric and a magnetic field that can be represented by oscillating vectors

$$\mathbf{E} = \mathbf{E}_0 \cos 2\pi\nu t \tag{5-1}$$

$$\mathbf{H} = \mathbf{H}_0 \cos 2\pi\nu t \tag{5-2}$$

where $\mathbf{E}_0$ and $\mathbf{H}_0$ are the maximum amplitudes, ν is the frequency, and t is the time, constrained to lie in planes that are perpendicular to each other, the oscillations being at right angles to the direction of propagation (see Fig. 5-1). The plane of polarization is usually taken to be that of the electric vector, although this has not always been the case.

Plane-polarized light can be considered the combination of two *circularly polarized* components of opposite chirality shown in Fig. 5-2, the electric vectors* of which trace out cylindrical helices with the same frequency as

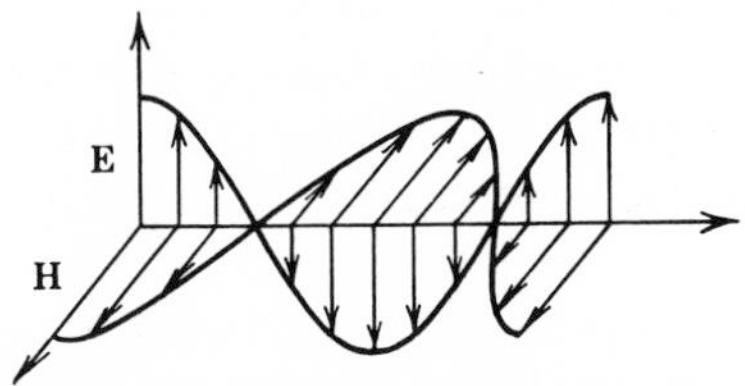

Fig. 5-1 The electric (E) and magnetic (H) fields in plane polarized light.

* As the electric and magnetic vectors are always at right angles, we need to consider only one of them.

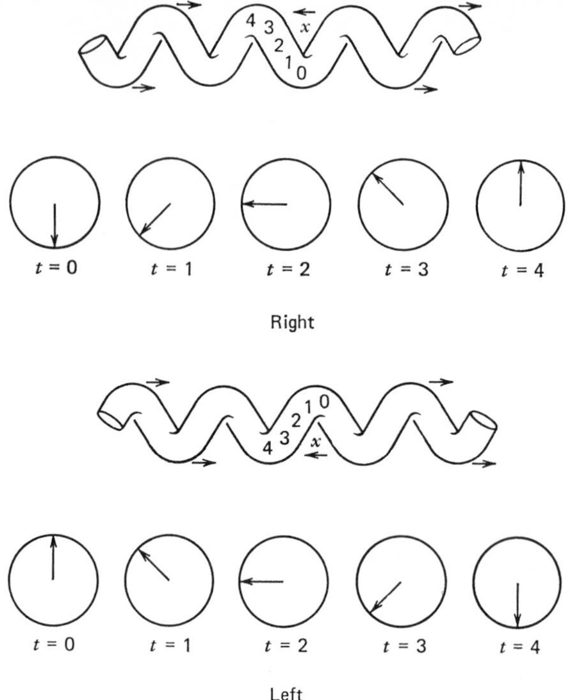

Fig. 5-2 Left and right circularly polarized light.

the plane-polarized radiation and have amplitudes equal to half that of the resultant radiation. The trace of the circularly polarized vector in a plane perpendicular to the direction of propagation of the light (its vibration form) is a circle. If the oncoming ray is viewed from a point x, the electric vector is in the position shown in Fig. 5-2 at the time $t = 0$. After times $t1$, $t2$, $t3$, and $t4$ the points 1, 2, 3, and 4 are aligned with x, and the electric vector is in the positions displayed in the figure. For right circularly polarized light the vector moves in a clockwise direction and for left, in an anticlockwise direction.

When plane-polarized light is passed through an optically active transparent medium, the refractive indices of the left and right circularly polarized components are different. If n_l is greater than n_r, the left is delayed relative to the right circularly polarized wave, and after traversing the medium, the two components are out of phase by an angle ϕ (see Fig. 5-3) and recombine to form plane-polarized light whose plane has been rotated

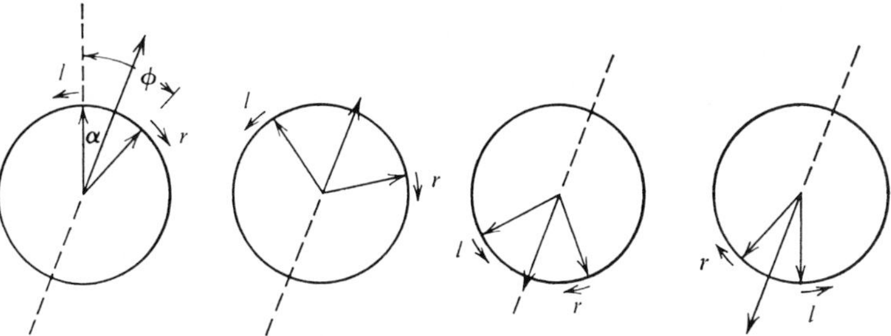

Fig. 5-3 Rotation of the plane of polarization by an optically active medium with $n_l > n_r$.

an angle α in the clockwise (positive) direction. The angle of rotation, in degrees per decimeter, is given by the Fresnel equation

$$\alpha = \frac{1800(n_l - n_r)}{\lambda_{\text{vac}}} \tag{5-3}$$

If the radiation has a frequency close to an absorption band, both the refractive indices and the extinction coefficients for the two forms of circularly polarized light are not identical. Thus, after traversing the medium, the two circularly polarized waves do not recombine to form plane-polarized light, but have an elliptically polarized vibration form, the major axis of which is tilted by an angle α from the original plane of polarization (see Fig. 5-4). The ellipticity (the ratio of the minor to the major axes) is a measure of the circular dichroism and is given by the expression

$$\psi \simeq (0.576)lc(\varepsilon_l - \varepsilon_r) \qquad \text{radians}$$

where l is the path length in centimeters, and c is the molar concentration.

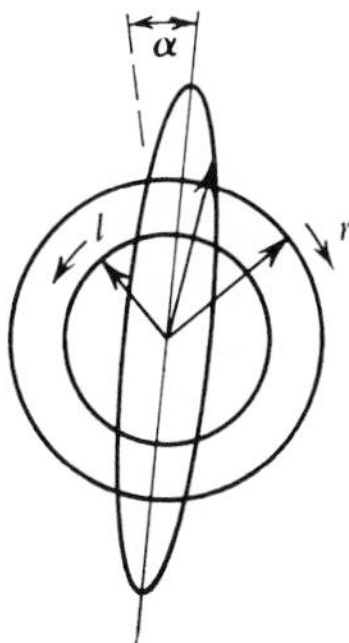

Fig. 5-4 Elliptical vibration form after a plane polarized ray has traversed an optically active medium with $n_l > n_r$ and $\varepsilon_l > \varepsilon_r$.

Let us now consider in greater detail what happens when electromagnetic radiation interacts with an isotropic medium. First, the value of $\mathbf{E}_0$ for a vacuum is reduced because of the induction of dipole moments in the molecules of the material. The induced moment per molecule $\mathbf{p}$ is given by

$$\mathbf{p} = \alpha \mathbf{E}' \tag{5-5}$$

where α is the molecular polarizability, and $\mathbf{E}'$, the instantaneous value of the field strength at the molecule, is expressed as

$$\mathbf{E}' = \mathbf{E} + \frac{4\pi \mathbf{P}}{3} \tag{5-6}$$

where $\mathbf{P} = N'\mathbf{p}$ for N' molecules per cm^3. At optical frequencies, $\mathbf{E}'$ is given approximately by the Lorentz field

$$\mathbf{E}' = \frac{(n^2 + 2)\mathbf{E}}{3} \tag{5-7}$$

and therefore, from Eqs. (5-1), (5-5), and (5-7),

$$\mathbf{p} = \frac{\alpha(n^2 + 2)}{3} \mathbf{E}_0' \cos 2\pi\nu t \tag{5-8}$$

where $\mathbf{E}_0'$ is the amplitude of the electric vector in the medium and is related to that in a vacuum by

$$\mathbf{E}_0' = \frac{\mathbf{E}_0}{\epsilon} \tag{5-9}$$

ϵ being the dielectric constant for the medium. For diamagnetic media, the induced magnetic moment $\mathbf{m}$ is negligible, and thus the effective magnetic field at the site of the molecule is equal to the external field.

For a medium containing optically active molecules these equations no longer hold, because there is an additional electric moment induced in the molecules because of the change in the magnetic field with time, and further, an induced magnetic moment is produced because of the change in the electric field with time:

$$\mathbf{p} = \alpha \mathbf{E}' - \frac{\beta}{c} \frac{\partial \mathbf{H}}{\partial t} \tag{5-10}$$

$$\mathbf{m} = \frac{\beta}{c} \frac{\partial \mathbf{E}}{\partial t} \tag{5-11}$$

where c is the velocity of light and β is an optical activity molecular parameter, which is a measure of the ability of a molecule to assume an electric moment in an alternating magnetic field and vice versa. According to the

Lorentz local electric field,

$$\mathbf{p} = \frac{(n^2 + 2)}{3} \alpha \mathbf{E}_0' \cos 2\pi\nu t + \frac{(n^2 + 2)}{3} \beta \frac{2\pi\nu}{c} \mathbf{H}_0 \sin 2\pi\nu t \qquad (5\text{-}12)$$

$$= \mathbf{p}_\alpha + \mathbf{p}_\beta$$

$$\mathbf{m} = -\frac{(n^2 + 2)}{3} \frac{\beta 2\pi\nu}{c} \mathbf{E}_0' \sin 2\pi\nu t \qquad (5\text{-}13)$$

$$= \mathbf{m}_\beta$$

Because $\mathbf{p}_\alpha$ is common to both isotropic and optically active media, the important parameters are $\mathbf{p}_\beta$ and $\mathbf{m}_\beta$. From these equations it can be shown that

$$\alpha = 1800\left(\frac{4\pi}{\lambda}\right)^2 \frac{(n^2 + 2)}{3} N'\beta \qquad (5\text{-}14)$$

$$[\alpha] = \frac{\alpha N}{N'M} \qquad (5\text{-}15)$$

where $[\alpha]$ is the specific rotation, N is Avogardo's number, and M is the molecular weight.

ROTATIONAL STRENGTH OF CHROMOPHORES

Rosenfeld [168] used quantum-mechanical perturbation theory to calculate the induced electric and magnetic moments. He showed that

$$\beta = \frac{c}{3\pi h} \sum_b \frac{R_{ba}}{\nu_{ba}{}^2 - \nu^2} \qquad (5\text{-}16)$$

where h is Planck's constant, ν is the frequency of the radiation, ν_{ba} is the frequency of a transition from a ground state a to an excited state b, and R_{ba}, the *rotational strength* of the transition $a \to b$, is given by

$$R_{ba} = \operatorname{Im}\{(a|\mathscr{P}|b)\cdot(b|\mathscr{M}|a)\} \qquad (5\text{-}17)$$

where Im means *the imaginary part of*—$\operatorname{Im}\{u + iv\} = v$, for example—and $(a|\mathscr{P}|b)$ and $(b|\mathscr{M}|a)$ are the matrix components of the electric and magnetic moments connecting the states a and b. This may be simplified to

$$R_{ba} = pm \cos \theta \qquad (5\text{-}18)$$

where p and m are the respective moments and θ the angle between them.

It can be shown that

$$\sum_b R_{ba} = 0 \qquad (5\text{-}19)$$

that is, the sum of the rotational strengths for all transitions from the ground state a is equal to zero; for example,

$$\sum_b R_{ba} = \text{Im}\left\{\sum_b [(a|\mathscr{P}|b)\cdot(b|\mathscr{M}|a)]\right\}$$
$$= \text{Im}\,(a|\mathscr{P}\cdot\mathscr{M}|a) \tag{5-20}$$

Because $(a|\mathscr{P}\cdot\mathscr{M}|a)$ is a diagonal element of a real observable and, therefore, must be real, the imaginary part of this real number is zero.

For a transition to be optically active, its rotational strength must be different from zero. This is possible only if the chromophore has no improper axis of rotation, including a plane and center of symmetry. These are the conditions which are also necessary for optical isomerism to be possible. Their justification can be relatively easily shown. For example, if there is a center of symmetry, the states of the molecule are either even or odd with respect to inversion. $\mathscr{P}$ changes sign on inversion as

$$\mathscr{P} = (\mathbf{i}x + \mathbf{j}y + \mathbf{k}z) \tag{5-21}$$

where **i**, **j**, and **k** are unit vectors along mutually perpendicular axes. If there is to be a nonvanishing value for $(a|\mathscr{P}|b)$, then either but not both a and b must be odd with respect to inversion. On the other hand, $\mathscr{M}$ does not change sign on inversion as

$$\mathscr{M} = \left[\mathbf{i}\left(y\frac{\partial}{\partial z} - z\frac{\partial}{\partial y}\right) + \mathbf{j}\left(z\frac{\partial}{\partial x} - x\frac{\partial}{\partial z}\right) + \mathbf{k}\left(x\frac{\partial}{\partial y} - y\frac{\partial}{\partial x}\right)\right] \tag{5-22}$$

and, therefore, $(b|\mathscr{M}|a)$ is nonvanishing only between states that are either both even or both odd. The conditions for $(a|\mathscr{P}|b)$ and $(b|\mathscr{M}|a)$ to be different from zero cannot hold simultaneously. Thus, if the chromophore has a center of symmetry, the rotational strength is zero.

Rosenfeld [168] also showed that

$$\alpha = \frac{2}{3h}\sum_b \frac{v_{ba}D_{ba}}{\nu_{ba}{}^2 - \nu^2} \tag{5-23}$$

where D_{ba}, the *dipole strength* of the transition, is given by

$$D_{ba} = |(a|\mathscr{P}|b)|^2 \tag{5-24}$$

THE COTTON EFFECT

The optical activity of a compound can be investigated by either circular birefringence or circular dichroism measurements. Traditionally the former is studied by measuring the rotation of the plane of polarization of plane-polarized light. When this is done as a function of the energy of the radiation, the technique is called *optical rotatory dispersion* (ORD). For an isolated nondegenerate absorption band the shape of the ORD curve mirrors that of the variation of refractive index with energy (see Fig. 5-5). It is to be noted that in practice the difference in the refractive indices of a compound for left and right circularly polarized light is very small (perhaps in the sixth place of decimals). Drude [54] first showed that the rotation for an isolated transition is proportional to $1/(\lambda^2 - \lambda_{ba}^2)$. This is not obeyed in the vicinity of the transition $a \rightarrow b$, because the rotation would be expected to go to $\pm\infty$. The observed curve results from radiation damping and can be fitted by an expression that has been modified to allow for the finite bandwidth of the absorption.

In this book, it is not intended to consider optical rotatory dispersion further. As a technique it is extremely important for the study of transitions in the ultraviolet, because, even at energies considerably removed from those of the actual electronic transitions, the rotation can be different from zero. Thus, some data can be gained for transitions that are themselves inaccessible because of instrumental limitations. However, the very nature of the shape of the ORD curve in the vicinity of the absorption band and the fact that the rotation is nonzero at energies remote from the excitation energy for an absorption make it exceptionally difficult to resolve the observed curves into components for individual transitions, especially for the *d–d*

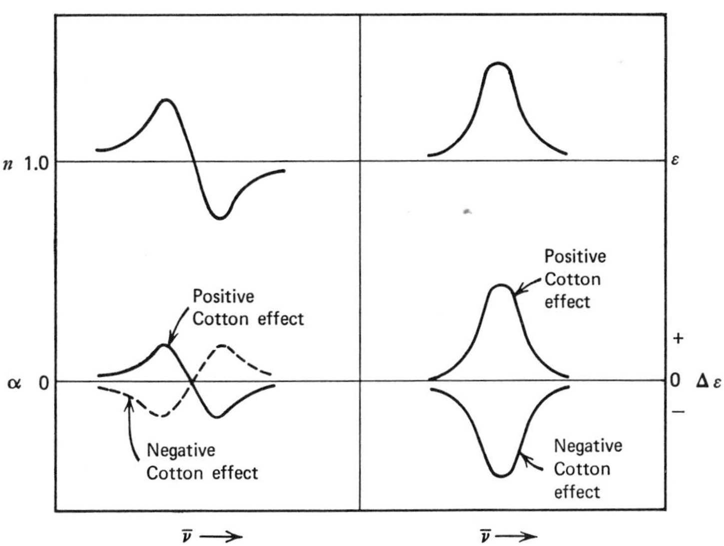

Fig. 5-5 Idealized curves for an isolated transition.

transitions of metal complexes, which are often very close in energy. This resolution is essential for the assignment of absolute configuration, because the crucial observation is the sign of the rotational strength for the individual transitions.

It is far preferable to study the circular dichroism (CD) of the compound: the CD bands are restricted to the region of the absorption, and their shape approximates that of a Gaussian band, thus simplifying the identification of the contributions of the transitions (see Fig. 5-5). In some cases, however, even with CD, it is extremely difficult to resolve an observed spectrum into the components for the individual transitions. In Fig. 5-6, four possible spectra are shown that have resulted from the mutual cancelation of two CD bands: in (*a*) the negative band has been completely swamped, a positive band of reduced intensity appearing; in (*b*) both positive and negative bands of considerably reduced intensity appear, the energies of their maxima differing considerably from the actual energies of the transitions; in (*c*) the negative band has partially canceled out the positive resulting in a double-humped positive band that could be incorrectly interpreted as two positive transitions; and in (*d*) the cancelation has led to the appearance of three bands (+, −, +). Nevertheless, as long as one is aware of this problem, the interpretation of the CD spectra is usually not too difficult.

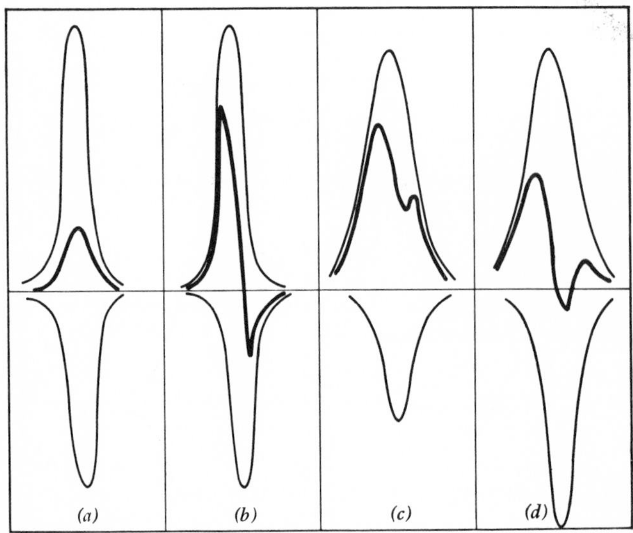

Fig. 5-6 Resultant CD curves from the mutual cancelation of a positive and a negative Cotton effect.

Cotton [37] was the first to observe a CD band, and its appearance as well as the appearance of the *anomalous rotatory dispersion* in the vicinity of an absorption band has been called a *Cotton effect*. These can be positive or negative, as shown in Fig. 5-5.

The rotational strength of an isolated transition can be obtained from the CD by the equation*

$$R_{ba} = \frac{3hc\, 10^3 \ln 10}{32\pi^3 N} \int \frac{\Delta\varepsilon}{\nu}\, d\nu \tag{5-25}$$

$$= 22.9 \times 10^{-40} \int \frac{\Delta\varepsilon}{\nu}\, d\nu \qquad \text{cgs units} \tag{5-26}$$

If the curve is Gaussian,

$$R_{ba} = 2.45 \times 10^{-40} \frac{\Delta\varepsilon_{\max} \Delta\nu_{1/2}}{\nu_{ba}} \tag{5-27}$$

As stated above, if there is serious overlapping between the CD bands of two or more adjacent transitions, it is extremely difficult objectively to resolve the observed spectrum into Gaussian components to calculate the actual rotational strengths of the individual transitions. This does not seem to restrict some workers who, apparently without hesitation, sketch up to six Gaussian bands under a complicated curve and work out the individual rotational strengths.

The dipole strength and the oscillator strength f_{ba} of a transition can similarly be determined from an absorption spectrum:

$$D_{ba} = \frac{3hc\, 10^3 \ln 10}{8\pi^3 N} \int \frac{\varepsilon}{\nu}\, d\nu \tag{5-28}$$

$$= 91.8 \times 10^{-40} \int \frac{\varepsilon}{\nu}\, d\nu \qquad \text{cgs units} \tag{5-29}$$

$$f_{ba} = \frac{8\pi^2 mc\nu}{3he^2} D_{ba} \tag{5-30}$$

$$= 0.476 \times 10^{30} \nu D_{ba} \qquad \text{cgs units} \tag{5-31}$$

The sum of the oscillator strengths of all the electronic transitions should equal the number of valence electrons.

ABSORPTION SPECTRA OF METAL COMPLEXES

In a book of this kind there is no necessity to go into great detail about the whole subject of absorption spectra of metal complexes. Suitable monographs are available on this subject (see the list of recommended texts at

* Units for $\Delta\varepsilon$ are cm^{-1} l mol^{-1}. Molecular ellipticity curves are sometimes published in place of $\Delta\varepsilon$. These are related by $[\theta]$ (deg cm^{-1} mol^{-1} ml × 10) = 3298 $\Delta\varepsilon$.

the end of the present chapter). There are some aspects, however, that are extremely pertinent to this discussion, and they will be dealt with here. A crucial issue in the determination of the absolute configuration from the measurement of CD is the correct assignment of the sign of the Cotton effect for particular transitions. This requires that the transitions are correctly identified. There are three steps to the solution of this problem for the *d–d* transitions of octahedral metal complexes:

1. The determination of the holohedrized* symmetry of the orthoaxial† chromophores and the splittings of the cubic levels under this field.
2. The evaluation of the effective molecular symmetry that gives rise to the splitting of degenerate levels from the holohedrized symmetry, that is, the prediction of the number of nondegenerate transitions that lie under the broad absorption bands.
3. The assessment of the relative energies of the transitions and their symmetry properties, including the determination of the relationship of these transitions to those of different symmetries.

In the ensuing discussion we consider only those electronic configurations which are dealt with in the CD discussions: mainly d^3 and spin-paired d^6, but also d^8 and d^9. The spin allowed transitions of interest for the d^3 and d^6 configurations have T_{1g} and T_{2g} symmetries‡ for d^3, ${}^4A_{2g} \rightarrow {}^4T_{2g}$ (lower in energy) and ${}^4A_{2g} \rightarrow {}^4T_{1g}$; and for d^6, ${}^1A_{1g} \rightarrow {}^1T_{1g}$ (lower) and ${}^1A_{1g} \rightarrow {}^1T_{2g}$. Under the octahedral ligand field both these transitions are electric-dipole-forbidden, and this holds essentially true even when the molecular symmetry of an octahedral complex is much lower than O_h. The electric-dipole-moment operator has T_{1u} symmetry, and $T_{1u} \times T_{1g}$ and $T_{1u} \times T_{2g}$ do not contain the totally symmetric representation, which is a requirement for an allowed transition. The *d–d* transitions gain electric-dipole intensity through a vibronic mechanism involving the vibrations of ML_6. In complexes of lower symmetry than O_h, where the improper axes of rotation, including the center and planes of symmetry, have been removed, the static low-symmetry field also

* *Holohedrized symmetry* is derived after any contribution occurring at x, y, z is divided into two halves, one half is left at x, y, z, the other is removed to $-x$, $-y$, $-z$ [178]. Thus, if a metal ion is feeling the field of two ligands a and b along a particular axis a—M—b, the interaction can be represented by c—M—c, where $c = \frac{1}{2}a + \frac{1}{2}b$. This means that for a basically octahedral chromophore the holohedrized symmetry cannot be less than D_{2h}.

† An *orthoaxial chromophore* is one in which, if the metal ion is placed at the origin of a Cartesian coordinate system, the complex can be oriented so that the axes are directed through the ligand nuclei [178]. For each Cartesian axis it is the sum of the contributions that is important in determining the energy of the *d* orbitals.

‡ The symmetry of a transition is given by the direct product of the irreducible representations of the ground and excited states.

contributes to the transition dipole moment. At present, however, little quantitative information is available regarding this contribution.

The T_{1g}, but not the T_{2g}, transition is magnetic-dipole-allowed because the magnetic-dipole-moment operator transforms as T_{1g}. Moffitt has calculated that the magnetic moment centered in this transition has the value $\sqrt{12}$ BM for the d^3 and $\sqrt{24}$ BM for the d^6 configurations [144]. The O_h selection rules are sufficiently powerful, even for the dissymmetric molecules, to ensure that the rotational strengths of the transitions derived from $T_{1g}(O_h)$ are of a higher order of magnitude than those from T_{2g}. For the T_{2g} transitions to become optically active they must borrow both electric and magnetic dipole moments, whereas those from T_{1g} require only electric dipole. This makes the T_{2g} components less reliable for assigning absolute configurations than their T_{1g} counterparts.

There are three fundamental holohedrized symmetries that we need to consider here: cubic, tetragonal, and rhombic. For the cubic field the contributions along the three axes are identical; for the tetragonal, the ligand fields along two of the axes x, y are the same but are different to the field along the third axis z; and for the rhombic symmetry the fields along the three axes are different from one another. It is this holohedrized symmetry that determines the gross features of an absorption spectrum. For example, the solution spectra of complexes such as $[Coen_3]^{3+}$ and *cis*-$[Cogly_3]$ show no splitting of the T_{1g} or T_{2g} absorption bands, because the contributions along the three axes are equivalent. The degeneracy of the transitions has been partly removed because of the lower D_3 and C_3 molecular fields, and this is apparent in their CD spectra, as will be discussed later. This trigonal splitting, however, is so small that it is not apparent in the absorption spectra.

For an octahedral complex ML_6 the energy of the T_{1g} transition is given by

$$\mathscr{E}_{^1T_{1g}} = \Delta_L - 35F_4 \qquad \text{for } d^6 \tag{5-32}$$

$$\mathscr{E}_{^4T_{1g}} = \Delta_L \qquad \text{for } d^3 \tag{5-33}$$

where Δ_L is the ligand-field splitting parameter for the complex ML_6, and F_4 is one of the Slater-Condon interelectronic repulsion parameters. If a number of the L groups are replaced by different ligands X, the approximate energy of the T_{1g} band is given by the *rule of average environment*, which states that the position of such a band is given by the weighted average of the energies of this transition for the complexes ML_6 and MX_6. For example, $\mathscr{E}_{^1T_{1g}}$ for $[Co(NH_3)_4ox]^+$ is given by $\frac{2}{3}\mathscr{E}_{^1T_{1g}}[Co(NH_3)_6]^{3+} + \frac{1}{3}\mathscr{E}_{^1T_{1g}}[Coox_3]^{3-}$, which equals 19.58 kK, in excellent agreement with the observed maximum of 19.53 kK.*

* The tetragonal splitting is apparently such that the absorption maximum corresponds to $\mathscr{E}_{T_{1g}}$.

When a tetragonal perturbation is applied to the cubic field, the T_{1g} and T_{2g} bands are each split to give two transitions with E_g and A_{2g}, and E_g and B_{2g} symmetries, respectively. For this study we are mainly interested in the relative energies of the tetragonal components for the T_{1g}. Many models have been proposed to calculate this, using both crystal-field and molecular-orbital approaches. They are not reviewed in detail here, but the reader is referred to two papers dealing with this subject by Yamatera [197] and by Schäffer and Jørgensen [178]. The tetragonal splitting can be expressed as

$$\Delta\mathscr{E}_{\text{tet}}(T_{1g}) = \mathscr{E}_{A_{2g}} - \mathscr{E}_{E_g} = \frac{1}{2}\left\{\frac{\sum_{xy} \Delta_{L_{xy}}}{4} - \frac{\sum_z \Delta_{L_z}}{2}\right\} \tag{5-34}$$

where $\sum_{xy} \Delta_{L_{xy}}$ and $\sum_z \Delta_{L_z}$ are the sum of the Δ values for octahedral complexes with each of the ligands in the *xy* plane and on the *z* axis; for example, for the oxalatotetramminecobalt(III) complex considered above

$$\begin{aligned}\Delta\mathscr{E}_{\text{tet}}(^1T_{1g}) &= \frac{1}{2}\left\{\frac{2\Delta_{\text{ox}} + 2\Delta_{\text{NH}_3}}{4} - \frac{2\Delta_{\text{NH}_3}}{2}\right\}\\ &= \tfrac{1}{4}(\Delta_{\text{ox}} - \Delta_{\text{NH}_3})\\ &= -1.3 \text{ kK}\end{aligned}$$

the $^1A_{2g}$ is lower in energy than the ^1E_g component by 1.3 kK. In Fig. 5-7 an energy-level diagram is presented, showing the relative energies of T_{1g} states and their components for the complexes ML_6, MX_6, ML_5X, *trans*-ML_4X_2, and *cis*-ML_4X_2.

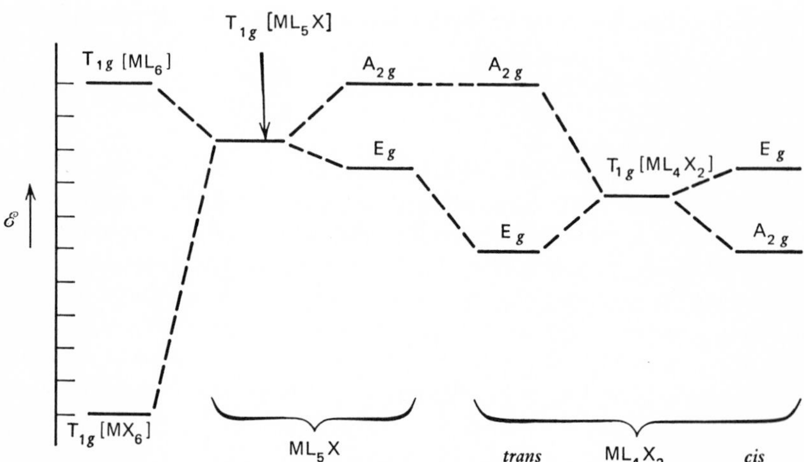

Fig. 5-7 Energy level diagram for the T_{1g} transition of some tetragonal complexes.

The expression above for the splitting ignores other terms, such as the difference in the configurational interaction energy for the E_g and A_{2g}, which, although they are relatively minor compared with the splittings above, can account for the observed discrepancies between the calculated and the experimentally found splittings [92]. For example, there is interaction between the E_g levels belonging to the T_{1g} and T_{2g} states, having the effect of increasing the tetragonal splitting for a compound with the E_g component lower in energy and of decreasing the splitting when the components are reversed, as shown in Fig. 5-8.

When the chromophore has a rhombic holohedrized symmetry, using considerations similar to the above, the various models predict that the order of the three components is the reverse of the order of the field strengths along the axes that transform as the component transitions. For *trans*-$[Co(NH_3)_3Cl_3]$ the components, which transform as the N–N, N–Cl, and Cl–Cl axes, increase in energy in that order.

After determining the holohedrized symmetry and estimating the tetragonal or rhombic splitting under this symmetry, we are in a position to interpret the solution absorption spectra. However, the holohedrized symmetry is too high for the interpretation of the CD spectra, because it contains a center and planes of symmetry. The degeneracy of the transitions is reduced because of the lower molecular symmetry. This secondary splitting has been the subject of some controversy, especially with regard to the trigonal splitting of the T_{1g} band.

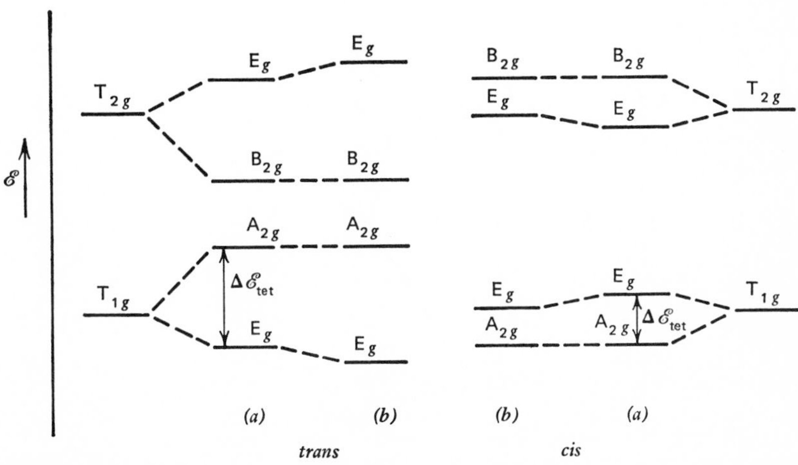

Fig. 5-8 Tetragonal splittings of T_{1g} and T_{2g} bands for *trans* and *cis*-$[Co(NH_3)_4Cl_2]^+$ (not to scale): (*a*) without and (*b*) with configurational interaction.

Schäffer interpreted the CD solution spectrum of d-$[Coen_3]^{3+}$ under the $^1A_{1g} \rightarrow {}^1T_{1g}$ absorption band as being composed of three Gaussian (in wavelength) bands, two degenerate positive components of $E(D_3)$ and a negative band corresponding to the $A_2(D_3)$ transition [175,176]. The splitting was estimated to be about 120 cm^{-1}, with the E transition lower in energy. Dingle, however, measured polarized crystal spectra of $[Coen_3]^{3+}$ in three different crystal lattices and found that the trigonal splitting of the 0,0 bands was close to zero, whereas the maxima for the E and A_2 components, which derive mainly from vibronic interactions, were separated by 140 cm^{-1}, the A_2 component being at the lower energy [43,45]. Dingle suggested that, because the rotational strength depends on the electric dipole moment from the static low-symmetry field and not from the vibronic contributions [145], the trigonal splitting of the 0,0 bands, or that of the bands constructed from totally symmetric vibrations based on the 0,0 bands, is the pertinent splitting as far as the CD is concerned. Schäffer has pointed out, however, that the transitions involved are Franck-Condon vertical transitions with the internuclear distances distributed as in the ground state and that the splitting is the energy difference between points

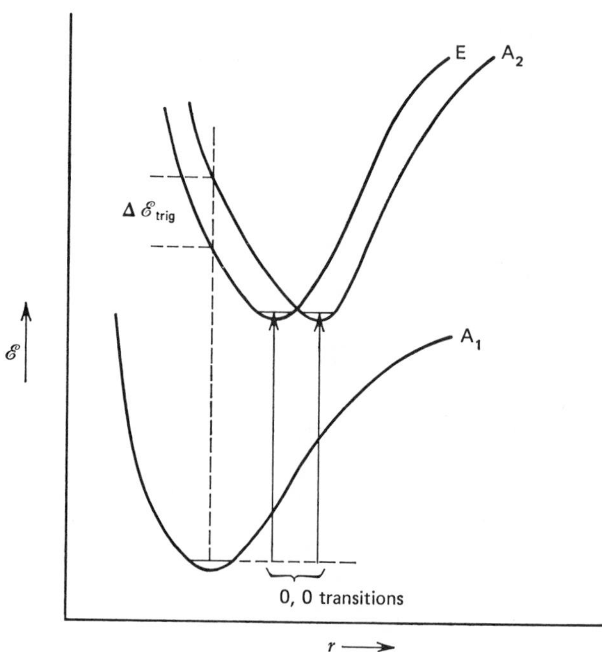

Fig. 5-9 Possible potential energy diagram for $M(aa)_3$ (D_3).

on the potential surfaces of the A_2 and E states far away from the minima [177]. This could be significant even if the 0,0 levels were degenerate (see Fig. 5-9). Schäffer further pointed out that even a totally symmetric vibration of a dissymmetric chromophore is able to mix ungerade functions into the gerade *d*-basis set and so make the transitions electric-dipole-allowed and also enable the D_3 selection rules to be applied, although there is a vibronic contribution to the intensity.

Denning has more recently proposed that a dynamic Jahn-Teller distortion of the excited state is responsible for the small trigonal field splitting and has quenched the orbital angular momentum of the excited state [41]. The latter was found from magnetic-circular-dichroism studies. Denning also studied the vibrational fine structure of the $^1A_1 \rightarrow {}^1E(^1T_{1g})$ CD band of *d*-$[Coen_3]^{3+}$ measured in *dl*-$([Rhen_3]Cl_3)_2 \cdot NaCl \cdot 6H_2O$ at 5°K. He found that the same vibrational structure appears in the CD and absorption spectra, showing that the same vibronic intensity mechanism is operating in both cases. He claimed that the energies of the vibrations agree well with the six skeletal frequencies of CoN_6 and that the relative intensities bear no apparent relationship to the *g* or *u* character of the vibrations due to Jahn-Teller distortion of the upper state. However, there seems to be no foundation for saying that the vibrations correspond to the six skeletal vibrations of CoN_6, especially because the vibrations are of the excited state. Denning finds vibrational modes with energies of 174 (T_{1g}), 246 (T_{2g}), 332 (T_{1u}), 408 (E_g), 445 (T_{1u}), and 491 (A_{1g}) cm^{-1} and assigns them as indicated.* IR and Raman studies of the ground state have led to the CoN_6 skeletal vibrations for $[Co(NH_3)_6]^{3+}$ being assigned to the 325 (T_{1u}), 440 (E_g), 495 (A_{1g}), 503 (T_{1u}) cm^{-1} bands, and the T_{2g} has been calculated to be about 370 cm^{-1}. The energy of the forbidden T_{2u} transition is not known [99]. Dingle and Ballhausen interpreted the vibrational fine structure in the T_{1g} absorption band of *dl*-$([Coen_3]Cl_3)_2 \cdot NaCl \cdot 6H_2O$ as being due to progressions of a 255 cm^{-1} vibration built on one quantum of the following modes: 185, 255, 345, and 400 cm^{-1} [45]. They proposed that these vibrations are associated with the D_3 components of the two T_{1u} and T_{2u} CoN_6 vibrations. It is interesting to compare them with the vibrational fine structure found by Wentworth for $[Co(NH_3)_6]^{3+}$: a *uniquantal progression* of a 420 cm^{-1} mode in combination with a 390 cm^{-1} mode [193]. Obviously more work needs to be done before the factors governing the splitting of the E and A_2 components are fully understood.

From the CD studies it is apparent that dissymmetric complexes with a tetragonal holohedrized symmetry have three nondegenerate bands with T_{1g} parentage, one corresponding to the $A_{2g}(D_{4h})$, the other two coming

* In the octahedral symmetry of CoN_6 there is no vibration with T_{1g} symmetry.

from the $E_g(D_{4h})$ transition. The observed splitting of the E_g level has been explained in a number of ways for complexes of the type *trans*-$[Co(l\text{-pn})_2Cl_2]^+$. Hawkins, Larsen, and Olsen attributed the two CD bands with $E_g(D_{4h})$ parentage to the D_2 effective molecular symmetry (the methyl groups were ignored) [87]. Dingle, however, has studied polarized crystal absorption spectra of this complex in the crystal $[Co(l\text{-pn})_2Cl_2]Cl \cdot HCl \cdot 2H_2O$ and also of the equivalent ethylenediamine complex as its perchlorate salt [44]. He assumed that the two conformations were enantiomeric in the ethylenediamine complex as was found in *trans*-$[Coen_2Cl_2]Cl \cdot HCl \cdot 2H_2O$ [150] in contrast to the propylenediamine complex, which has two λ conformations [172]. Because he found that the two measured spectra were very similar, he concluded that the conformations were unimportant in determining the features of the spectra.* The polarized-crystal spectra show two transitions in the vicinity of the E_g band, one polarized parallel to the Cl—Co—Cl axis, the other perpendicular to it, the two separated by about 100 to 250 cm^{-1}, depending on the temperature. The polarization results could be equally explained by either the D_{4h}-vibronic† or D_2-vibronic mechanism of spectral intensity. Unfortunately, no vibrational fine structure was observed, and thus it is not known if the 0,0 bands for the two components are degenerate. Nevertheless, whether the splitting arises from a low-symmetry molecular field or from vibronic interactions in the D_{4h} or D_2 fields, two CD bands of opposite sign are observed under the E_g absorption band.

In an octahedral complex d^8 metal ions have three spin-allowed transitions of interest: ${}^3A_{2g} \rightarrow {}^3T_{2g}$, ${}^3A_{2g} \rightarrow {}^3T_{1g}(F)$ and ${}^3A_{2g} \rightarrow {}^3T_{1g}(P)$ (see Fig. 5-10). Only the first of these has T_{1g} symmetry and is thus magnetic-dipole-allowed. Under the D_3 molecular field it should yield E and A_2 components as found for d^3 and d^6 systems. Dingle and Palmer, who studied the polarized-crystal spectra of $[Nien_3](NO_3)_2$ (see Fig. 5-11), have found that this splitting is very small (~ 0 cm^{-1}) [46]. The spectral intensity was again found to be predominantly vibronic, although very little vibrational fine structure was observed, thus frustrating any attempt to study the allowing vibrations for the ${}^3A_{2g} \rightarrow {}^3T_{2g}$ transition.

An energy-level diagram for tetragonally distorted copper(II) complexes (d^9) is given in Fig. 5-12, along with the electronic configurations for these states. There are three transitions under the D_{4h} symmetry: ${}^2B_{1g} \rightarrow {}^2A_{1g}$

* Wentworth has published the polarized-crystal spectra for *trans*-$[Coen_2Cl_2]Cl \cdot HCl \cdot 2H_2O$ [194]. The intensities and energies of the transitions differ from those published by Dingle [44] for the perchlorate salt, and there are some discrepancies in the 24kK region. Wentworth's spectra are very similar to those published earlier by Yamada and coworkers [196].

† Ballhausen and Moffitt had previously interpreted the ethylenediamine complex spectra by way of a D_{4h}-vibronic model [10].

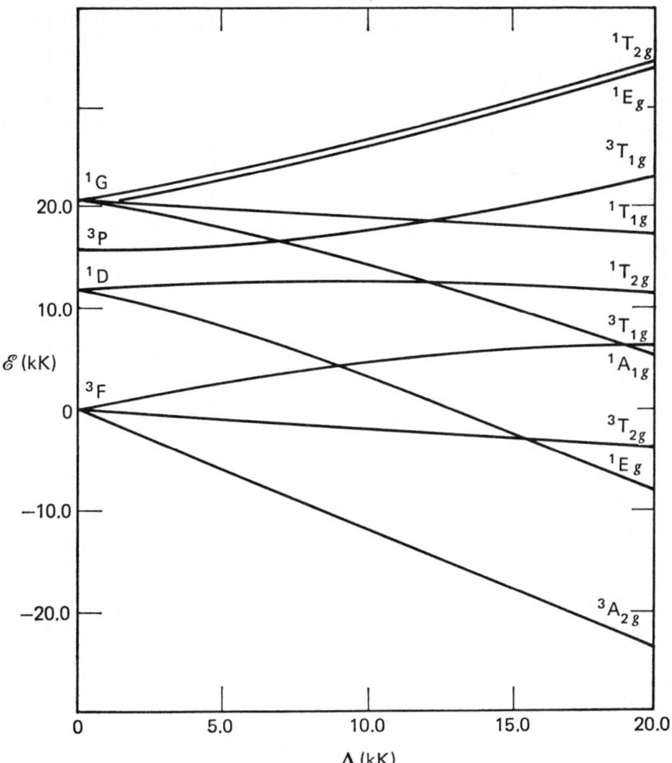

Fig. 5-10 Orgel diagram for octahedral nickel(II) (d^8) [153].

(B_{1g} symmetry), $^2B_{1g} \rightarrow {}^2B_{2g}$ (A_{2g} symmetry), and $^2B_{1g} \rightarrow {}^2E_g$ (E_g symmetry). The first of these is magnetic-dipole-forbidden,* the second transforms as R_z, and the third as R_x, R_y. Roos [166] has studied the energy levels in some copper(II) complexes using the Pariser, Parr, Pople model [156,164], which is a semiempirical self-consistent field molecular-orbital method, with a modification for the open shell system due to Roothaan [167]. He found that, after a correction for configurational interaction, the $^2A_{1g}$ state lies lower in energy than the $^2B_{2g}$ for the hexammine irrespective of the size of the tetragonal distortion. For the hexaqua complex the $^2A_{1g}$ and $^2B_{2g}$ states become approximately degenerate when the distortion is so large that the complex is almost square-planar. Otherwise the $^2A_{1g}$ has the lower energy.

* The transition to the $^2A_{1g}(D_{4h})$ level would gain some magnetic dipole moment through mixing with the ^2E_g level by spin-orbit coupling, but the mixing coefficient has been estimated to be small [93].

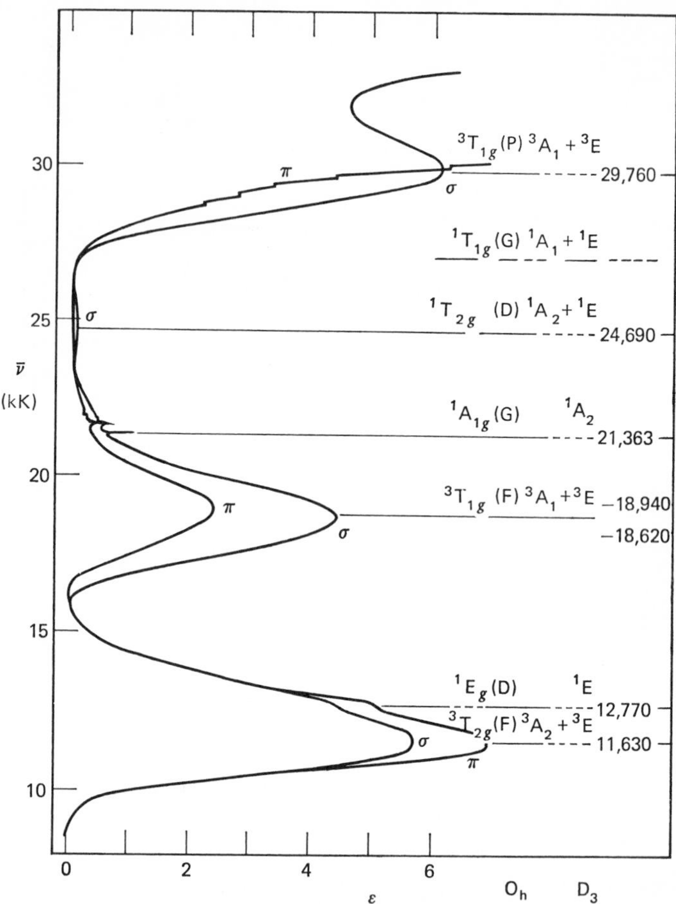

Fig. 5-11 Polarized crystal absorption spectra of $[Nien_3](NO_3)_2$ with the electric vector parallel (π) and perpendicular (σ) to the C_3 molecular axis [46].

When π bonding is negligible—in the ammine complexes, for example—the $^2B_{2g}$ and ^2E_g states are almost degenerate. According to Roos, however, π bonding leads to the levels being separated, with the $^2B_{2g}$ state lying lower in energy. For example, Roos calculated that for metal-ligand distances of 2.05 Å (plane) and 2.5 Å (axial) the transition energies for $[Cu(H_2O)_6]^{2+}$ are 10.50 kK ($^2B_{1g} \rightarrow {}^2A_{1g}$), 12.20 kK ($^2B_{1g} \rightarrow {}^2B_{2g}$), and 14.00 kK ($^2B_{1g} \rightarrow {}^2E_g$). It has been proposed that the doubly degenerate level is split by spin-orbit coupling [163]. The low molecular field of the optically active tetragonally distorted copper(II) complexes supplement this splitting, thus giving rise to four transitions under the $^2E_g \rightarrow {}^2T_{2g}$ cubic band.

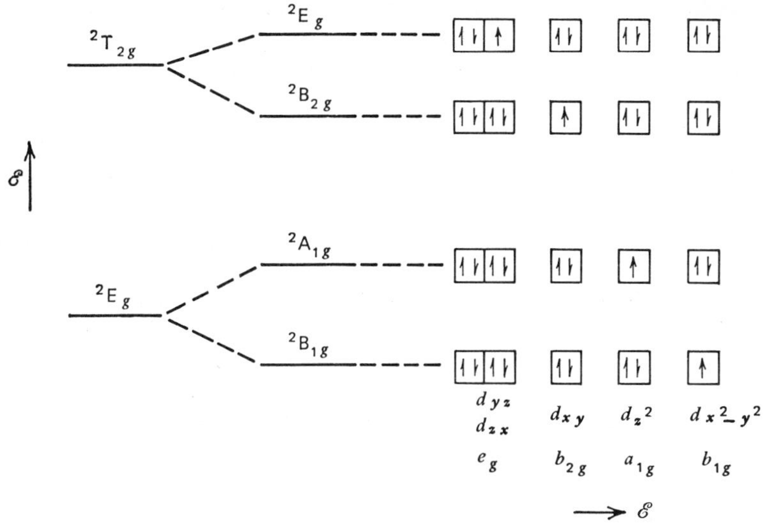

Fig. 5-12 Energy level diagram for d^9 tetragonally distorted complexes (D_{4h}).

We have seen how some *d–d* transitions are magnetic-dipole-allowed and, although parity-forbidden, do gain some electric dipole intensity by way of vibronic coupling and low-symmetry molecular fields. If we are to understand the optical activity of these transitions, we must know the factors that govern the extent to which the dipole moments contribute to the rotational strength and especially the factors that govern the sign of the rotational strength.

GENERAL THEORIES OF OPTICAL ACTIVITY

Since the early observations of the optical activity of solutions containing dissymmetric molecules, it has been realized that the phenomenon is closely associated with the way in which atoms and groups are distributed in space within the molecules. This close association between optical activity and molecular absolute configuration has yielded a potentially very powerful tool for the study of stereochemistry. However, if its full potential is to be realized, the molecular factors that govern the sign and perhaps the relative size of the rotational strength of the individual electronic transitions must be fully understood. It has been common practice to attempt to devise a model in order to calculate values of the molecular rotation parameter β and the angle of rotation α but this seems to be less important and even more difficult than the determination of the sign of the Cotton effect for each of the transitions of interest. At least the sign of the rotational strength for a particular transition does not depend on solvent, temperature, and so on, as does α, nor does it depend on the energy of the various other transitions.

The earliest attempts to propose a theory of optical activity concentrated on the asymmetric carbon atom as the source of dissymmetry. Crum Brown [39] and Guye [74] proposed that the activity is related to the *product of asymmetry* of the form

$$K = (a - b)(b - c)(c - d)(a - c)(a - d)(b - d) \qquad (5\text{-}35)$$

where a, b, c, and d are scalar quantities associated with the four different groups about an asymmetric carbon atom. The product has the correct properties of a pseudoscalar: it reduces to zero if any two of the parameters are equal, and it changes sign on inversion. However, they associated the parameters with the mass of the groups, and this was soon shown to be incorrect, because sizable rotations were observed for compounds that had groups of equal mass but different structure [188].

It is apparent now that at least two different kinds of models are required: one for chromophores that are inherently dissymmetric, the other for chromophores that, in the first approximation, are not but gain dissymmetry because of the *vicinal effect* of the atoms and groups that are dissymmetrically distributed about the chromophores. For example, the skewed dienes (see Fig. 5-13) and the coupled ligand transitions in complexes of the type d-[Niphen$_3$]$^{2+}$ are optically active, because the transitions, *per se*, have a chiral symmetry, whereas in d-[Coen$_3$]$^{3+}$ the central metal ion chromophore (CoN_6) has, in the first approximation, an octahedral symmetry that is made chiral by the secondary effect of the distribution and conformation of the chelate rings. The electrons of a chromophore essentially feel the

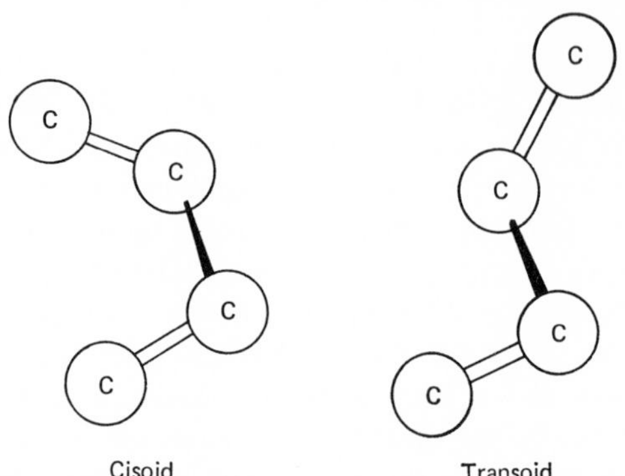

Fig. 5-13 Skewed dienes with a right-hand chirality.

force field of the atoms to which they primarily belong, and hence there is a tendency for the electrons to be governed by the selection rules that would hold rigorously if there were no external dissymmetric field.

No completely satisfactory model has been devised to predict successfully from first principles the sign of the Cotton effect for the symmetric chromophores. Nevertheless, it is instructive to review briefly the models that have been proposed. They can be divided into two fundamentally different kinds: those based on the coupled motion of two or more electrons, such as the theories of Born [15,16], Oseen [155], Kuhn [103,104,105], de Mallemann [130,131,132], Gray [73], Boys [22], and Kirkwood [102]; and the one-electron theories of Condon, Altar, and Eyring [36] and others.

Kuhn's classical coupled-oscillator theory is perhaps the easiest to understand. Having proved that Drude's model of an electron moving in a helical path [54] did not lead to optical activity, Kuhn concluded that it is necessary to have at least two oscillators whose motions are coupled. Let us consider the situation in Fig. 5-14. The two oscillators 1 and 2 of mass

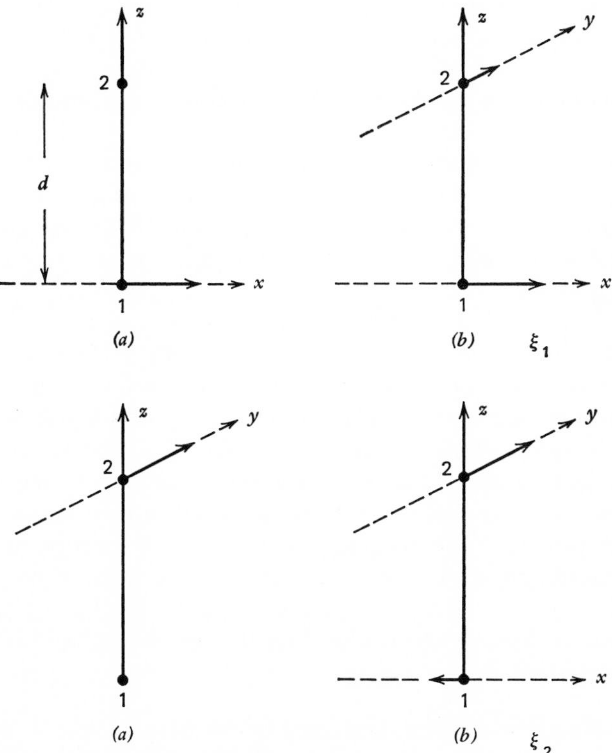

Fig. 5.14 Two resonators (*a*) without and (*b*) with coupling.

m_1 and m_2, representing two different chromophores in a molecule, are displaced a distance d from each other along the z axis. If we first assume that they each vibrate independently of the other, their potential energies U_1 and U_2 are given by

$$U_1 = \tfrac{1}{2}k_1 x_1^2 \qquad \text{and} \qquad U_2 = \tfrac{1}{2}k_2 y_2^2 \tag{5-36}$$

where k_1 and k_2 are the force constants for the vibrations, whose frequencies are

$$\nu_1 = \frac{1}{2\pi}\left(\frac{k_1}{m_1}\right)^{1/2} \qquad \text{and} \qquad \nu_2 = \frac{1}{2\pi}\left(\frac{k_2}{m_2}\right)^{1/2} \tag{5-37}$$

If the resonators are coupled, the total potential energy of the system is given by

$$\begin{aligned} U &= U_1 + U_2 + U_{12} \\ &= \tfrac{1}{2}k_1 x_1^2 + \tfrac{1}{2}k_2 y_2^2 + k_{12} x_1 y_2 \end{aligned} \tag{5-38}$$

where k_{12} is the coupling constant. Under this field, the force acting on oscillator 1, given by

$$F_{x_1} = -\frac{\partial U}{\partial x_1} = -k_1 x_1 - k_{12} y_2 \tag{5-39}$$

depends not only on the position of 1 but also on the position of 2. If one resonator moves, the other cannot remain at rest: their motions are coupled. The resulting motion can be resolved into two normal vibrations $\xi_1(\nu_1')$ and $\xi_2(\nu_2')$ by transformation to a new set of axes that are related to the old set by a rotation through an angle φ that is small if the coupling is small. The frequencies ν_1' and ν_2' are only slightly different from ν_1 and ν_2 if the coupling is weak.

When right circularly polarized light propagated in the z direction with $\lambda = 4d$ (for convenience) interacts with ξ_1 (see Fig. 5-15), maximum energy is transferred to the system in contrast to the situation with left circularly polarized light, where the oscillation along the y axis is out of phase with the electric vector of the radiation. Thus for ξ_1, n_r and ε_r are larger than n_l and ε_l, and the associated Cotton effect is negative. It can be similarly shown that the rotational strength for ξ_2 is positive. This result is independent of whether light is propagated in the z or $-z$ direction or even if the system is randomly oriented with respect to the direction of propagation.

The classical coupled-oscillator theory of Kuhn [108] and also quantum-mechanical methods of calculation [35] for this model yield the following expression for α:

$$\alpha = \frac{2\pi N_1}{3}\lambda^{-2}\frac{(n^2+2)}{3} d \sin\varphi \cos\varphi \frac{e_1 e_2}{(m_1 m_2)^{1/2}}\left\{\frac{1}{(\nu_2'^2 - \nu^2)} - \frac{1}{(\nu_1'^2 - \nu^2)}\right\} \tag{5-40}$$

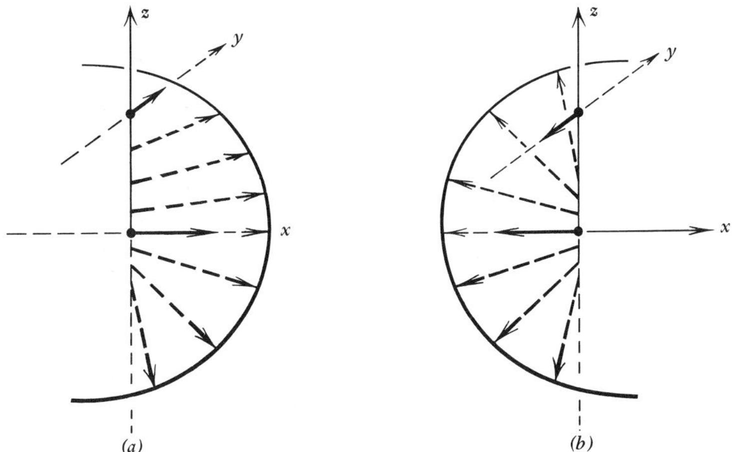

Fig. 5-15 Interaction of (*a*) right and (*b*) left circularly polarized light with ξ_1.

where ν is the frequency of the light at which the measurement is being made. For the system to be optically active, the oscillators must be coupled, that is, $\varphi \neq 0$, and the resonators must be separated by a distance $d \neq 0$. From Eq. (5-40) it is also seen that if ν_1' and ν_2' are degenerate, the system is inactive.

The sign of the Cotton effect for ξ_1 and ξ_2 can be arrived at very simply if we consider the electric and magnetic dipole vectors associated with them. Consider two electric-dipole-allowed transitions localized at points 1 and 2, polarized along the x and y axes, respectively, and coupled to give two transitions ξ_1 and ξ_2. The displacement current

$$\mathbf{j} = \frac{1}{c}\frac{\partial \mathbf{p}_1}{\partial t} \tag{5-41}$$

associated with the oscillating electric dipole transition moment $\mathbf{p}_1$ gives rise to a magnetic dipole moment at point 2 [129]:

$$\mathbf{m}_2 = \frac{1}{2c}\left(\mathbf{d}_{12} \times \frac{\partial \mathbf{p}_1}{\partial t}\right) \tag{5-42}$$

Similarly at point 1 there is a magnetic moment due to the oscillating dipole $\mathbf{p}_2$:

$$\mathbf{m}_1 = \frac{1}{2c}\left(\mathbf{d}_{21} \times \frac{\partial \mathbf{p}_2}{\partial t}\right) \tag{5-43}$$

The directions of the magnetic moments are given by the right-hand thumb rule and are shown in Fig. 5-16. The oscillating electric dipole $\mathbf{p}_1$ does not lead to optical activity in oscillator 1 by itself, because the magnetic moment

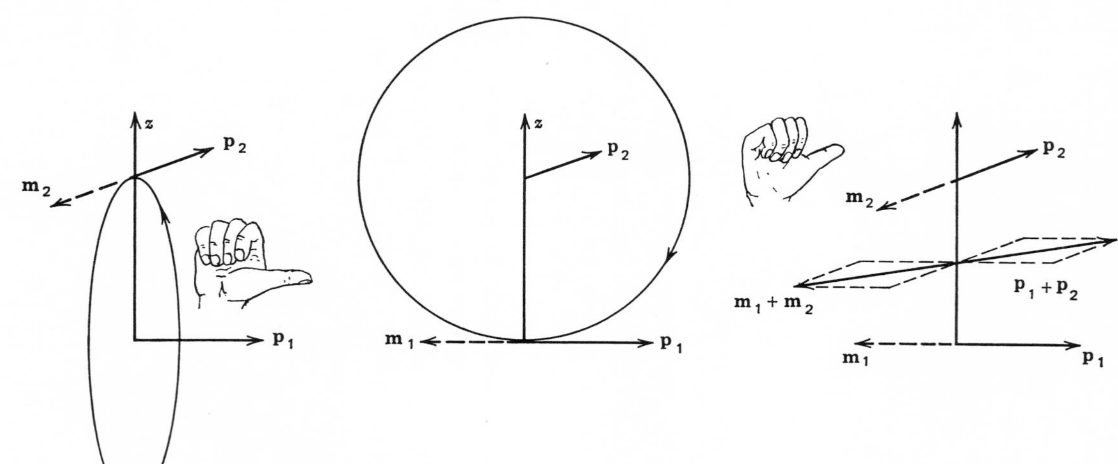

Fig. 5-16 Magnetic moments $\mathbf{m}_2$ and $\mathbf{m}_1$ derived from the oscillating electric dipoles $\mathbf{p}_1$ and $\mathbf{p}_2$ and the resultant electric and magnetic moments for ξ_1.

$\mathbf{m}_2$ is at right angles to $\mathbf{p}_1$. When the transitions at 1 and 2 are coupled, however, the resultant magnetic and electric dipole moments for ξ_1 are found to be antiparallel, and therefore, from Eq. (5-18), the rotational strength is negative. Similarly ξ_2 is found to have a positive Cotton effect.

This procedure is applied to a number of complexes later. However, it is only useful for systems that have coupled electric-dipole-allowed transitions of known polarization that are skewed in a definite manner with respect to each other. In other words, it is able to determine the sign of the Cotton effect for certain kinds of inherently dissymmetric chromophores.

The coupled-oscillator theory is not normally applicable to the *d–d* transitions of metal complexes, because they are electric-dipole-forbidden. Kuhn, however, attempted to apply the classical coupled-oscillator theory to $[Coen_3]^{3+}$ and $[Coox_3]^{3-}$ [106,107]. He postulated that the metal ion transitions gained optical activity by coupling with oscillations on the ligands, which were themselves coupled together. The oscillations he considered had their electric moments directed along the edges of the octahedron spanned by the chelates. These yielded two perturbed metal ion transitions of almost the same energy, corresponding to the in-phase and out-of-phase coupling of the ligand resonators shown in Fig. 5-17. Kuhn concluded that (a) when there is weak coupling between the metal ion and the ligands, complexes with the L configuration have negative A_2 and positive E metal ion transitions*; (b) when the coupling is of medium strength, both the lower-energy A_2 and the E transition have a positive Cotton effect; and (c) when the coupling is strong, A_2 is positive and E is negative. He suggested that in $[Coox_3]^{3-}$ the coupling is of medium strength, and, therefore, you would expect one Cotton effect as is found. His prediction, however, that *l*-$[Coox_3]^{3-}$ has the L configuration is incorrect [121]. Further, A_2 and E are not always in the same order, and the signs of their Cotton effects do not depend on the size of any coupling.

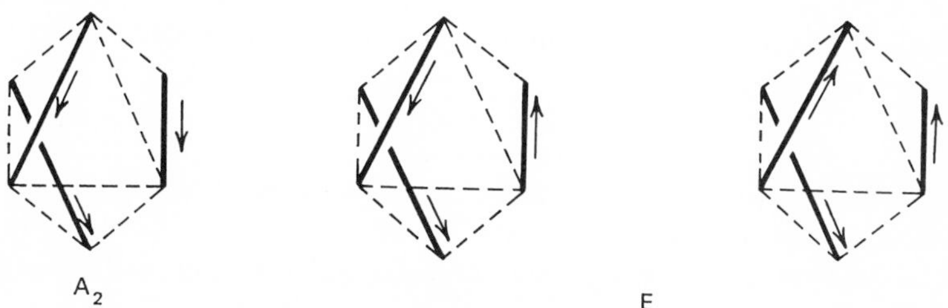

Fig. 5-17 Coupling of ligand transitions.

* Kuhn also claimed that when there is only a small energy difference between the transitions, the positive band is dominant.

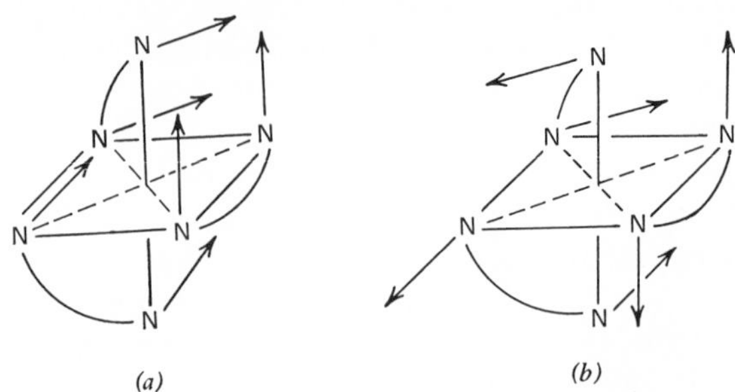

Fig. 5-18 Coupling of ligand transitions polarized perpendicular to the planes of chelates: (*a*) one component of $E(T_{1u})$ and (*b*) one component of $E(T_{2u})$.

McCaffery and Mason [119] have since proposed that in $[Coen_3]^{3+}$ the *d–d* transitions gain intensity by coupling with charge-transfer transitions centered on about 50.0 kK that are derived from component excitations polarized perpendicular to the planes of the chelate rings, as shown in Fig. 5-18. This proposal was based on their discovery from circular-dichroism studies of crystals that a charge-transfer transition with E symmetry has a negative Cotton effect for the D configuration. In terms of the coupled-oscillator theory this result could only be rationalized if the component excitations were polarized perpendicular to the planes of the chelate rings and not if they were polarized in the planes of the chelates, as proposed by Kuhn.

Another model that depends on the coupled motion of two or more electrons is the polarizability theory of Kirkwood [102]. This has two fundamental assumptions:

1. Electrons may be considered assignable to definite groups between which there is no exchange;
2. Each transition is localized to a first approximation in a distinct group.

According to Kirkwood, the molecular parameter β is made up of three parts:

$$\beta = \beta_0 + \sum_{i \neq j} \beta_{ij} + \sum_{i} \beta_{ii} \tag{5-44}$$

where $\sum_i \beta_{ii}$ is the sum of the intrinsic rotations of the groups, $\sum_{i \neq j} \beta_{ij}$ is the contribution due to the coupling of the magnetic moment on one group with an electric moment on another, and β_0 is similar in origin to the coupling of oscillations in the theories of Kuhn, Born, and others. The terms $\sum_i \beta_{ii}$ and $\sum_{i \neq j} \beta_{ij}$ were thought to be insignificant relative to β_0 and so were discarded. The model led to the expression

$$\beta \simeq \beta_0 = -\frac{1}{6} \sum_{j > i} \alpha_j \alpha_i a_j a_i G_{ji} \mathbf{R}_{ji} \cdot (\mathbf{b}_j \times \mathbf{b}_i) \tag{5-45}$$

where α_i, α_j are the mean polarizabilities of the groups i and j; a_i, a_j are the anisotropies of the groups i and j defined by

$$a = \frac{\alpha_{\parallel} - \alpha_{\perp}}{\alpha} \tag{5-46}$$

$\alpha_{\parallel}$ being the polarizability along the axis of symmetry of the groups and $\alpha_{\perp}$ the polarizability perpendicular to this axis; $\mathbf{R}_{ji}$ is the vector from the center of gravity of the group j to that of i; $\mathbf{b}_i$ and $\mathbf{b}_j$ are unit vectors along the cylindrical axes of the cylindrical groups; and G_{ji} is a dipole–dipole interaction term between the groups i and j:

$$G_{ji} = \frac{\mathbf{b}_j \cdot \mathbf{b}_i}{R_{ji}^{\,3}} - \frac{3(\mathbf{b}_j \cdot \mathbf{R}_{ji})(\mathbf{b}_i \cdot \mathbf{R}_{ji})}{R_{ji}^{\,5}} \tag{5-47}$$

The parameters in Eq. (5-45) are experimentally determinable: the α's can be computed from atomic refractions, the anisotropies from studies of the Kerr effect, and the **b**'s and $\mathbf{R}_{ji}$'s from the structure of the molecule.

The one-electron theory of Condon, Altar, and Eyring is based on the assumption that the coupling of electrons is relatively unimportant compared with terms arising from the motion of an electron in a dissymmetric field [36]. The model follows from the Hartree approximation that an electron moves in an average field because of the average action on it (the *vicinal effect*) of all the other electrons and nuclei in the molecule. The potential governing this motion has the form

$$V = \tfrac{1}{2}(k_1x^2 + k_2y^2 + k_3z^2) + Axyz \tag{5-48}$$

where k_1, k_2, k_3, and A are largely determined by the average charge on the different atoms of the molecule. If the molecule is dissymmetric, A is different from zero, and Eq. (5-48) describes an equipotential surface that has the form of a twisted ellipsoid.

In 1956 Moffitt proposed that the *d–d* transitions of metal complexes could become electric-dipole-allowed under a static trigonal perturbing crystal field, such as found in tris(bidentate) complexes, by mixing some *p* character into the *d* wave functions [144]. Because of an error of sign in the values of the electronic angular-momentum change involved in the *d–d* transitions, however, Moffitt concluded that the A_2 and E components of the T_{1g} band for cobalt(III) and chromium(III) complexes had the same sign. This error was pointed out by Sugano, who showed that no rotational strength would result from the model [185].

Piper considered an admixture of both 4*p* and 4*f* orbitals under a trigonal crystal field and predicted that tris(bidentate) complexes with the L configuration and a chelate ring angle ∠LML less than 90° would have a negative A_{2g} transition at higher energy than the E component [160,161]. Measurements have shown that, at least in solution, both these conclusions are incorrect. Moreover, Piper emphasized that the sign of the Cotton effect depended on whether ∠LML was greater or less than 90°. This has also been shown to be incorrect [71]*. Subsequently, Piper put forward a molecular-orbital model based on a trigonal splitting of the T_{1g} band, but this was also unsuccessful in accounting for the experimental data [101].

A number of other proposals have been published [81,115,177,183] without any marked advance, although Schäffer's angular-overlap model [177] seems to show some promise, and the recent study of the vibronic fine structure in the CD of metal complexes could lead to a more rapid advance in this subject.

The amount of experimental information in the literature on the circular dichroism of metal complexes has reached such a level that it is possible to determine the absolute configuration of these compounds using empirical relationships. Nevertheless, a unifying model is highly desirable, and it is hoped that this will not be long in coming.

EMPIRICAL METHODS BASED ON THE *d–d* TRANSITIONS

Empirical methods have been developed for studying the various sources of optical activity in a molecule:

1. The vicinal effect due to an asymmetric carbon atom in an optically active ligand
2. The vicinal effect due to an asymmetric donor atom
3. The conformations of the chelate rings
4. The distribution of chelate rings
5. The distribution of unidentate ligands

The contributions of these to the overall rotational strength of the *d–d* transitions is thought to be additive.

Vicinal Effects due to an Asymmetric Carbon Atom

Let us consider a complex of the kind shown below where R is a group containing an asymmetric carbon.

* Note added in proof: In a recent note Judkins and Royer reported the single crystal CD spectrum of *d*-$[Cotn_3]Cl_3 \cdot 4H_2O$ with light propagated parallel to the C_3 molecular axis—identified by a single crystal ^{59}Co nmr study. (R. R. Judkins and D. J. Royer, *Inorg. Nucl. Chem. Letters*, **6**:305 (1970)). The authors claimed that for this polarization only the $^1A_1 \rightarrow {}^1E$ (D_3) transition should be active under the T_{1g} band. The observed CD band was positive. It was proposed that this supported Piper's theory that the sign of the Cotton effect was dependent on ∠LML. It is difficult to evaluate the result because no experimental detail was given for the nmr assignment of the C_3 axis, and because light was not directed down an optic axis of the biaxial crystal.

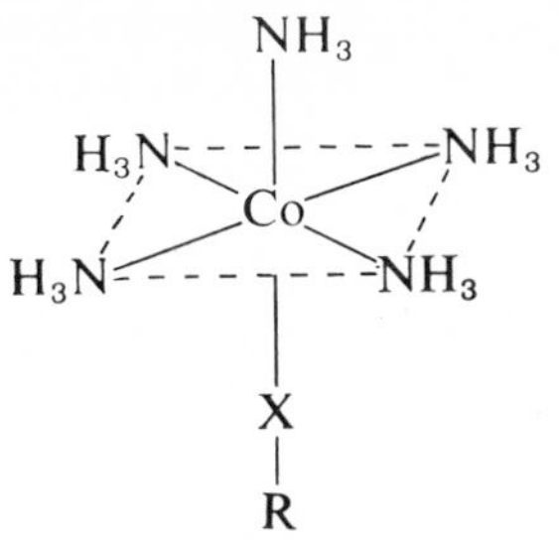

NH_3

H_3N — Co — NH_3

H_3N — — NH_3

X

R

It is well known in organic chemistry that the asymmetric distribution of groups about a central carbon atom is able to impose rotational strength on an otherwise symmetric chromophore that is physically removed some distance from the source of asymmetry. This effect, which is thought to be transmitted through space and by way of the bonds linking the asymmetric center to the chromophore, is known to depend on the number of atoms between the chromophore and the asymmetric carbon. In compounds such as CH_3—CH_2—$CH(CH_3)$—$(CH_2)_n$—CHO and CH_3—CH_2—$CH(CH_3)$—$(CH_2)_n$—CO—CH_3, however, the carbonyl chromophore at 3000 Å exhibits a Cotton effect even when n is 2 or 3 [47]. In the same way, a basically octahedral central metal ion chromophore can show optical activity under the influence of a ligand containing an asymmetric carbon. Larsen and Olsen first studied this problem and tentatively concluded that chelation was necessary if the imposed Cotton effect was to be measurable [111]. This has since been shown to be incorrect. Although only a small number of systems have been studied, it would appear that the size of the induced rotational strength depends on the donor group of the ligand, and the sign of the Cotton effect is determined by the absolute configuration of the ligand.

Take, for example, the series of compounds $[Co(NH_3)_5$—O—CO—CH-$(R)NH_3]^{3+}$ containing the optically active amino acids acting as unidentate ligands through their carboxylate groups. Circular-dichroism data for these compounds are given in Table 5-1. The actual symmetry of the coordination sphere is C_{4v} (CoN_5O), although the complex has a D_{4h} holohedrized symmetry. The T_{1g} band, which is centered at 20.0 kK for all these compounds, is split into E_g and A_{2g} components. The former is at lower energy (by approximately 1.4 kK), according to Eq. (5-34), and, from a Gaussian analysis of the absorption band, the E_g is thought to lie at about 19.70 kK and the A_{2g} at 21.85 kK.

The experimental data are best rationalized by assuming that the degeneracy of the E_g transition has been removed by the low molecular field; the relative rotational strengths of the components depend on the groups about the asymmetric carbon. An energy-level diagram is given in Fig. 5-19. For the L configuration of the amino acid, Γ_a and Γ_b, the components of the $E_g(D_{4h})$ band,

Table 5-1 Absorption and CD data for the T_{1g} band of $[Co(NH_3)_5L\text{-amH}](ClO_4)_3$

	Absorption		CD in H_2O		CD in 0.05 *M* Na_2SO_4		CD in 1 *M* NH_3		
Amino acid	$\bar{\nu}$, kK	ε	$\bar{\nu}$, kK	$\Delta\varepsilon$	$\bar{\nu}$, kK	$\Delta\varepsilon$	$\bar{\nu}$, kK	$\Delta\varepsilon$	Ref.
Alanine	19.96	68.4	18.45	+0.003	19.23	+0.031	18.52	+0.011	88
			20.70	−0.015			21.14	−0.008	
Phenylalanine	19.94	67.5	19.70	−0.086	19.70	+0.048	20.00	−0.040	88
Methionine	20.00	68.0	19.70	−0.137	20.08	−0.018	20.12	−0.035	88
Tyrosine	19.96	70.9	19.70	−0.070	19.70	+0.054	19.61	−0.064	88
Histidine	19.96	66.9	20.00	−0.015	19.53	+0.023	20.20	−0.019	88
Proline	19.94	67.4	19.05	+0.026	19.42	+0.080	19.49	+0.041	88
Tryptophan	19.96	69.8	19.70	+0.196	19.70	+0.321	19.61	+0.209	88
Asparagine	19.98	69.2	19.80	−0.120	19.80	−0.077	19.70	−0.091	88
Leucine	19.94	69.9	19.70	−0.063	18.87	+0.014	20.12	−0.035	88
Phenylglycine[a]	20.00	70.1	20.12	+0.074	19.15	+0.069	18.78	−0.042	88
Serine[b]	20.00	67.6	19.80	−0.05					200
Valine[b]	20.00	64.6	19.80	−0.05					200
Threonine[b]	20.00	66.1	19.69	−0.10					200

[a] D Isomer was studied; the signs of the Cotton effects have been inverted.
[b] As chloride.

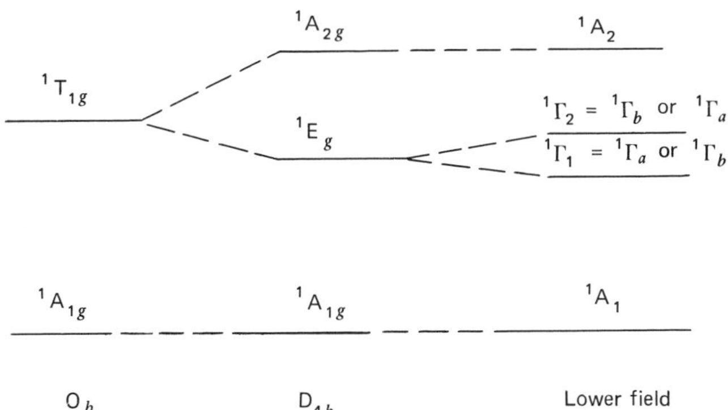

Fig. 5-19 Energy levels for $[Co(NH_3)_5O—CO—CH(NH_3)—R]^{3+}$.

have positive and negative Cotton effects, whereas the $A_{2g}(D_{4h})$, hereafter called the A_2 component for the sake of brevity, transition is negative [88]. For all the amino acids studied, Γ_a is the lowest energy component (Γ_1) and $\Gamma_b = \Gamma_2$. For most of the perchlorate salts Γ_b dominates the CD spectra,

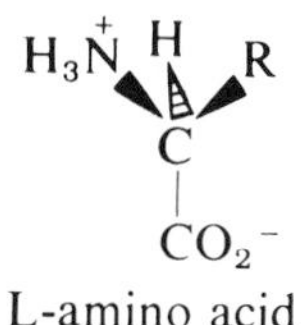

L-amino acid

A_2 being hidden under the residual Γ_b band and Γ_a being canceled completely. For alanine, however, Γ_a and Γ_b have approximately equivalent rotational strengths, and for proline and tryptophan, Γ_a dominates the CD. As found in other systems, the addition of polarizable anions, such as SO_4^{2-}, PO_4^{3-}, and SeO_3^{2-}, has a large effect on the observed CD [71,137,138]. The addition of sulfate increases the rotational strength of the Γ_a component relative to the other two components. When the $—NH_3^+$ group is deprotonated, there is a considerable change in the CD spectra, which can be explained by assuming that the rotational strengths of Γ_a and Γ_b become more equivalent. Furthermore, the observed CD is not so dependent on the added anions as it was for the protonated complex.

Dunlop and Gillard postulated that the sign of the dominant Cotton effect for the transitions under the T_{1g} band indicates the absolute configuration of the coordinated carboxylate: if the dominant Cotton effect "is negative, then the acid has the L-configuration related to that of L(+)-lactic acid" [55]. Such a broad generalization seems completely ill founded; even for the above series of closely related complexes the "rule" is not obeyed. The observed variation in the relative rotational strength of the transitions with the coordinated amino acid, with the pH of the solution, and with the presence of various anions renders any rule based solely on observed dominating bands highly suspect. Further, it seems unjustified to postulate such a general rule as the above for α carboxylates without considerable experimental data covering a large number of α substituents. There is no experimental evidence available to suggest that all carboxylates with the absolute configuration shown, which is possessed by L(*d*)-lactate with X = OH, impose the same sign of Cotton effect on a particular chromophore independently of the nature of X—NH_3^+, F, Cl, Br, C_2H_5, C_6H_5, for example:

X H CH_3
C
|
CO_2^-

From the data given in Table 5-1 and Figs. 5-20 to 5-22, and from the above discussion, it is possible to propose the following empirical rule: for complexes of the type $[Co(NH_3)_5\text{L-amH}]^{3+}$, L amino acids, which are coordinated

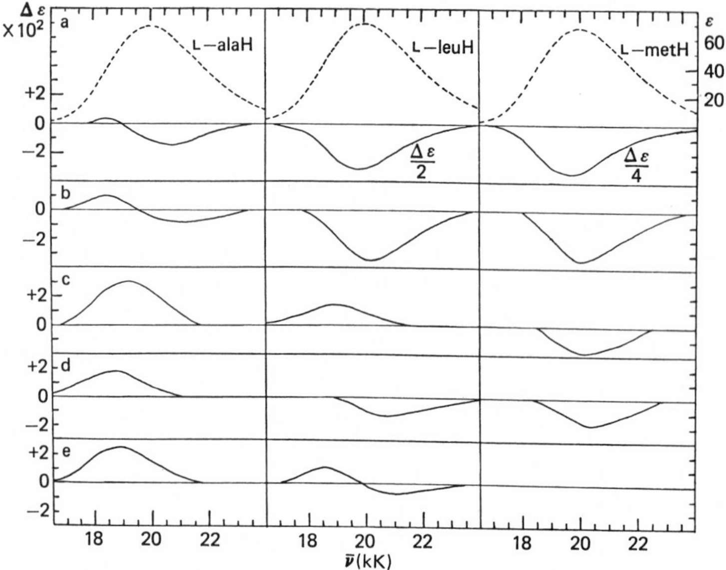

Fig. 5-20 Absorption (– – –) and circular dichroism (——) spectra of $[Co(NH_3)_5(amH)]^{3+}$ in (*a*) water, (*b*) 1 *M* aqueous ammonia, (*c*) 0.05 *M* sodium sulfate, (*d*) 0.05 *M* sodium phosphate, and (*e*) 0.05 *M* potassium selenite [88].

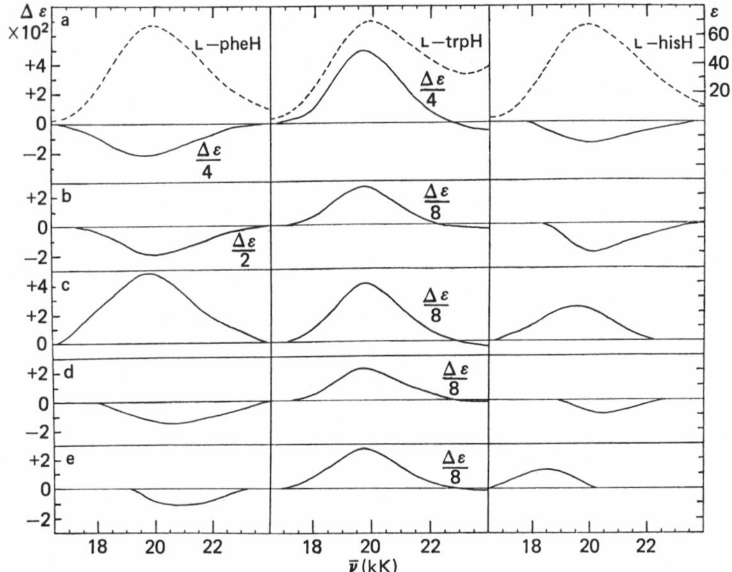

Fig. 5-21 Absorption (- - -) and circular dichroism (——) spectra of $[Co(NH_3)_5(amH)]^{3+}$ in (*a*) water, (*b*) 1*M* aqueous ammonia, (*c*) 0.05 *M* sodium sulfate, (*d*) 0.05 *M* sodium phosphate, and (*e*) 0.05 *M* potassium selenite [88].

as unidentate carboxylate ligands, impose a negative Cotton effect on the $^1A_{1g} \rightarrow {}^1A_{2g}(D_{4h})$ transition and give rise to two nondegenerate transitions of opposite sign corresponding to the $E_g(D_{4h})$ band [88].

When two carboxylates are coordinated trans to each other, as in *trans*-$[Coen_2(O—CO—R)_2]^{n+}$, the splitting of the E_g and A_{2g} components is twice that for the mono complex. The absorption spectra show the two transitions as separate bands [200], and this makes it much easier to assign the observed Cotton effects to particular transitions. In Fig. 5-23, the CD spectra are given for a series of L-amino acid complexes of this kind. The Cotton effects of Γ_b (= Γ_2) and A_2, which are clearly discernible, have the same signs as found for the mono complexes.

When an amino acid is chelated, the central metal ion derives its optical activity from the vicinal effect, the conformation of the chelate ring, and also, if there are two or more chelates dissymmetrically arrayed, from the distribution of the chelates. It is very difficult to separate the contributions from the first two. The conformations of five-membered amino acid chelate rings were discussed in Chap. 3. It was found that the rings are extremely flexible, the

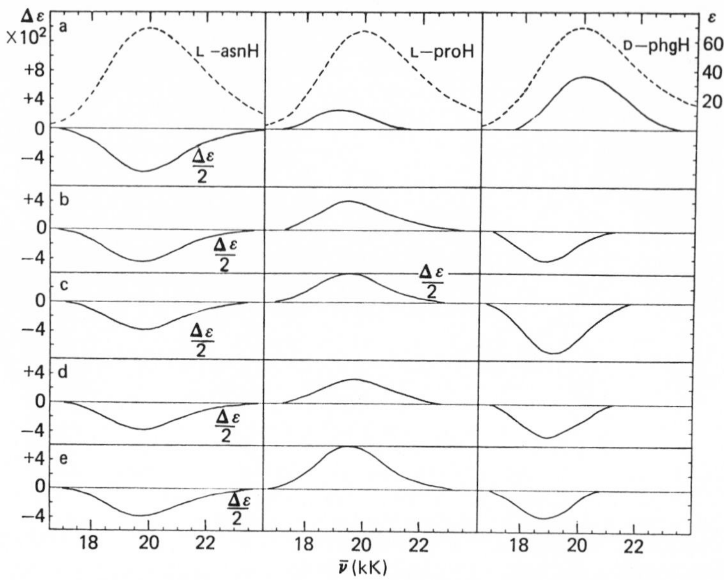

Fig. 5-22 Absorption (– – –) and circular dichroism (——) spectra of $[Co(NH_3)_5(amH)]^{3+}$ in (*a*) water, (*b*) aqueous ammonia, (*c*) 0.05 *M* sodium sulfate, (*d*) 0.05 *M* sodium phosphate, and (*e*) 0.05 *M* potassium selenite [88].

energy differences between a whole range of equatorial and axial conformations being very small. Thus the conformation is not likely to have a marked effect on the CD spectra of complexes of the type $[Co(NH_3)_4$L-am$]^{2+}$.

You would not expect to find the vicinal effect for the chelated amino acids to be identical in every respect to that for the unidentate. The position of the asymmetric grouping with respect to the chromophore is different in the two (see Fig. 5-24); the effect is transferred to the central metal ion through both the carboxylate and amino groups in the chelate; and although the unidentate is not completely free to rotate about its bonds and the chelate ring tends to be flexible, there is far more freedom for rotation in the unidentate than in the chelate. Nevertheless, the chelated L amino acids enforce the same signs for the Cotton effects as the unidentate acids in tetragonal complexes of the type $[Co(NH_3)_4$L-am$]^{2+}$, $[Coen_2$L-am$]^{2+}$, *trans*-[CoEDDA L-am], *trans*-[Co(L-am)$_3$], and *trans*-[Coox(L-am)$_2$]$^-$.

The amino acid tetrammine series has been extensively studied [89,200]. The coordination chromophore CoN_5O is identical to that for the unidentates. The E_g component, which is at the lower energy, is split to give two transitions of opposite sign because of the low molecular symmetry. Absorption and CD data are given in Table 5-2, and some spectra are presented in

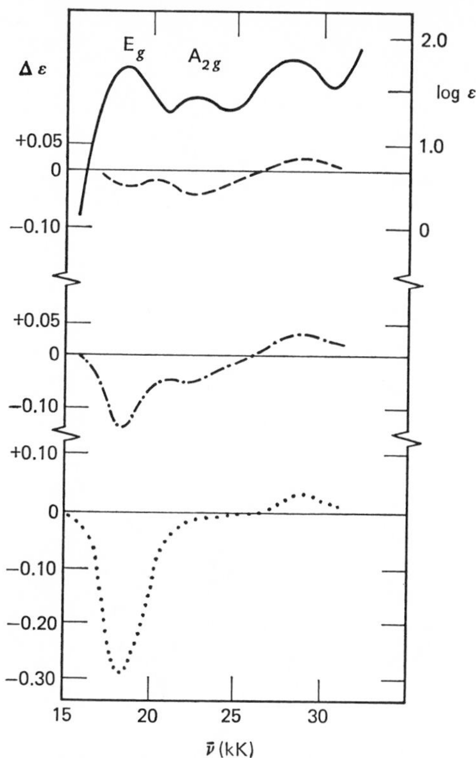

Fig. 5-23 Absorption (——) and CD spectra of *trans*-$[Coen_2X_2]^{3+}$, where X is L-proline (– – –), L-alanine (–·–·), and L-serine (· · · ·) (data from [200]).

Fig. 5-25. Generally two negative Cotton effects dominate the T_{1g} region, although there are some exceptions—with L-proline the central positive band is dominant, for example.* For all the complexes studied the transitions maintain the same relative energies—Γ_b, Γ_a, and A_2.

Liu and Douglas were the first to suggest that the rotational strengths due to the configurational effect (distribution of chelates) and the vicinal effects were additive in complexes of the type $[Coen_2$ L-am$]^{2+}$ [118]. They attributed the circular dichroism of *dl*-$[Coen_2$ L-am$]^{2+}$ (see Table 5-6) solely to the vicinal effect and subtracted this from the CD of the *d* and *l* isomers to obtain a measure of the configurational effect. The "racemic" isomer showed a −, +, − pattern under the T_{1g} band. The chromophore is again CoN_5O,

* The CD of this complex is discussed again in the section on the vicinal effect due to an asymmetric donor atom.

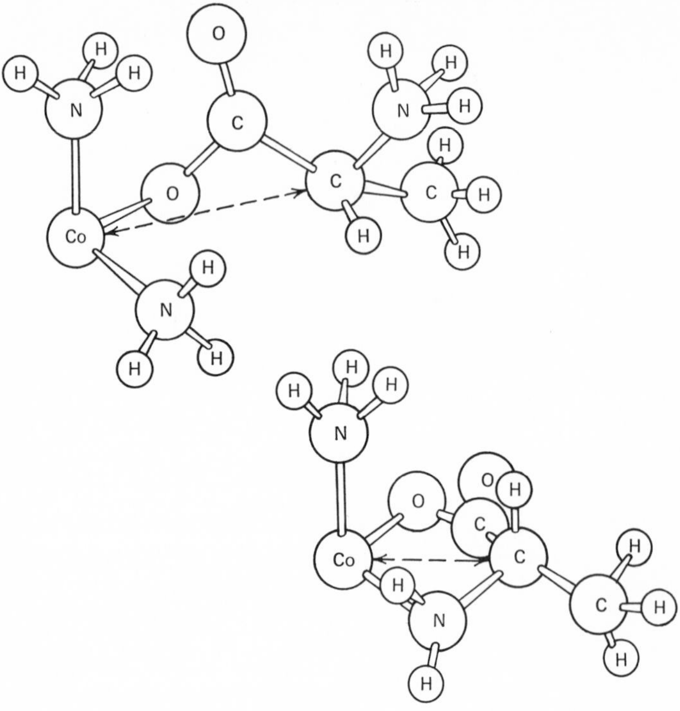

Fig. 5-24 Relationship of the asymmetric carbon to the cobalt in unidentate (carboxylate) and chelated alaninecobalt complexes.

and thus the A_2 is negative, as found for the systems discussed above. The vicinal effect for L-alanine in the complexes *trans-N,N'*-ethylenediamine-diacetato-L-alaninatocobalt(III) [113] and *trans*-tris(L-alaninato)cobalt(III) [51] has been obtained by adding the observed CD for the L and D isomers. The resultant spectra correspond to twice the vicinal effect and six times the vicinal effect, respectively. Both complexes have the rhombic symmetry *trans*-$[CoN_3O_3]$, as shown in Fig. 5-26, and the observed CD components can be accounted for by such a field. Γ_{NN} and Γ_{OO} are lowest and highest in energy, respectively. As previously found, the L-amino acid gave rise to one positive and two negative Cotton effects for the T_{1g} band (Γ_{NN} and Γ_{OO} negative and Γ_{NO} positive). Hidaka and Shimura have applied the same principle as proposed by Douglas and his coworkers to determine the vicinal effect for L-alanine in the complex *trans*-$[Coox(L\text{-}ala)_2]^-$ [94]. The vicinal effect for four

Table 5-2 Absorption and CD data for the T_{1g} band of $[Co(NH_3)_4$L-am$]SO_4$

	Absorption		CD		
Amino acid	$\bar{\nu}$, kK	ε	$\bar{\nu}$, kK	$\Delta\varepsilon$	Ref.
Alanine	20.28	77.6	18.28	−0.02	62
			19.53	+0.07	
			20.20	−0.22	
Alanine	20.24	84.8	18.25	−0.023	89
			19.72	+0.058	
			21.93	−0.200	
Leucine	20.33	79.4	18.38	−0.05	200
			19.80	+0.04	
			21.83	−0.30	
Isoleucine	20.28	79.4	18.52	−0.12	200
			21.74	−0.32	
Valine	20.20	81.3	18.52	−0.16	200
			21.65	−0.34	
Phenylalanine	20.20	81.3	18.32	−0.06	200
			19.69	+0.02	
			21.83	−0.34	
Phenylalanine	20.24	89.0	18.47	−0.080	89
			19.51	+0.012	
			21.74	−0.401	
Phenylalanine[a]	20.28	79.4	18.28	−0.026	118
			19.61	+0.003	
			21.83	−0.114	
Proline	20.20	77.6	17.99	−0.07	200
			19.69	+0.42	
			22.22	−0.21	
Proline	20.16	78.6	17.92	−0.073	89
			19.69	+0.570	
			22.03	−0.295	
Methionine	20.24	80.0	18.08	−0.025	89
			19.72	+0.166	
			21.88	−0.260	
Tryptophan[b]	20.33	78.3	18.80	−0.137	89
			21.60	−0.371	
Serine	20.33	75.9	21.01	−0.32	200
Threonine	20.33	79.4	21.32	−0.30	200

[a] As iodide.

[b] As perchlorate.

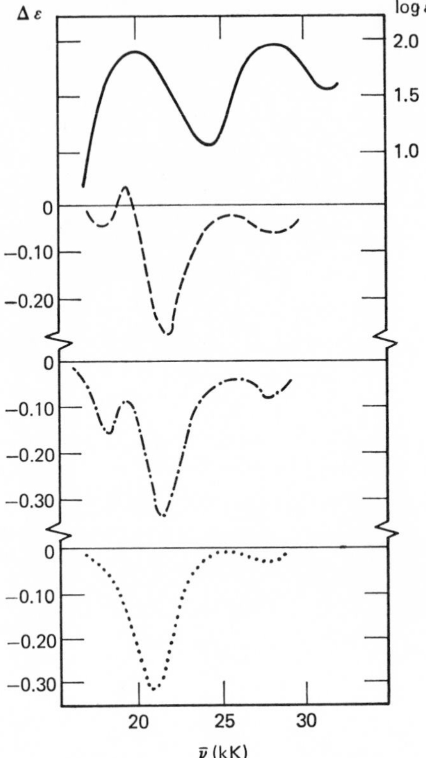

Fig. 5-25 Absorption (——) and CD spectra of $[Co(NH_3)_4(\text{L-am})]^{2+}$, where am is leucine (– – –), valine (–·–·), and serine (· · · ·) (data from [200]).

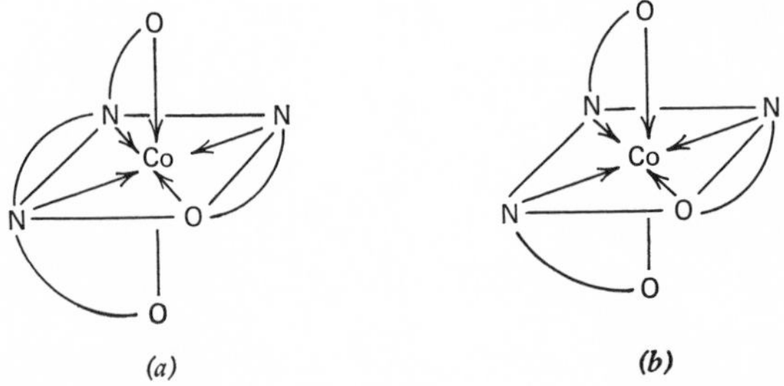

Fig. 5-26 (*a*) *Trans*-N,N′-ethylenediaminediacetato-L-alaninatocobalt(III) and (*b*) *trans*-tris(L-alaninato)cobalt(III).

molecules of the amino acid was obtained by adding the CD spectra of the D and L isomers (see Fig. 5-27). This complex has a *trans*-[CoO_4N_2] chromophore; the $A_{2g}(D_{4h})$ component at lower energy than E_g has a negative Cotton effect, as expected.

From these studies it is possible to extend the empirical rule above: "In tetragonal or rhombic complexes, amino acids with the L configuration impose a negative Cotton effect on the transition with $A_{2g}(D_{4h})$ parentage and give rise to two nondegenerate transitions of opposite sign corresponding to the $E_g(D_{4h})$ band."

The vicinal effect of an L-amino acid in a complex with cubic symmetry has been studied only for *cis*-[Co(L-ala)$_3$] [51]. A negative Cotton effect is observed, but insufficient evidence is available to assign it to either the E or A_2 components of the C_3 molecular field.

The vicinal effect of L-amino acids in their copper(II) complexes has also been studied extensively. [93,159,189,190,191,192,198]. Wellman and his coworkers have claimed that the observed ORD in the *d–d* region is due to a conformational effect [189,190,191,192]. They pointed out that the curves for [Cu(L-am)gly] increase in amplitude in the order R = CH_3— < CH_3CH— < $CH_3CH_2CH_2$— < $(CH_3)_2CH$— < ϕ—CH_2— < ϕ, which, they proposed, is the same order as for the energy differences between the axial and equatorial conformations. In fact, the axial-equatorial energy differences for all the

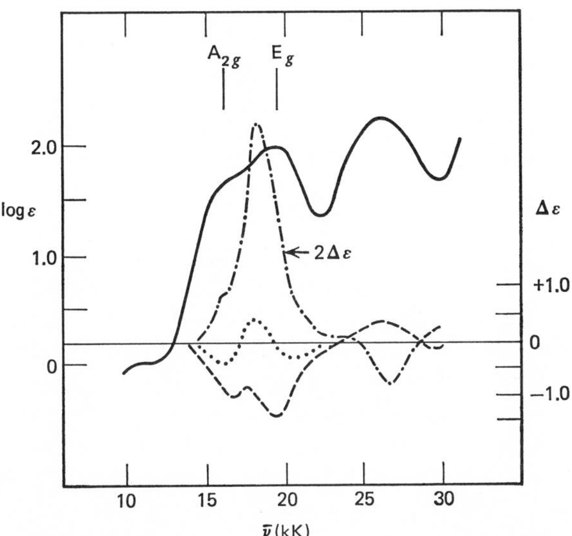

Fig. 5-27 Absorption spectrum of d_{546}-*trans*-[Coox(L-ala)$_2$]$^-$ (——) and the CD spectra of the d_{546} (– – –) and l_{546} (–·–·) isomers and of the vicinal contribution (. . .) (data from [94]).

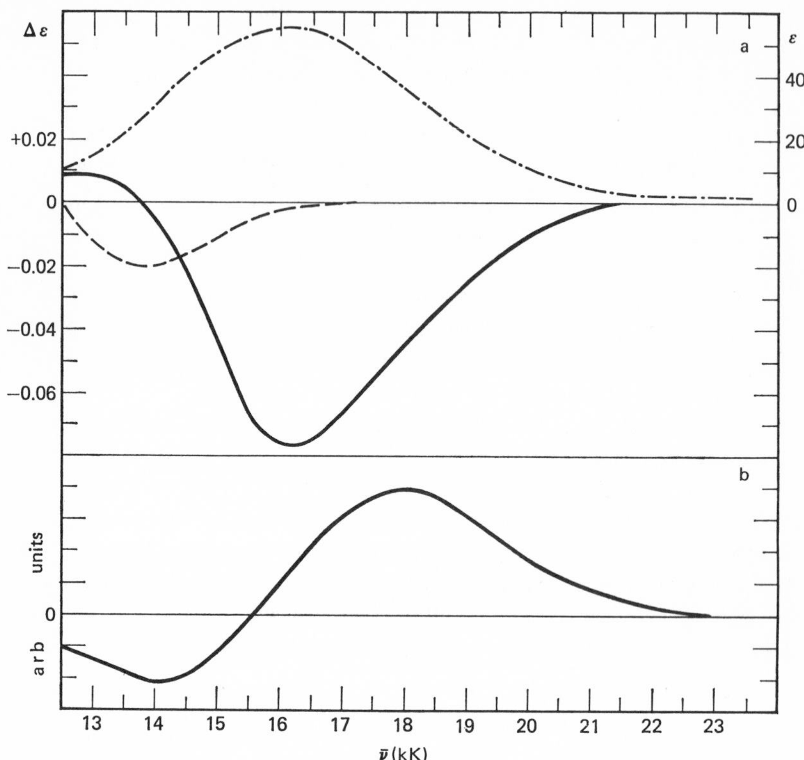

Fig. 5-28 Absorption (–·–·) and circular dichroism (——) spectra of Cu(L-ala)$_2$: (*a*) computed CD curves for 1:1 (– – –) and 1:2 species and absorption spectrum of 1:2 complex dissolved in water; (*b*) KBr disk [93].

R″(R′)CH— groups above are almost identical, because the groups can orient themselves with R′ and R″ away from any serious nonbonded interactions. Wellman and his coworkers also measured the ORD of L-histidinatoglycinatocopper(II), L-3-methyl- and L-1-methyl-histidinatoglycinatocopper-(II). The curves for the first two were dominated by a positive Cotton effect, in contrast to the l-methyl derivative and the majority of the equivalent L-amino acid complexes. This, they said, was due to the imidazole nucleus coordinating in the L-histidine and L-3-methylhistidine complexes to make these ligands terdentate and enforcing a λ conformation. The apical coordination was impossible in the l-methylhistidine complex, and, therefore, it coordinated as a bidentate with the δ conformation. Perhaps a conformational effect is important for the terdentate ligands and causes the observed changes. Nevertheless, there is no evidence to suggest that the observed Cotton effects can be accounted for simply by a conformational effect. There are four transitions of similar energy under the absorption band, and the observed dominating Cotton effect depends on the relative rotational strength of the four components, which you might expect to alter as the chelates change from bidentate to terdentate. Until there is direct evidence to the contrary, it is perhaps better to consider the Cotton effects as arising from a vicinal effect.

Yasui has published CD data for copper(II) amino acid complexes and suggests that there are four observable CD bands with the sign pattern +, +, −, −, the negative peaks dominating the spectra [198,199]. The instrument used by Yasui for his measurements was relatively insensitive, and subsequent workers were unable to reproduce the detail of his spectra. Hawkins and Wong [93] have determined the CD spectra of some 1:1 and 1:2 copper(II)

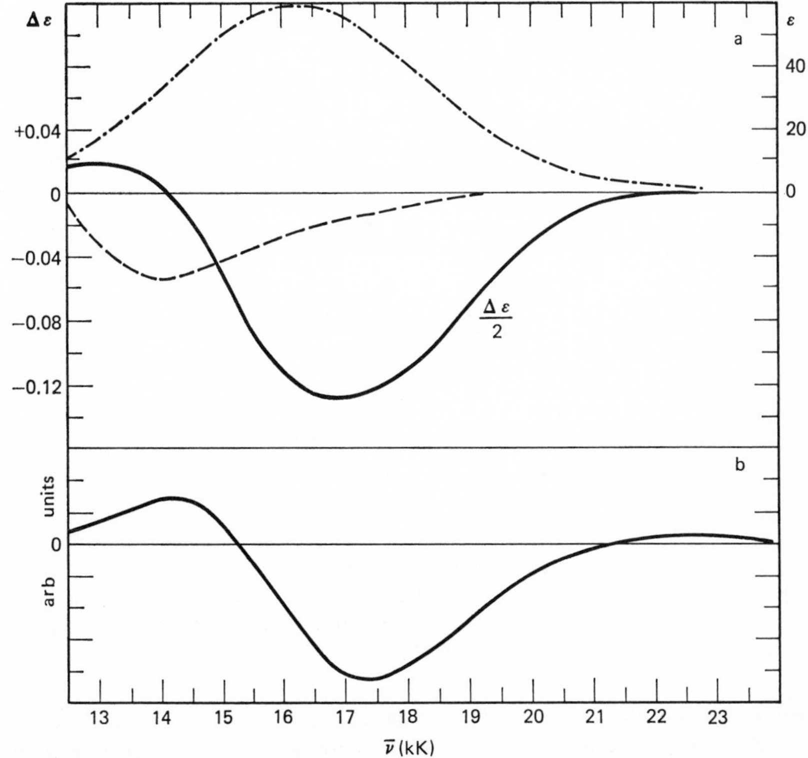

Fig. 5-29 Absorption (–·–·) and circular dichroism (——) spectra of Cu(L-val)$_2$: (*a*) computed CD curves for 1:1 (– – –) and 1:2 species and absorption spectrum of 1:2 complex dissolved in water; (*b*) (*c*) KBr disk [93].

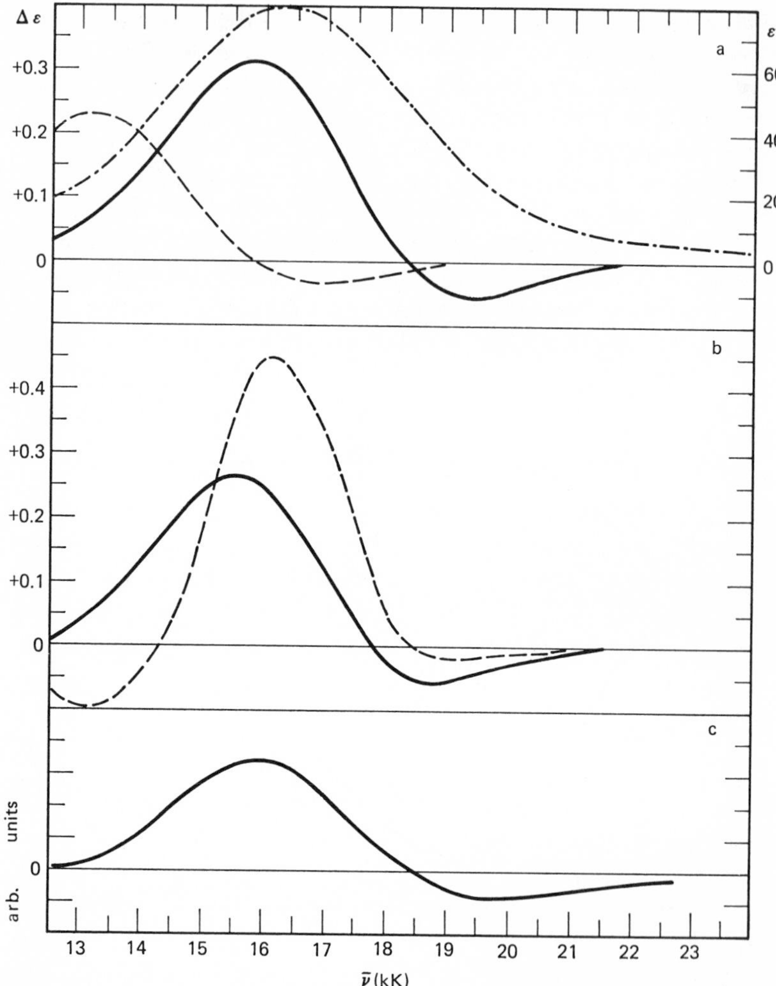

Fig. 5-30 Absorption (-·-·) and circular dichroism (——) spectra of Cu(L-pro)$_2$: (*a*) computed CD curves for 1:1 (- - -) and 1:2 species and absorption spectrum of 1:2 complex dissolved in water; (*b*) complex dissolved in DMSO—H_2O (molar ratio 1:2) at 20° and at $-90°$ (- - -); (*c*) KBr disk [93].

L-amino acid complexes and have observed three Cotton effects under the *d–d* absorption band with a $+$, $-$, $+$ sign pattern, except for L-proline which exhibited a $-$, $+$, $-$ pattern (see Figs. 5-28 to 5-32). Spectroscopic data from their studies are presented in Tables 5-3 and 5-4. The lowest energy transition ($^2B_{1g} \rightarrow {}^2A_{1g}$) is magnetic-dipole-forbidden in both tetragonal and rhombic fields, in contrast to the other transitions. Even in the low molecular

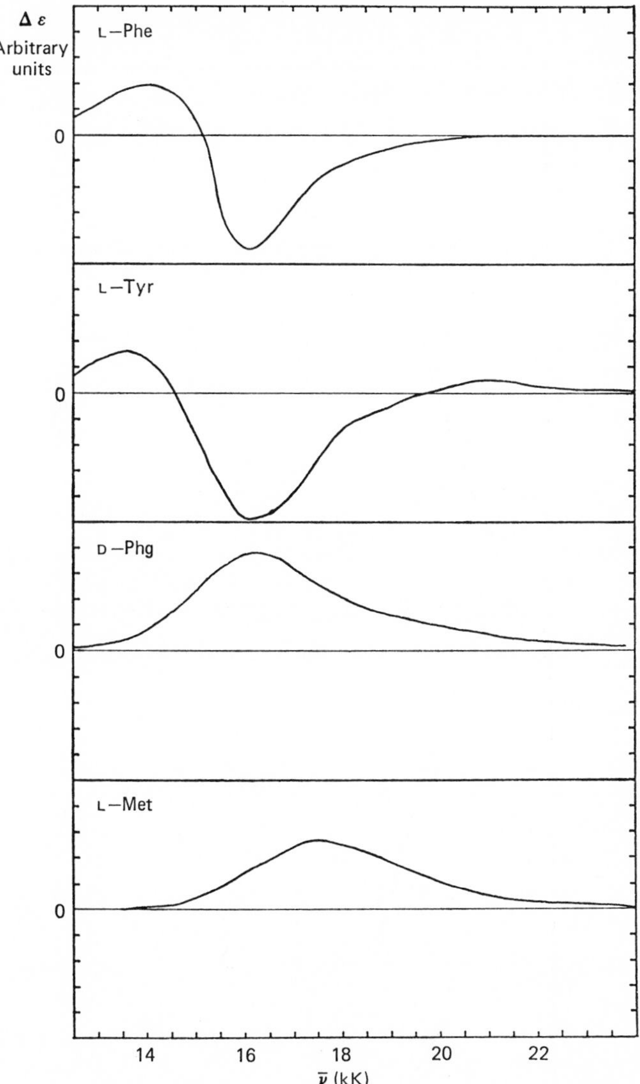

Fig. 5-31 CD spectra of $Cu(am)_2$ in KBr at 20°, except for L-tyrosine, which is at −190° [93].

symmetry and after allowing for spin-orbit coupling between the A_{1g} and E_g levels, this transition is expected to have a much lower rotational strength than the other transitions. Because of this Hawkins and Wong assigned the three observed Cotton effects to $^2B_{1g} \rightarrow {}^2B_{2g}(D_{4h})$ and to the two components of the tetragonal transition to the ^2E_g state (Γ_a and Γ_b). For all the

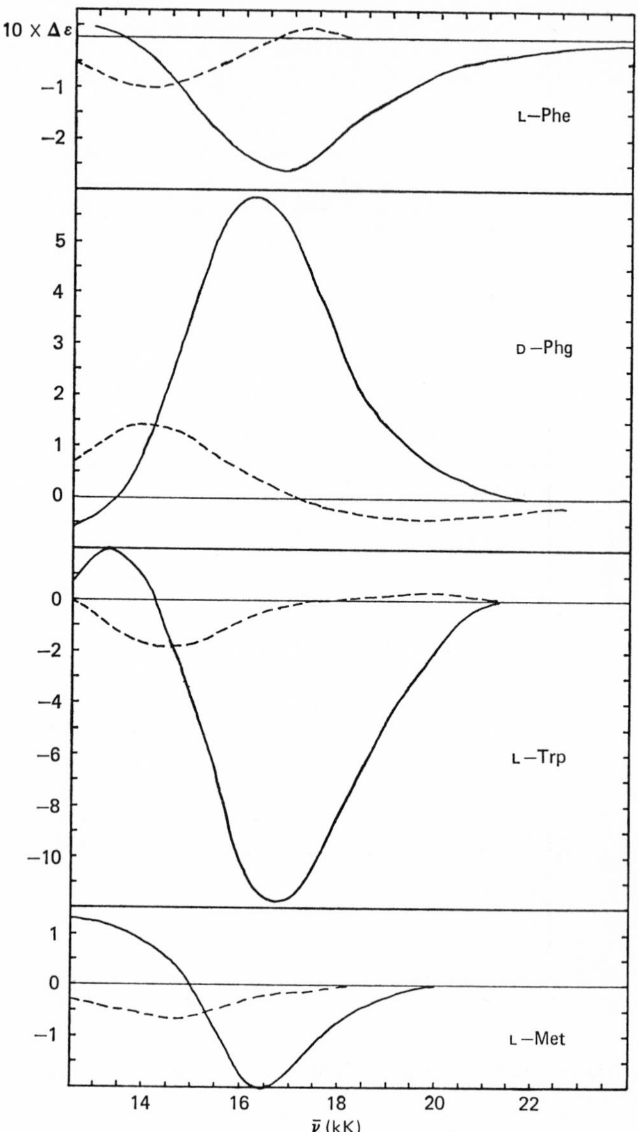

Fig. 5-32 Computed CD spectra for 1:1 (– – –) and 1:2 (——) copper(II) amino acid complexes [93].

Table 5-3 Absorption and CD data for bis(α-aminocarboxylato)copper(II) complexes[a] [93]

	H_2O				DMSO-H_2O[d]				
	Absorption		CD		Absorption		CD		
	$\bar{\nu}$, kK	ε	$\bar{\nu}$, kK	$\Delta\varepsilon$	$\bar{\nu}$, kK	ε	$\bar{\nu}$, kK	$\Delta\varepsilon$	Assignment[c]
L-Ala	16.25	55.0	13.00	+0.008	16.30	63.3	~12.5	+0.025	$^2B_{2g}$
			16.15	−0.075			16.80	−0.096	$^2E_g(\Gamma_a)$
L-Val	16.41	58.9	13.00	+0.034	16.30	63.9	13.45	+0.041	$^2B_{2g}$
			17.10	−0.197			17.25	−0.221	$^2E_g(\Gamma_a)$
L-Pro	16.40	79.9	15.53	+0.255	16.46	83.4	15.45	+0.265	$^2E_g(\Gamma_a)$
			19.30	−0.072			18.80	−0.065	$^2E_g(\Gamma_b)$
L-Pro[b]					16.55	89.2	13.16	−0.100	$^2B_{2g}$
							16.10	+0.449	$^2E_g(\Gamma_a)$
							19.25	−0.020	$^2E_g(\Gamma_b)$

[a] Isolated complexes dissolved in solvent.
[b] At −90°, no correction for contraction.
[c] Tentative assignment: excited state under D_{4h} symmetry.
[d] Dimethylsulfoxide solvent containing 0.67 mole fraction of water.

Table 5-4 CD data for mono- and bis(α-aminocarboxylato)copper(II) complexes at 20° [93]

Amino acid	CuL		CuL_2		Assignment[a]
	$\bar{\nu}$,kK	$\Delta\varepsilon_1$, max	$\bar{\nu}$, kK	$\Delta\varepsilon_2$, max	
L-Ala	13.80	−0.021	12.80	+0.008	$^2B_{2g}$
			16.15	−0.078	$^2E_g(\Gamma_a)$
L-Phe	14.10	−0.095	<12.5	~+0.027	$^2B_{2g}$
	17.40	+0.021	16.75	−0.255	$^2E_g(\Gamma_a)$
D-Phg	14.00	+0.141	<12.5	~−0.06	$^2B_{2g}$
	19.65	−0.042	16.22	+0.587	$^2E_g(\Gamma_a)$
L-Trp	14.50	−0.182	13.20	+0.199	$^2B_{2g}$
	19.90	+0.030	16.75	−1.175	$^2E_g(\Gamma_a)$
L-Val	14.08	−0.055	13.05	+0.036	$^2B_{2g}$
			16.90	−0.260	$^2E_g(\Gamma_a)$
L-Pro	13.11	+0.230	15.83	+0.312	$^2E_g(\Gamma_a)$
	16.85	−0.037	19.40	−0.060	$^2E_g(\Gamma_b)$
L-Met	14.60	−0.070	<12.5	~+0.14	$^2B_{2g}$
			16.45	−0.202	$^2E_g(\Gamma_a)$

[a] Tentative assignment for CuL_2: excited state under D_{4h} symmetry.

observed solution CD spectra the Γ_a transition was dominant and had a negative Cotton effect for the L-amino acids other than proline.

Throughout the discussion above we have concentrated on the vicinal effect from carboxylate coordinated unidentate and from chelated amino acids. Because the vicinal effect is thought to be transmitted to the central metal ion by way of space and through the bonds linking the asymmetric center to the chromophore, we must also consider for the chelate the Cotton effect that is imposed by way of the —NH_2 group. The asymmetric center is adjacent to the nitrogen atom but is one atom removed from the coordinated oxygen. It has been shown that the rotational strength of a chromophore decreases as the number of atoms between it and the asymmetric center increases [47,198]. Thus it might be expected that the vicinal effect through the nitrogen would be greater than that through the oxygen. This does not seem to be the case. Complexes with α-substituted ethylamines have shown only very small Cotton effects in the *d–d* region [90]. Perhaps the carboxylate is so successful in transmitting rotational strength, because its chromophore is coupled with that of the central metal, whereas any chromophore involving the amine group has too high an energy to interact successfully. Vicinal effects for *N*-bonded ligands have previously been measured for a copper(II) complex with *l*-*N*,*N*′-bis(1-chloromethylpropyl)ethylenediamine [83] and for platinum(II) olefin amine complexes [40].

Martin and his coworkers have proposed a double-octant rule to account for their observations of the vicinal effects imposed by polypeptides coordinated to copper(II) and nickel(II) [133]. The Cotton effects imposed by gly-gly-ala, gly-ala-gly, and ala-gly-gly are different but additive to almost reproduce the observed Cotton effect for ala-ala-ala. The tripeptide complex, when placed in the double octant, as shown in Fig. 5-33, has the methyl groups all in negative space, and, therefore, the effects should be additive. It is very difficult to assess this proposition, especially because you are viewing resultant unassigned broad CD bands.

CD data have been reported for a series of complexes of the type shown below with R = S-α-methylbenzyl-3-R-menthyl-, S-*sec*-butyl, and S-2-methylbutyl, and R′ = H and CH_3 [100]:

A significant vicinal effect was observed for the complexes that had a group with an asymmetric α carbon, but there was no measurable CD for the S-2-methylbutyl complex, which has an asymmetric β carbon atom. The groups with the R configuration imposed a dominating positive Cotton effect, and those with S, a negative Cotton effect.

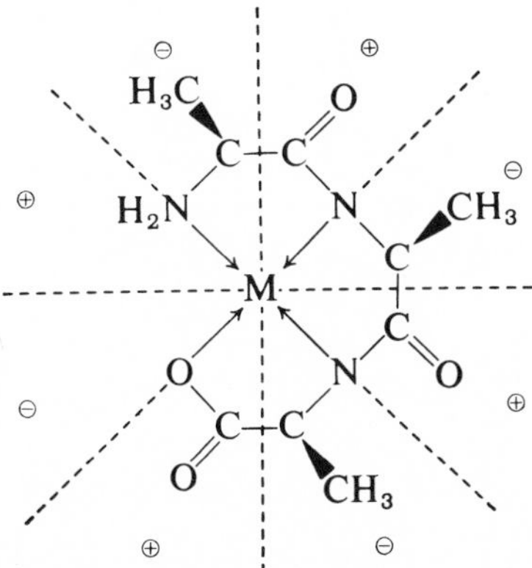

Fig. 5-33 Copper(II) complex of ala-ala-ala in Martin's double octant with the signs for the space above the plane [133].

Vicinal Effects Due to an Asymmetric Donor Atom

Although the rotational strength induced in a central metal ion chromophore by an amine containing an asymmetric carbon is very small, it is particularly significant when the donor atom is itself asymmetric. Sarcosine (*N*-methylglycine) [80] and *N*-methylethylenediamine [26,27] have been resolved while coordinated to cobalt(III) in complexes of the type $[Co(NH_3)_4L]^{n+}$ and related complexes.

The conformational effect in the sarcosine complexes should not be important. X-ray analysis of *l*-$[Coen_2sar]^{2+}$ [14] and d_{436}-$[Co(NH_3)_4sar]^{2+}$ [112] has shown that, at least in the crystals, the sarcosine chelate ring is slightly puckered. Nevertheless, in solution the energy difference between the axial and equatorial conformations is probably small. The observed CD shown in Fig. 5-34 for l_{436}-$[Co(NH_3)_4sar]^{2+}$ is thus due mainly to the asymmetric nitrogen with the R configuration. In the complex $[Co(NH_3)_4N$-Me-L-leu$]^{2+}$ the *N*-methyl-L-leucinate coordinates stereospecifically with the R-nitrogen configuration [171]. In Fig. 5-35 its CD is compared with that of L-leucinatotetramminecobalt(III). If the CD of the latter is subtracted from that of the *N*-methyl derivative, the result does not give an exact measure of the vicinal effect because (a) the ligand fields in the two complexes are different ($\Delta_{RNH(Me)} \neq \Delta_{RNH_2}$) and so the positions of the E and $A_2(D_{4h})$ components are different in the two; (b) the size of the C* vicinal effects are probably different for the two complexes, because the groups around

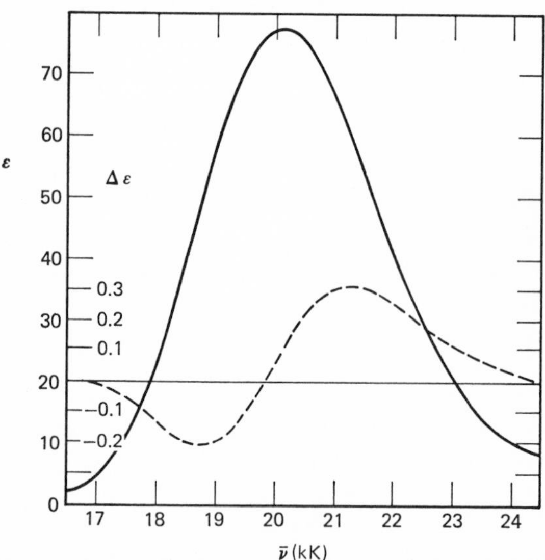

Fig. 5-34 Absorption (——) and CD (– – –) spectra of l_{436}-$[Co(NH_3)_4sar]^{2+}$.

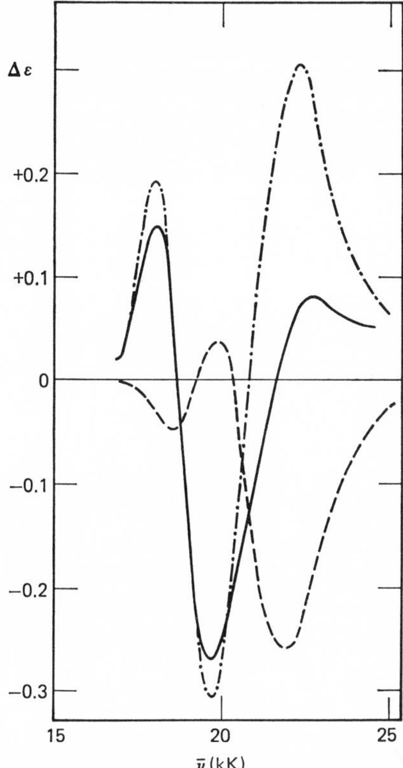

Fig. 5-35 Cotton effects under the first ligand field band for $[Co(NH_3)_4$R-N-Me-L-leu$]^{2+}$ (——) and $[Co(NH_3)_4$L-leu$]^{2+}$ (– – –) and a curve (–·–·) representing the difference between the two spectra (data from [171] and [200]).

the asymmetric carbon are not identical; and (c) any conformational effect, no matter how small, is different for the two complexes. Nevertheless, the resultant CD gives us some idea of the signs and approximate relative sizes of the rotational strengths of the T_{1g} components from the asymmetric nitrogen: Γ_b(E) and $A_2(D_{4h})$ are positive and Γ_a(E) is negative, with Γ_a and A_2 large compared with Γ_b. From this it would appear that the two observed CD bands under T_{1g} in *l*-$[Co(NH_3)_4sar]^{2+}$ and also in $[Co(NH_3)_4$-(*N*-Me-L-ala)$]^{2+}$ are Γ_a (−ve) and A_2 (+ve). Saburi and Yoshikawa also studied the CD of L-pipecolinatotetramminecobalt(III), whose structure is shown in Fig. 5-36 [171]. The rigid-chair conformation of the six-membered piperidine ring enforces some rigidity on the amino acid chelate ring and stabilizes one chirality for the chelate ring. Further, the ligand coordinates stereospecifically with the nitrogen having the R configuration. Because the chirality

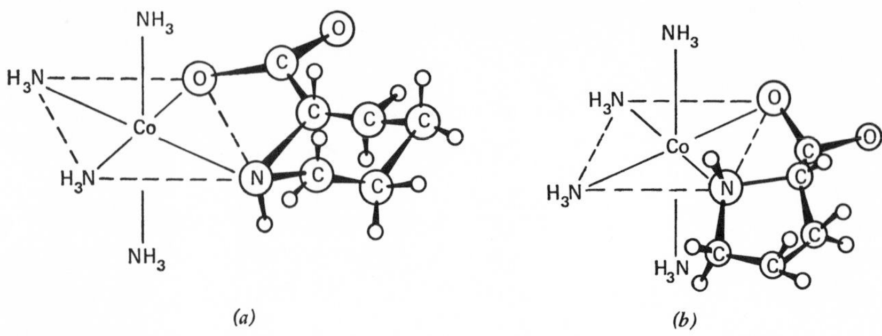

Fig. 5-36 Structure of (*a*) $[Co(NH_3)_4$(L-pipecolinato)$]^{2+}$ and (*b*) $[Co(NH_3)_4$L-pro$]^{2+}$.

of the conformation is relatively fixed, one might expect a conformational effect to be important. The measured CD, however, is not much different from that of the *N*-methyl-L-alaninate complex. From Dreiding stereomodels it would appear that a range of conformations is available to the amino acid ring, varying from a conformation with the carboxylate carbon in the NMO plane to a series of envelope conformations. For all these the dihedral angle ω is smaller than for the diamines, and for some it approaches zero. This could possibly account for the conformational effect not significantly contributing to the observed CD.

L-Proline is another interesting example in which the nitrogen coordinates stereospecifically but, in contrast to L-pipecolinate, with the S configuration (see Fig. 5-36). The CD of $[Co(NH_3)_4$L-pro$]^{2+}$ (see Fig. 5-25) shows the sign pattern $-$, $+$, $-$ with the central positive band Γ_a the largest. The rotational strengths from C* and N* reinforce one another as each imposes Γ_b ($-$ve), Γ_a ($+$ve), and A_2 ($-$ve). This is not the case for Cu(L-pro)$_2$, in which the vicinal effect from the asymmetric nitrogen is opposite to that from the carbon. The asymmetric nitrogen vicinal effect imposes a negative Cotton effect on the $^2B_{1g} \rightarrow {}^2B_{2g}$, and $^2E_g(\Gamma_b)(D_{4h})$ transitions and a positive Cotton effect on the $^2B_{1g} \rightarrow {}^2E_g(\Gamma_a)(D_{4h})$ transition. The amino acid chelate-ring conformation is again relatively rigid, but the displacement from the CoN_3O plane is rather small, thus making any conformational effect unimportant.

The *N*-substituted ethylenediamine complexes have been studied by Buckingham, Marzilli, and Sargeson [26,27]. The interpretation of the published CD spectra will be left until the conformational effect has been discussed.

Asymmetric donor atoms are often found in multidentate systems; for example, *trans*-triethylenetetramine complexes can be resolved, and the CD of the isomers must be governed to some extent by the presence of the asymmetric secondary nitrogen atoms. The exact contribution of the N* vicinal effect is difficult to judge, however, especially because the asymmetric nitrogen atoms have two groups that differ only at the third atom displaced from the nitrogen (see Fig. 5-37).

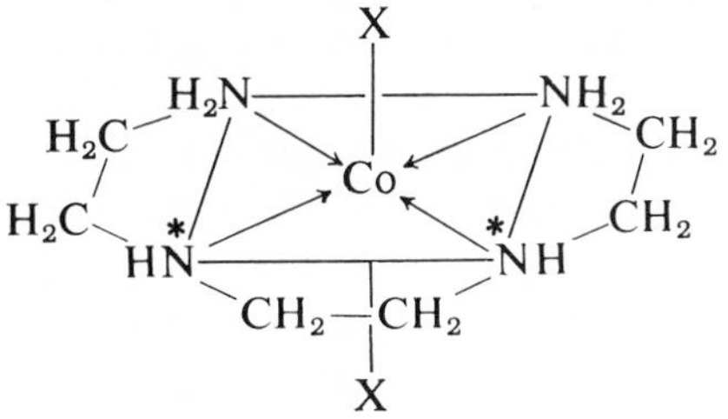

Fig. 5-37 *Trans*-triethylenetetraminecobalt(III).

In Chapter 1 it was mentioned that unsymmetrical multidentates are potential sources of dissymmetry in complexes. Take, for example, the complex triammine(*N*-2′-aminoethyl)-2-aminoethanolcobalt(III). The facial isomers shown in Fig. 5-38 can be resolved. The two chelate rings are not chirally disposed, and the conformations are enantiomeric. The molecule's asymmetry

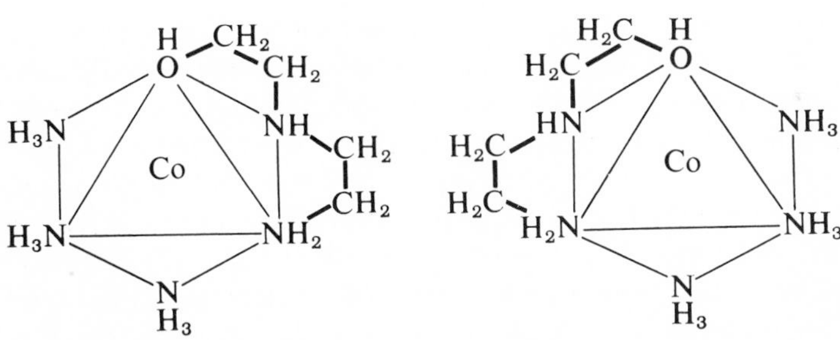

Fig. 5-38 Optical isomers of *cis*-triammine{N-(2′-aminoethyl)-2-aminoethanol}-cobalt(III).

derives from the asymmetric nitrogen atom, which is the source of any circular dichroism.

At present experimental work is being undertaken to prepare a complex with a unidentate asymmetric nitrogen-donor ligand [173]. This work will be most useful in providing information on the rotational strength imposed by asymmetric donor atoms.

The Conformational Effect

The chiral conformations of chelate rings impose optical activity on the central metal ion. In practice it is usually difficult to separate the conformational effect from vicinal effects. Nevertheless, for diamines, aminoalcohols, and other strongly puckered ring systems, the experimental data strongly suggest that it is the more important source of rotational strength. The exact features of the conformation that govern the size of the rotational strength are unknown at present. It seems reasonable to expect, however, that the degree of puckering is important. The dihedral angle ω (see Chap. 3) defines this puckering, as does the relative position of the two carbon atoms that are adjacent to the donor atoms. This is the basis of an octant rule proposed in 1965 by Hawkins and Larsen for determining the conformations from the signs of the Cotton effects of the T_{1g} components of d^6 and d^3 complexes [85]. It can be stated as follows:

Conformations with negative octant signs (λ) in complexes with tetragonal symmetry impose a negative Cotton effect on the $A_{2g}(D_{4h})$ component of the T_{1g} cubic band.

Let us examine the experimental data that were the basis of this empirical rule. Unfortunately most of the published work has been centered on closely related ligands, mostly five-membered diamines. The number of different systems being studied, however, is increasing rapidly, and in the near future sufficient information will be available to provide a rigorous test for the octant rule.

A series of compounds with the general formula $[Co(\text{R-pn})(NH_3)_2X_2]^{n+}$ has been studied [87,91,92]. The λ conformation with the methyl group equatorial, which has a negative octant sign, is favored for the three geometrical isomers, which all have an approximate tetragonal symmetry (see Fig. 5-39). There are four diastereoisomers of the cis-cis isomer, but for the moment we shall consider the mixed species where the conformation, and perhaps the vicinal effect, is the only source of optical activity. The CD for these compounds with $X = Cl^-$ are shown in Fig. 5-40. The *trans*-diammine and the cis-cis isomers have the $A_2(D_{4h})$ component at lower energy than the E by approximately 2.2 kK [92], whereas in the *trans*-dichloro compound the A_2 is at higher energy by about 5.0 kK. The A_2 component of the latter compound is obviously negative; the E component has a dominant positive Cotton effect and, by comparison with *trans*-dichlorobis(R-propylenediamine)cobalt-(III) [87], there is probably a negative component of the E at lower energy which is swamped by the dominating positive band. The sign of the Cotton effect for the individual transitions is independent of *X*. Gaussian analyses of the absorption spectra of the *cis-cis* and *trans*-diammine complexes show that the spectra can be accounted for by two bands, one at approximately 16.5 kK (A_2), and the other at about 18.7 kK (E). Although the two components of the

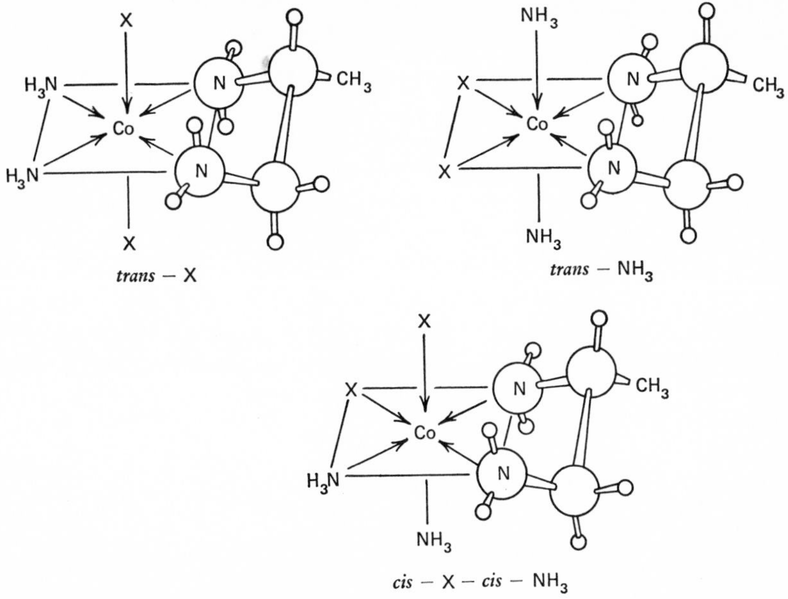

Fig. 5-39 Three geometrical isomers of $[Co(R\text{-}pn)(NH_3)_2X_2]^{n+}$.

E are not degenerate, the actual splittings are thought to be of the order of 0.15 kK or less. For the *trans*-diammine complex the positive component of the E is dominant, as found for the *trans*-dichloro compound. The negative band is found at higher energy than the absorption maximum of the A_2, and, because any cancellation by the positive should push the A_2 CD band well below the position of the absorption maximum, the observed negative is probably composed of contributions from the A_2 and the negative component of the E, both of which are dominated by the positive transition. For the *cis-cis* complex two positive bands are observed under the T_{1g} band.* The shape and positions of the bands could be interpreted in a number of ways. The spectrum is consistent with a dominant high-energy positive component with a lower energy negative band, which leads to partial cancellation of the positive band. For all three complexes a positive CD band with E parentage dominates the T_{1g} region. From these studies it would appear that the signs of the Cotton effects depend on the conformation of the chelate ring.

A number of complexes of the type *trans*-$[Co(R\text{-}diamine)_2X_2]^{n+}$ have been studied for *l*-propylenediamine and *l*-*trans*-1,2-diaminocyclohexane [87,195].

* An earlier published CD spectrum [91] could not be reproduced when greater care was taken in maintaining equal concentrations of the four diastereoisomers and in preventing decomposition in the solid state.

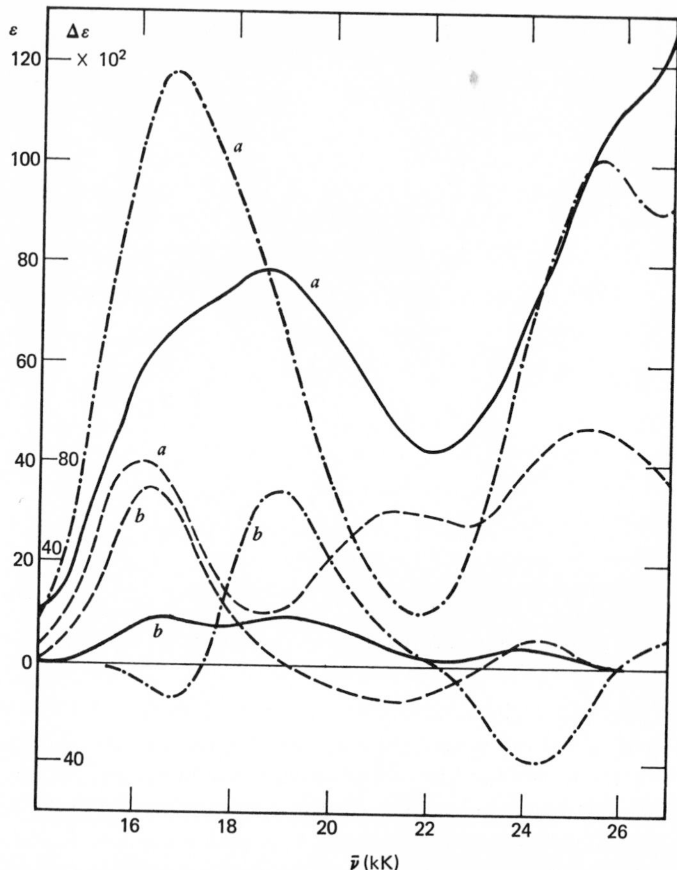

Fig. 5-40 Absorption (*a*) and circular dichroism (*b*) spectra of *cis**- $[Co(R\text{-}pn)(NH_3)_2^*Cl_2^*]^+$ (——) and *trans**-$[Co(R\text{-}pn)(NH_3)_2^*Cl_2]^+$ (–·–·–) in dimethylsulfoxide [92] and of *trans**-$[Co(R\text{-}pn)(NH_3)_2Cl_2^*]^+$ (– – –) in methanol [87].

The latter bidentate is completely stereospecific as far as the conformation is concerned. The signs of the Cotton effects of equivalent transitions are the same for the two ligands: A_2 is negative, and E has a dominant positive band. The CD for the dichloro compounds with the two diamines are shown in Fig. 5-41. The rotational strengths are seen to be approximately twice those of the mono(propylenediamine) complex, as expected from the ratio of the number of puckered chelate rings. It is important to note that the rotational strength for the bis(propylenediamine) complex is only slightly less than for the bis(*trans*-1,2-diaminocyclohexane). In the latter there are two asymmetric carbons per ligand compared with one in the former. This suggests that the

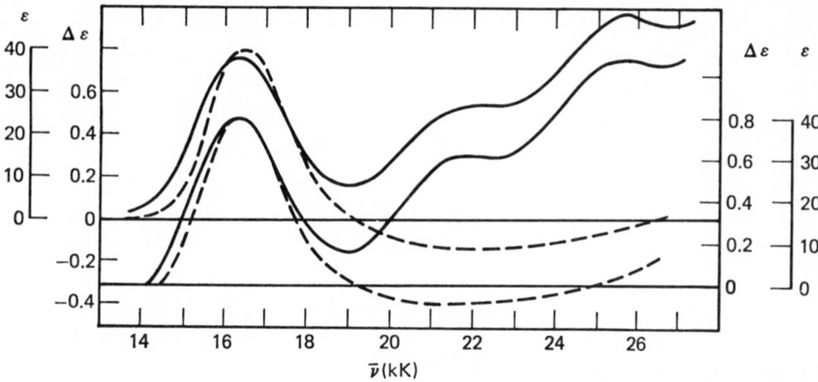

Fig. 5-41 Absorption (——) and CD (– – –) spectra of *trans*-$[Co(R\text{-}pn)_2Cl_2]^+$ (*upper curves and left scale*) and *trans*-$[Co(R\text{-}chxn)_2Cl_2]^+$ (*lower curves and right scale*) in methanol (data from [87]).

rotational strength does arise from the conformations. Further, because the cyclohexane derivative is stereospecific, the fact that the propylenediamine complex has a rotational strength for, say, its positive E component only slightly less than that for the diaminocyclohexane compound implies that in the former the ratio of equatorial to axial conformations is greater than 9:1.

Hall and Douglas have studied a similar system with rhodium(III) and the diamines R-propylenediamine and R,R-2,3-diaminobutane [77]. The A_2 transition is again found to be negative. Further, the rotational strengths for the equivalent complexes are almost identical, with the butane derivative imposing slightly larger rotational strengths. From this it can again be concluded that the rotational strength is derived from the conformation and not the vicinal effect and that, in these complexes, the equatorial conformation is favored over the axial by a significant energy difference.

The discussion above has been restricted to complexes that have a pseudo-octahedral stereochemistry, in which the equatorial conformation is favored over the axial because of the interaction of the axial methyl groups with the ligands in the apical positions. In square-planar complexes, the energy difference between the two kinds of conformations approaches zero. In solution, however, solvent molecules oriented in the fifth and sixth positions would lead to a preference for the equatorial. This probably is the explanation for the observations of Ito, Fujita, and Saito, who studied some square-planar platinum(II) and gold(III) complexes of propylenediamine and cyclohexanediamine [97]. It was found, for example, for platinum(II), that for, say, band 4, assigned to the $^1A \rightarrow {}^1E$ $(d_{zx}, d_{yz} \rightarrow d_{x^2-y^2})$ transition, the rotational strength for the bis(R-*trans*-1,2-diaminocyclohexane) complex was approximately twice that for diammine(R-*trans*-1,2-diaminocyclohexane) and bis(R-propylenediamine) and four times that for ethylenediamine(R-propylenediamine) and diammine(R-propylenediamine). This suggests for R-propylenediamine that the ratio of equatorial to axial is of the order of 3:1, which corresponds to an energy difference of 0.6 kcal mol^{-1}.

Ito and his coworkers offered a different explanation for their observations [97]. They suggested that, because C^4 and C^5 of the cyclohexane lie in negative octants, as do the C^1 and C^2 atoms, whereas C^3 and C^6 almost lie in the plane of the complex, any rotational strength derived from C^4 and C^5 reinforces that from C^1 and C^2. They further suggest that this effect might be more significant for these complexes than for those involving other metal ions because of the increased covalency of these compounds.

An additional alternative to account for the observations is that the vicinal effect is all important, and the rotational strength depends on the number of asymmetric carbon atoms. In view of the results from the study of the cobalt(III) and rhodium(III) complexes, however, this alternative suggestion is not favored.

The CD spectra of $[Co(R\text{-}pn)(NH_3)_4]^{3+}$ [87] and d_{400}-$[Co(Meen)(NH_3)_4]^{3+}$ [26]† have similar overall features (see Fig. 5-42). The chromophore is approximately cubic, although there is a tetragonal perturbation because Δ_{pn} and Δ_{Meen} are different from Δ_{NH_3}. The tetragonal splitting would be about 0.1 kK, which is of the same order of magnitude as that expected for the two components of the E due to the low molecular field. The $A_2(D_{4h})$ Cotton effect would be at higher energy but would be completely swamped by the higher of the E components. If the methyl group is equatorial for the *N*-substituted complex, the S-N* and R-N* configurations correspond to λ and δ conformations, respectively. The ratio of the rotational strengths of the two observed Cotton effects are different for the *C*-methyl and *N*-methyl complexes, and the rotational strength of the dominant positive band for the latter is approximately half that for the former. This suggests that the conformational and N* vicinal effects are not reinforcing one another in the *N*-methyl complex. The factor that dominates, however, is uncertain.

The CD spectrum for *l*-*trans*-$[Co(Meen)_2Cl_2]^+$ (see Fig. 5-43) has a different form from that of *trans*-$[Co(R\text{-}pn)_2Cl_2]^+$ (see Fig. 5-41). In the latter, the chelates have the λ conformation, and this enforces Γ_b (−ve), Γ_a (+ve), and A_2 (−ve), Γ_b lying at lower energy than Γ_a and being dominated by it. The CD of the *N*-methyl complex has been interpreted by Buckingham, Marzilli, and Sargeson as arising from a dominant positive E component

† The frequency scale for the published spectrum was in error. The correct scale is shown in Fig. 5-42.

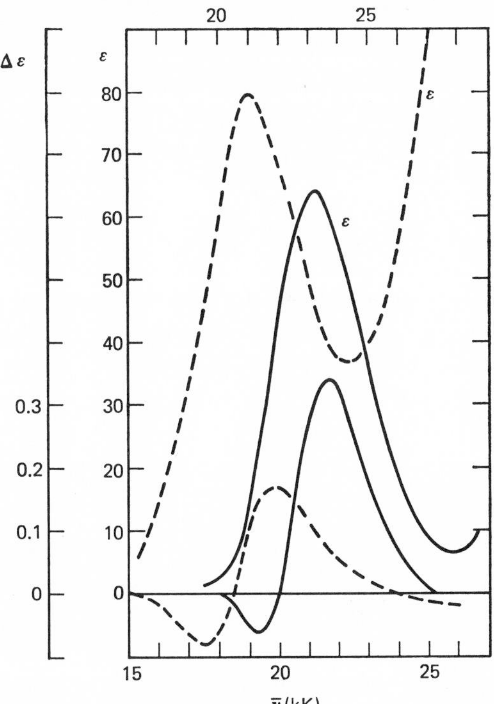

Fig. 5-42 Absorption and CD spectra of $[Co(R\text{-}pn)(NH_3)_4]^{3+}$ (——) (*lower frequency scale*) [87] and *d*-$[Co(Meen)(NH_3)_4]^{3+}$ (– – –) (*upper frequency scale*) [26]. The frequency scale in the published spectrum of the latter complex was in error; the correct scale is shown here.

(Γ_a) with a negative band partially canceling Γ_a to yield two positive peaks followed by a negative A_2 band [27]. This, they said, was due to the ligand's having the S configuration and a λ conformation. X-ray studies, however, have shown that the conformation is in fact δ, and the asymmetric nitrogen has the R configuration [165]. As the δ conformation would impose a positive A_2 Cotton effect, the CD spectrum is dominated by the asymmetric nitrogen. An explanation of the CD is presented in Fig. 5-43, where the conformational effect is represented by the CD of *trans*-$[Co(S\text{-}pn)_2Cl_2]^{+}$* [84]. Very recently Saburi, Tsujito, and Yoshikawa reported the CD spectrum of *trans*-dichlorobis(*N*-methy-S-propylenediamine)cobalt(III), which has the δ conformation

* The energy scale for the spectra of *trans*-$[Co(S\text{-}pn)_2Cl_2]^{+}$ was adjusted until the absorption maximum for the E_g band corresponded to that for *trans*-$[Co(Meen)_2Cl_2]^{+}$.

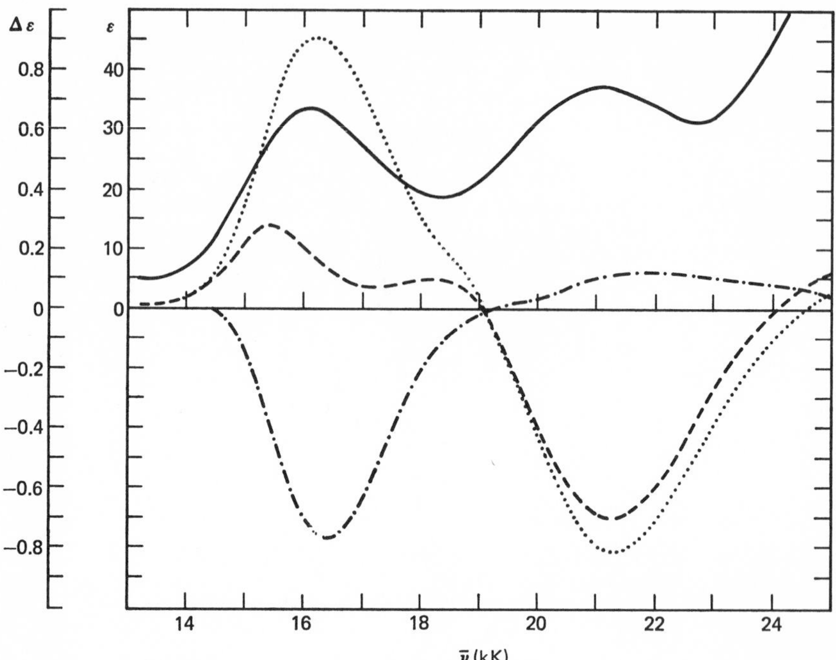

Fig. 5-43 Absorption (——) and circular dichroism (– – –) spectra of *l-trans*-[Co(Meen)$_2$Cl$_2$]$^+$ in methanol showing the proposed contributions from the conformation (–·–·–) and asymmetric nitrogen (· · · ·) [84].

and R-N* preferred, and compared the CD with that† of *l-trans,trans*-[Co(Meen)$_2$Cl$_2$]$^+$ [170]. They correctly concluded that the latter had the same configuration as the N-methyl-S-propylenediamine complex.

Resolved *trans*-dichlorotriethylenetetraminecobalt(III) acquires its CD from the conformations of the chelate rings and from the asymmetric secondary N atoms. The two possible optically active conformers are shown in Fig. 5-44 and the CD spectra of the *d* isomer and *d-trans*-dichloro(S-3,8-dimethyltriethylenetetramine)cobalt(III) [30,201] in Fig. 5-45. Because the vicinal effect from the asymmetric carbon is thought to be unimportant for such a system, it is obvious from the CD spectra that the two compounds have the same configuration. In the trans complex, S-3,8-dimethyltriethylenetetramine coordinates stereospecifically with the methyl groups equatorial, with the S configuration for the asymmetric nitrogens, and with conformations that yield an overall positive octant sign. If we ignore the asymmetric nitro-

† The CD of *l-trans,trans*-[Co(Meen)$_2$Cl$_2$]$^+$ reported differs in some respects from that published by Buckingham, Marzilli, and Sargeson [27] and from that measured in this laboratory.

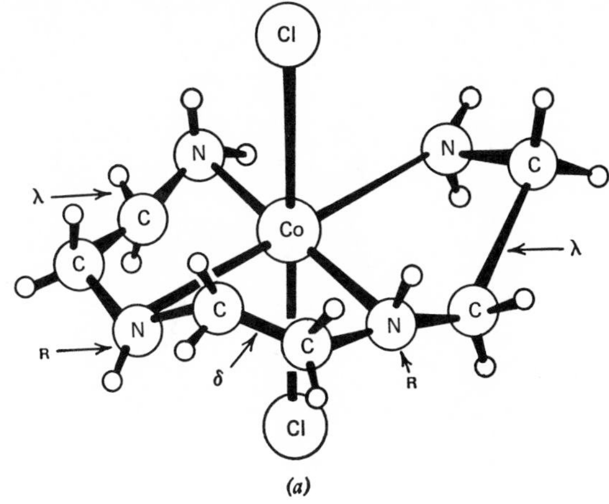

(*a*)

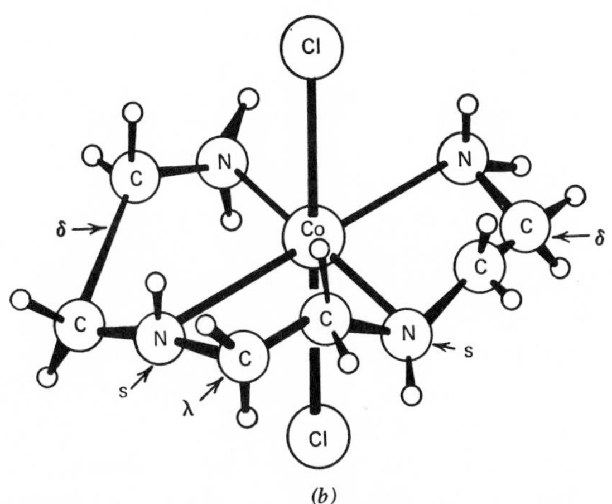

(*b*)

Fig. 5-44 The dissymmetric conformers of *trans*-dichlorotriethylenetetraminecobalt-(III): (*a*) negative octant sign; (*b*) positive octant sign.

gens, the positive octant sign is expected to give rise to a positive Cotton effect for the $A_2(D_{4h})$ transition. This is found. We are unable to predict what sign the asymmetric nitrogens will impose, but it is thought that the rotational strength derived from this source will be small here, because the asymmetric nitrogens have two groups that are almost identical, ($—CH_2CH_2NH_2$ and $—CH_2CH_2NHR$). The assigned absolute configuration is supported by acid-hydrolysis studies because the trien complex undergoes hydrolysis

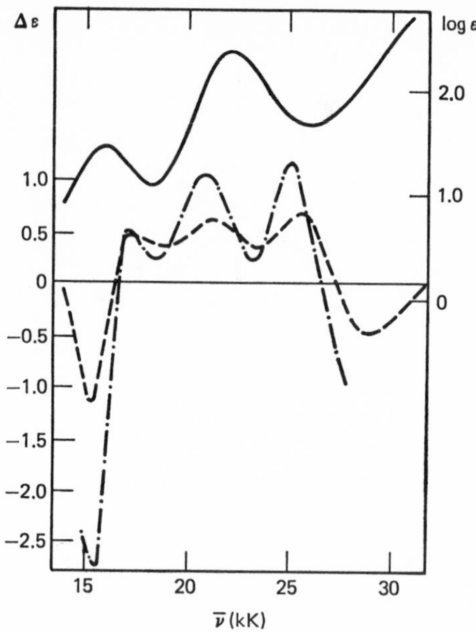

Fig. 5-45 Absorption (——) and CD (– – –) spectra of d-*trans*-[Cotrien Cl_2]$^+$, and the CD spectrum (–·–·) of d-*trans*-[Co(s-3,8-dimetrien)Cl_2]$^+$ [30].

stereospecifically to form D(*d*)-*cis*-β-chloroaquatriethylenetetraminecobalt-(III) [28].

Gillard and Wooton have published CD spectra for alkaline solutions of copper(II) and the ligands s-2-aminopropanol and s-ψ-ephedrine (C_6H_5CH-(OH)$CH(CH_3)NH(CH_3)$) with a metal-ligand ratio of 1:2 [68]. They observed three CD bands with the signs +, −, +, the CD for the ephedrine complex being much larger than that for 2-aminopropanol. The water molecules occupying the fifth and sixth positions would stabilize the δ conformation for both ligands. In the ephedrine complex this conformation has the nitrogen coordinated stereospecifically as the R configuration, and this, the authors suggest, is the reason for the much larger Cotton effects observed for this complex. Unfortunately the exact species present in solution have not been established. The ligands do not form very stable complexes even as chelates, and because there was only a metal-ligand ratio of 1:2 and the solution was alkaline there is the possibility that polynuclear μ-hydroxo bridged complexes were present in solution. It seems very unwise to publish CD of such systems without first having carefully studied the equilibria that exist in solution. Gillard has also published the CD of an aqueous solution containing copper(II) and *l*-propylenediamine [66]. Again the exact molecular composition of the solution is not known. Two positive CD bands are observed under the *d–d* absorption band due primarily to the conformations of the chelate rings.

A number of attempts have been made to determine the conformational effects of optically active diamines in tris(bidentate)cobalt(III) complexes. CD spectral data for the diastereomeric pair *d*- and *l*-[Co(R-chxn)$_3$]Cl$_3$ have been published [162]. The two spectra were not enantiomeric, and the authors accounted for this by suggesting that the trigonal splitting of the E and $A_2(D_3)$ components of the T_{1g} band was different for the two complexes. Douglas explained a similar difference for the diastereomers *d*- and *l*-[Co(R-pn)$_3$]$^{3+}$, *d*- and *l*-[Co(R-pn)$_2$en]$^{3+}$, and *d*- and *l*-[Co(R-pn)en$_2$]$^{3+}$ by proposing that the configurational contributions for the two were equivalent but enantiomeric, whereas the conformational effect was identical for the two diastereomers [49]. If the CD spectra for the two are added, the resultant spectrum is a multiple of the conformational effect* of a propylenediamine ring with the λ conformation. It was found that the λ conformation gave a negative $E(D_3)$ component and a positive $A_2(D_3)$ transition, the latter being the dominant component in contrast to the conformational effect. Following this work by Douglas, two other groups published similar results for some of these compounds and arrived at the same conclusions as Douglas [126,151].

Mason and his coworkers have proposed that the sign and magnitude of the CD associated with the charge transfer at about 2150 Å in diamine cobalt(III) complexes are governed by both the configurational effect and the conformational effect [123,126]. From their published data they extracted quantitative configurational and conformational contributions to the charge-transfer band for a series of ethylenediamine and propylenediamine complexes: D and L configurations contribute $\Delta\varepsilon = -20$ and $+20$, and δ and λ conformations $\Delta\varepsilon = -8$ and $+8$ respectively. Their results, however, especially the extremely good agreement between the computed and experimental values of $\Delta\varepsilon$ for the whole range of complexes, must be viewed with some caution, because it would be difficult to obtain accurate measurements with the particular instrument used at 2150 Å for transitions with $\Delta\varepsilon/\varepsilon$ equal to 0.002 or less. Ogino, Murano, and Fujita have also studied the charge-transfer region for some of the complexes studied by Mason and his coworkers [151]. For the L($\delta\delta\delta$) configuration, however, they found two CD bands under the charge-transfer absorption band of interest, the most dominant component centered at about 2150 Å being opposite in sign to that published by Mason. Circular dichroism instruments with optics for measuring down to 1850 Å, which are currently available commercially, will enable this region to be studied much more successfully.

* Douglas classified it as a vicinal effect.

The Distribution of Chelate Rings—The Configurational Effect

The CD spectra of a large number of complexes that have dissymmetric arrangements of chelates have been published. Unfortunately, for a great many of these, it has been impossible to assign the observed Cotton effects to particular transitions. Nevertheless, because the absolute configurations of a number of representative complexes are known from other sources, it has been possible to develop empirical rules that have been most successful in predicting the chirality of certain kinds of chelates.

Complexes with cubic symmetry. In this classification we consider two series of complexes: $M(aa)_3^{n+}$, in which aa is a symmetrical bidentate, and *cis*-$M(ab)_3^{n+}$, in which ab is an unsymmetrical bidentate—an amino acid, for example. In complexes with d^3, d^8, and spin-paired d^6 metal ions, the low-energy T_{1g} cubic absorption band is found to have two CD components. As discussed previously, the exact reason for this splitting is not known. The D_3 and C_3 molecular symmetries, however, somehow impose themselves on the chromophore. The two bands observed for *d*-$[Coen_3]^{3+}$ are not due to the presence of the D(λλλ) isomer, as recently proposed by Dingle and Ballhausen [45], because complexes of the type *d*-$[Co(R\text{-}chxn)_3]^{3+}$ show two bands, although there is only one possible configuration, L(λλλ).* The splitting is small, leading to gross cancellation of the two components, and the sign of the splitting is presently not predictable, making the assignment of the transitions sometimes very difficult.

Three procedures have been devised for identifying the components: (a) oriented crystal CD studies; (b) the effect of polarizable oxy anions, such as sulfate, phosphate, and selenite; and (c) the empirical observation that the E component often dominates the region of the T_{1g} band.

The oriented-crystal approach developed for metal complexes by the Mason school [119,121]† has as a prerequisite that the structure of the crystal must be known. Further, although it is possible to obtain spectra in biaxial crystals, it is desirable to study the complex in a uniaxial crystal in which the polarizations of the transitions have a direct relationship to the unique axis. Light propagated along the unique axis of a uniaxial crystal that has the molecular C_3 axes parallel to the unique axis has its electric vector at right angles to the C_3 axes and therefore excites only the transition with E symmetry. The CD spectrum of *d*-$[Coen_3]^{3+}$ in the uniaxial crystal (*d*-$[Coen_3]$-

* Piper and Karipides referred to the isomer with this configuration as (−). This is thought to refer to the sign of the dominant Cotton effect under the T_{1g} band, because the sign of the rotation at the Na_D line is positive.

† Drouard and Mathieu had previously measured the CD of *d*-$[Coen_3]^{3+}$ oriented in a crystal ($XBr_3 \cdot 2H_2O$) [53].

$Cl_3)_2 \cdot NaCl \cdot 6H_2O$ is shown in Fig. 5-46. The $E(T_{1g})$ and $E(T_{2g})$ bands are both positive, and thus, in the solution spectrum of *d*-$[Coen_3]^{3+}$, the positive band can be assigned to the E component and the negative to the A_2.

It has been found that various anions have a measurable effect on the CD spectra of complexes in solution [184]. The polarizable oxy anions, sulphate, phosphate, and selenite, have a dramatic effect on the spectra of some tris-(diamine) complexes, giving rise to a marked increase in the rotational strength of the A_2 component at the expense of the E component [71,137,138]. This effect is observed where hydrogen atoms attached to the nitrogens of the diamines are oriented approximately parallel to the C_3 molecular axis. The distance between these atoms, for example, in L-$[Co(R\text{-}pn)_3]^{3+}$ and $[Cotn_3]^{3+}$, is approximately 2.5 Å, which corresponds to the O—O distances in the oxy anions. Whenever the hydrogens are absent, for example, in tris(1,10-phenanthroline) complexes, and where the hydrogens cannot be oriented in the above way, the effect described above is not observed. This suggests that the effect arises because of H bonding by the anions to these particular hydrogen atoms.*

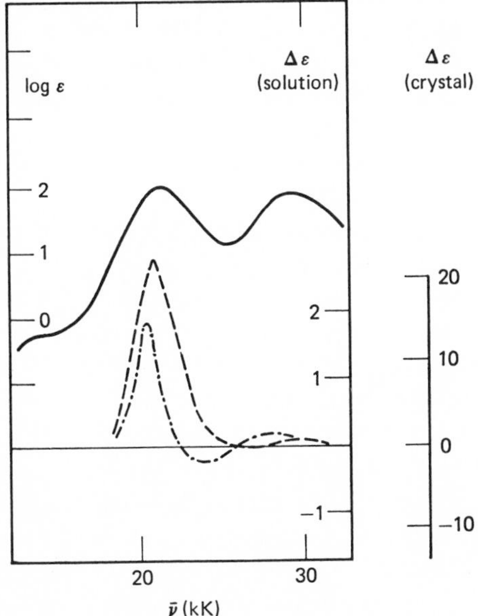

Fig. 5-46 Absorption (——) and CD (–·–·) spectra of d-$[Coen_3]^{3+}$ in aqueous solution, and the axial CD spectrum of (d-$[Coen_3]Cl_3)_2$. $NaCl.6H_20$ (– – –) [119].

* Note added in proof: A recent publication (R. R. Judkins and D. J. Royer, *Inorg. Nucl. Chem. Letters*, **6**: 305 (1970)) has thrown some doubt onto this method of identifying the components of the T_{1g} band, but, as stated above, until further experimental data are published it is difficult to evaluate the evidence.

The most common method of assigning the trigonal components has been to rely on the observations that in many complexes the E component is dominant over the A_2. Although it has been most useful, this is a particularly dangerous procedure, especially if the degree to which one component is larger than the other is only slight and if CD bands of transitions of similar energy to the T_{1g} band overlap its region.

CD data for the T_{1g} band of $M(aa)_3{}^{n+}$ and *cis*-$M(ab)_3{}^{n+}$ are given in Table 5-5. The table includes the method used to assign the transitions and, where possible, the method used to determine the absolute configuration. This is not meant to cover every CD spectrum published, but sufficient information is given to show that "for complexes with the D configuration, the E and A_2 transitions have positive and negative Cotton effects, respectively, and, for complexes with the enantiomeric chirality, the signs are reversed."

It must be remembered that in some of these complexes vicinal and conformational effects are also present and they must be taken into account when assessing the configurational effect. Douglas [49] has shown that when the conformational contribution is subtracted, the configurational effect of the two isomers of tris(R-propylenediamine)cobalt(III) and of the mixed R-propylenediamine ethylenediamine complexes is similar to the CD curves for the isomers of $[Coen_3]^{3+}$, showing that the preference for the lel configurations of $[Coen_3]^{3+}$ is not very marked. When the vicinal effect is subtracted from complexes of the type *cis*-[Co(L-am)$_3$], the resultant spectrum has one dominant band, as found in *cis*-[Cogly$_3$].

For all the complexes listed in Table 5-5, in which the dominance of E has been used to assign the transitions, one component obviously has the larger rotational strength, and this assignment is supported by a comparison of the relative energies of the components with those for closely related compounds, where the assignment has been made by one of the other methods. Unfortunately, there is a tendency to apply "the dominating E" assignment even in situations where both the observed $\Delta\varepsilon$ values are small and almost equivalent. This practice should be discouraged. It should also be noted that the order of the two components in solution need not necessarily be the same as found in a crystal, and, therefore, one cannot base an assignment on oriented-crystal absorption studies. For example, Dingle and Ballhausen have shown that the absorption maximum of the A_2 component is at lower energy than that of the E for *d*-$[Coen_3]^{3+}$ in some crystal lattices [45]. In solution [119], however, and in a solid isotropic medium, such as a KBr disk [125], the E component is at the lower energy. Mason and his coworkers [121] have used the polarized crystal absorption spectra for $[Crmal_3]^{3-}$ [82] to assign the E transition to the lower-energy positive CD band for *l*-$[Crmal_3]^{3-}$, although it has the smaller rotational strength. Such an assignment, and the consequent absolute configuration, must be disregarded.

Table 5-5 Absorption and CD data for the T_{1g} band of $M(aa)_3^{n+}$ and *cis*-$M(ab)_3^{n+}$

Complex	Absorption $\bar{\nu}$, kK	Absorption ε	CD $\bar{\nu}$, kK	CD $\Delta\varepsilon$	Transition	Method of assignment	Absolute configuration	Ref.[a]
d-$[Coen_3]^{3+}$	21.28	84	20.28	+1.89	E	Crys.	D^{XR}	121
			23.37	−0.17	A_2			
l-$[Co(R\text{-}pn)en_2]^{3+}$	[b]		20.5	−2.0	E	Dom.	L^{E}	49
			23.2	+0.3	A_2			
l-$[Co(R\text{-}pn)_2en]^{3+}$	[b]		20.4	−2.85	E	Dom.	L^{E}	49
			24.4	+0.72	A_2			
l-$[Co(R\text{-}pn)_3]^{3+}$	21.37	96	20.28	−1.95	E	An.	L^{XR}	121
			22.78	+0.58	A_2			
d-$[Co(R\text{-}pn)_3]^{3+}$	[b]		21.0	+2.47	E	Dom.	$D^{(XR)}$	49
d-$[Co(R\text{-}chxn)_3]^{3+}$	[b]		20.0	−2.28	E	Dom.	L^{CA}	162
			22.5	+0.69	A_2			
l-$[Co(R\text{-}chxn)_3]^{3+}$[c]	[b]		20.8	+3.9	E	Dom.	D^{CA}	162
l-$[Rhen_3]^{3+}$	32.6	251	31.3	+2.0	E	Crys.-an.	D^{E}	121
			34.9	−0.1	A_2			
d-$[Rh(R\text{-}pn)_3]^{3+}$	33.33	270	31.25	−2.12	E	An.	L^{CA}	77
			34.97	+0.61	A_2			
d-$[Rh(R\text{-}bn)_3]^{3+}$	33.33	270	31.25	−2.31	E	Dom.	L^{CA}	77
			34.97	+0.61	A_2			
d-$[Cotn_3]^{3+}$	20.49	75	18.69	−0.08	E	An.	L^{XR}	71
			29.96	+0.17	A_2			
l-$[Crtn_3]^{3+}$	[b]		21.05	+0.28	E	An.	D^{E}	12
l-$[Co(bgH)_3]^{3+}$	21.1	204	19.5	−3.49	A_2	Dom.	D^{E}	143
			22.0	+4.68	E			

Table 5.5—*Continued.*

Complex	Absorption		CD		Transition	Method of assignment	Absolute configuration	Ref.[a]
	$\bar{\nu}$, kK	ε	$\bar{\nu}$, kK	$\Delta\varepsilon$				
l-$[Cr(bgH)_3]^{3+}$	20.8	103	19.2	−2.73	A_2	Dom.	D^E	143
			21.8	+4.14	E			
l-$[Ni(R\text{-chxn})_3]^{2+}$	11.4	8	10.7	−3.2	E	Dom.	L^{CA}	186
			13.0	+0.9	A_2			
l-$[Coox_3]^{3-}$	16.61	153	16.20	+3.30	E	Crys.	D^E	121
d-$[Crox_3]^{3-}$	17.51	74	15.87	−0.58	A_2	Crys.	D^E	121
			18.11	+2.83	E			
d-$[Rhox_3]^{3-}$	24.87	330	25.00	+2.85	E	Crys.	D^E	121
d-$[Cren_3]^{3+}$	21.74	74	21.93	+1.36	E	An.	D^E	121
l-$[Cr(R\text{-pn})_3]^{3+}$	21.74	71	21.28	−1.30	E	An.	L^{CA}	121
			24.51	+0.06	A_2			
l-$[Cobipy_3]^{3+}$	22.20	72	19.90	−0.30	A_2	Dom.	D^E	58
			22.28	+2.90	E			
l-$[Co(L\text{-pro})_3]$	18.6	112	18.0	−2.6	E	Dom.	L^{CA}	42
			20.5	+0.15	A_2			
d-$[Co(L\text{-pro})_3]$	18.5	157	17.5	+4.1	E	Dom.	D^{CA}	42
			20.0	−0.5	A_2			
l-$[Co(L\text{-ala})_3]$	19.5	175	19.0	−2.5	E	Dom.	L^E	42
d-$[Co(L\text{-ala})_3]$	19.5	195	18.5	+1.3	E	Dom.	D^E	42
			21.0	−0.2	A_2			
d-$H_3[Co(L\text{-asp})_3]$	[d]		18.0	+1.5	E	Dom.	D^E	180
			21.0	−0.2	A_2			

Complex	Absorption $\bar{\nu}$, kK	Absorption ε	CD $\bar{\nu}$, kK	CD $\Delta\varepsilon$	Transition	Method of assignment	Absolute configuration	Ref.[a]
l-H_3[Co(L-asp)$_3$]	[d]		19.2	−1.5	E	Dom.	L^{E}	180
d-[Cogly$_2$(L-val)]	[b]		18.5	+1.30	E	Dom.	D^{E}	182
			20.9	−0.20	A_2			
l-[Cogly$_2$(L-val)]	[b]		18.8	−2.0	E	Dom.	L^{E}	182
d-[Co(L-val)$_2$gly]	[b]		18.5	+1.08	E	Dom.	D^{E}	182
			20.9	−0.03	A_2			
d-[Co(L-val)$_2$(L-ala)]	[b]		18.5	+0.92	E	Dom.	D^{E}	182
			20.9	−0.02	A_2			
l-[Co(L-val)$_2$(D-ala)]	[b]		18.9	−2.00	E	Dom.	L^{E}	182
d-[Cogly$_2$(L-ala)]	[b]		18.5	+1.36	E	Dom.	D^{E}	182
			20.9	−0.02	A_2			
d-[Co(L-ala)$_2$(D-ala)]	[b]		18.5	+1.33	E	Dom.	D^{E}	182
			20.9	−0.03	A_2			

[a] References refer to absorption and CD data.
[b] Data not given.
[c] Compounds listed as *d* and *l*-[Co(R-chxn)$_3$]$^{3+}$ were referred to in [162] as (−) and (+), respectively.
[d] Difficult to estimate from figure.

Abbreviations: Crys.—oriented crystal CD; Dom.—dominant E component; An.—added polarizable anions; XR—X-ray method; E—empirical method; CA—conformational analysis.

Complexes with tetragonal and rhombic symmetries. Within this general classification three different groups of compounds are considered: $[M(aa)_2X_2]^{n+}$, where aa is a bidentate ligand and X_2 represents two unidentate ligands or a second bidentate; *trans*-$[M(ab)_3]^{n+}$, where ab is an unsymmetrical bidentate; and finally complexes with multidentate ligands. In theory the assignment of the transitions under the T_{1g} band should be facilitated for this classification, because the tetragonal splitting is often obvious even in the absorption spectra, and various models are able to predict the order of the tetragonal components. However, the secondary splitting of the $E_g(D_{4h})$ band is not thoroughly understood, and it is impossible at the moment to predict the sign of this splitting, unless there is a known rhombic perturbation. In practice the A_2 band is only unambiguously assignable when the three T_{1g} components are observable or when the tetragonal splitting is so large that there can be no confusion.

In order to provide a backbone for the discussion of the CD of complexes in this classification, an empirical rule is given before the supporting evidence has been presented. This rule is equivalent to the octant rule proposed by Hawkins and Larsen [86]:

In complexes with a tetragonal metal ion chromophore a left-hand chiral distribution of chelate rings imposes a negative Cotton effect on the $A_2(D_{4h})$ component, and positive and negative Cotton effects on the two components of the $E(D_4h)$ band, the chirality being determined by the octant sign [85] or ring-pair [114] method.

$[Coen_2(am)]^{n+}$. This series of complexes has been studied extensively. The absolute configurations have been determined by X ray for two members of the series, with L-glutamate [56] and sarcosinate [14]; the *d* isomers have the D configuration. Because the CD spectra are very similar for the *l* series and the *d* series [31,79,118,169], it is possible to assign the L and D configurations, respectively, to these complexes. Where two CD bands are observed under the T_{1g} band, the dominant component is at the lower energy and is followed by a Cotton effect of opposite sign. For this tetragonal chromophore the A_2 transition is higher in energy than the E. The observed Cotton effects, however, are thought to correspond to the two E components, the A_2 band being canceled by the higher energy E component of opposite sign. This conclusion is supported by the results of a study of the corresponding rhodium(III) complexes [78]. For these compounds, the CD spectra of the *l* isomers have a dominant positive Cotton effect, and the general features of the spectra closely resemble those of the equivalent dextrorotatory cobalt(III) isomers. Thus the *l* rhodium complexes could be assigned the D configuration. The CD spectrum for *d*-$[Rhen_2(\text{L-met})]^{2+}$ has two negative Cotton effects under the T_{1g} band. The minor one, which is at the higher energy, is assignable to the A_2 transition, because the $E(D_{4h})$ band has not previously been observed

to give two components of the same sign. This result supports the above empirical rule, because the L configuration gives rise to two components with negative Cotton effects, one of which is the A_2 transition.

Cis-[Coen$_2$X$_2$]$^{n+}$ and related complexes. CD data have been published for complexes of this kind with X = Cl^-, H_2O, NH_3, N_3^-, CN^-, and NO_2^- [124]. Under the C_2 molecular field, the T_{1g} band gives rise to two transitions with B symmetry and one with A that has E trigonal parentage and is polarized along the C_2 axis. According to Mason and his coworkers, the two B transitions are mixed, whereas the upper state of the A (E,D_3) transition has a wave function that is invariant through the series $[Coen_3]^{3+}$, *cis*-$[Coen_2L_2]^{2+}$, and *cis*-$[Co(NH_3)_4L_2]^{n+}$, and, therefore, the A transition is the most reliable guide for assigning absolute configurations. When Δ_X is approximately equal to Δ_{en}, it was proposed that the major CD band is due to the A(E,D_3) and the B(E,D_3) transitions, whereas if Δ_X is different from Δ_{en}, the symmetry is approximately C_{2v} and the major CD band is now composed of contributions from the two transitions with B symmetry. They further proposed that, for the D configuration, the major band is always positive. Schäffer has questioned this method of analyzing the spectra and has put forward an alternative procedure using the angular-overlap model [177].

Here the chromophores are treated as being tetragonal. The order of the E and A_2 components depends on the relative size of Δ_{en} and Δ_X. If Δ_{en} is greater than Δ_X, then the $A_2(D_{4h})$ component is at lower energy, but if Δ_{en} is less than Δ_X, the $E(D_{4h})$ component is at the lower energy—see Eq. (5-34). The size of the splitting is given by $\frac{1}{4}(\Delta_{en} - \Delta_X)$. In the molecular field the symmetry is lower than D_{4h}, and the E transition is no longer degenerate. The splitting of the E level is no doubt small but could be of the order of 100 cm^{-1}.

With these proposals in mind, let us examine the CD data (see Table 5-6) available for complexes of this kind. Unfortunately, for all the complexes listed, the three components are not observable. For *d*-$[Coen_2(CN)_2]^+$, however, two positive Cotton effects are found under the T_{1g} band. Because cyanide has a stronger field than ethylenediamine, the higher energy component corresponds to the A_2 transition. According to the empirical rule above, the observation of two positive Cotton effects would suggest the D configuration. This has been verified by X-ray analysis [142]. The absolute configuration of *d*-*cis*-$[Co(l\text{-pn})_2(NO_2)_2]^+$ has also been determined by X-ray studies [11], and its CD closely resembles that of *l*-$[Coen_2(NO_2)_2]^+$, which must also have the L configuration. Because Δ_{NO_2} is approximately equal to Δ_{en} for these complexes, however, it is impossible at the moment to assign the observed transitions.

According to Eq. (5-34), the A_2 component for *d*-$[Coen_2(NH_3)_2]^{3+}$ is at lower energy by about 100 cm^{-1}, and because the splitting of the E level is

Table 5-6 Absorption and CD data for the T_{1g} band of *cis*-$[M(aa)_2X_2]^{n+}$

Complex	Absorption		CD		Ref.
	$\bar{\nu}$, kK	ε	$\bar{\nu}$, kK	$\Delta\varepsilon$	
d-$[Coen_2(NH_3)_2]^{3+}$	21.28	73	20.32	+0.42	124
			23.26	−0.04	
d-$[Coen_2(H_2O)_2]^{3+}$	20.20	83	17.86	−0.30	124
			20.62	+1.05	
d-$[Coen_2(CN)_2]^{+}$	25.00	99	22.67	+0.30	124
			27.25	+0.17	
d-$[Coen_2(NO_2)_2]^{+}$	22.73	224	21.74	+1.4	124
			25.00	−0.65	
d-[Co(R-pn)$_2$(NO_2)$_2$]$^+$	a	a	21.7	−1.1	11[b]
			24.5	+0.6	
l-$[Coen_2(N_3)_2]^{+}$	19.23	340	17.54	−0.65	124
			19.88	+1.2	
d-$[Coen_2Cl_2]^{+}$	18.69	69	16.26	−0.6	124
			18.58	+0.7	
d-$[Cren_2Cl_2]^{+}$	18.52	74	16.95	−0.5	124
			19.23	+0.6	
l_{546}-$[Coen_2(NH_3)CN]^{2+}$	22.93	78	21.6	−0.42	152
			25.3	+0.08	
d-$[Coen_2(H_2O)CN]^{2+}$	22.5	55	19.63	+0.40	152
			24.4	−0.08	
d_{546}-$[Coen_2(OH)CN]^{+}$	21.87	79	19.03	+0.51	152
			23.0	−0.57	
d_{546}-$[Coen_2(NO_2)CN]^{+}$	22.9	118	21.97	+0.28	152
			25.0	−0.14	
l-$[Coen_2(CN)I]^{+}$	18.83	155	16.87	+0.23	152
			19.27	−0.33	
d-$[Coen_2(NH_3)Cl]^{2+}$	a	a	18.05	+0.32	63
			20.92	+0.02(?)	
d_{546}-$[Coen_2(NCS)Cl]^{2+}$	a	a	17.30	−0.43	63
			19.80	+0.41	
d-$[Coen_2(NO_2)Cl]^{+}$	a	a	18.35	+0.35	63
			22.47	+0.20	
d-$[Coen_2(H_2O)Cl]^{2+}$	a	a	17.66	+0.74	63
l-$Co[en_2bipy]^{3+}$	21.93	114	20.83	−0.95	124
l-$[Coen_2phen]^{3+}$	21.51	117	20.79	−0.78	124
d-$[Coen_2ox]^{+}$	20.00	103	19.23	+2.6	124
d-$[Cren_2ox]^{+}$	20.83	90	20.83	+1.9	124
d-[Co(R-pn)$_2$ox]$^+$	20.00	125	19.42	+3.1	124
l-[Co(R-pn)$_2$ox]$^+$	20.00	125	19.23	−2.75	124
d-$[Coen_2acac]^{2+}$	19.80	159	19.61	+2.6	124
d-$[Coen_2CO_3]^{2+}$	19.49	143	18.87	+3.7	124
d-$[Crphen_2ox]^{+}$	20.00	62	20.05	+2.43	58
l-$[Cophen_2ox]^{+}$	19.70	77	19.23	−0.99	58

Complex	Absorption		CD		Ref.
	$\bar{\nu}$, kK	ε	$\bar{\nu}$, kK	$\Delta\varepsilon$	
l-$[Crbipy_2ox]^+$	20.10	61	17.92	+0.06	58
			20.55	−1.27 (?)	
l-$[Cobipy_2ox]^+$	20.05	81	19.05	−2.15	58
			21.28	+0.69	
d_{546}-$[Coox_2en]^+$	18.50	109	17.20	−2.27	50[b]
			20.20	−0.82	
l_{546}-$[Comal_2en]^+$	18.50	95	16.95	+2.8	50[b]
			18.55	−2.3	
			20.55	+1.0	
l-$[Coen_2sal]^+$	19.38	170	16.26	−0.04[b]	64
			17.70	+0.13	
			19.76	−0.84	
l-$[Coen_2salH]^{2+}$	19.65	173	19.42	−1.28	64
d-$[Coen_2sar]^{2+}$	20.5	100	19.40	+1.8	31[b]
l_{546}-$[Coen_2gly]^{2+}$	20.53	98.5	19.80	−2.10	118
d-$[Rhen_2gly]^{2+}$	31.65	250	29.85	−2.31	78
dl-$[Coen_2(\text{L-ala})]^{2+}$	20.52	110	18.67	−0.07	118
			20.35	+0.14	
			22.47	−0.30	
d_{546}-$[Coen_2(\text{L-ala})]^{2+}$	20.52	110	19.90	+2.38	118
			22.90	−0.23	
l_{546}-$[Coen_2(\text{L-ala})]^{2+}$	20.52	110	19.70	−2.19	118
l-$[Rhen_2(\text{L-ala})]^{2+}$	31.65	330	29.85	+2.42	78
d-$[Rhen_2(\text{L-ala})]^{2+}$	31.65	330	29.85	−2.31	78
d-$[Coen_2(\text{L-glu})]^{2+}$	[a]	[a]	19.6	+2.5	56[b]
			22.7	−0.1	
d-$[Rhen_2(\text{L-val})]^{2+}$	31.65	298	29.85	−2.34	78
l-$[Rhen_2(\text{L-val})]^{2+}$	31.65	298	29.85	+2.43	78
dl-$[Coen_2(\text{L-leu})]^{2+}$	20.53	109	18.74	−0.10	118
			20.24	+0.12	
			22.42	−0.29	
d_{546}-$[Coen_2(\text{L-leu})]^{2+}$	20.53	109	19.92	+2.38	118
			22.88	−0.24	
l_{546}-$[Coen_2(\text{L-leu})]^{2+}$	20.53	109	19.72	−2.24	118
d-$[Rhen_2(\text{L-leu})]^{2+}$	31.65	286	29.85	−2.35	78
l-$[Rhen_2(\text{L-leu})]^{2+}$	31.65	286	29.85	+2.43	78
d-$[Coen_2(\text{L-ser})]^{2+}$	20.50	118	19.60	+2.34	79
			23.00	−0.08	
l-$[Coen_2(\text{L-ser})]^{2+}$	20.50	118	19.90	−2.26	79
d-$[Rhen_2(\text{L-ser})]^{2+}$	31.65	335	29.85	−2.36	78
l-$[Rhen_2(\text{L-ser})]^{2+}$	31.65	335	29.85	+2.48	78
d-$[Coen_2(\text{L-thr})]^{2+}$	20.50	116	19.60	+2.35	79
			23.00	−0.09	
l-$[Coen_2(\text{L-thr})]^{2+}$	20.50	116	19.90	−2.28	79

Table 5-6—*Continued.*

Complex	Absorption		CD		Ref.
	$\bar{\nu}$, kK	ε	$\bar{\nu}$, kK	$\Delta\varepsilon$	
d-$[Rhen_2(\text{L-thr})]^{2+}$	31.65	320	29.85	−2.37	78
l-$[Rhen_2(\text{L-thr})]^{2+}$	31.65	320	29.85	+2.49	78
d-$[Coen_2(\text{L-met})]^{2+}$	20.50	105	19.90	+2.33	79
			23.20	−0.13	
l-$[Coen_2(\text{L-met})]^{2+}$	20.50	105	19.50	−2.24	79
d-$[Rhen_2(\text{L-met})]^{2+}$	31.55	285	29.24	−1.64	78
			34.13	−0.16	
l-$[Rhen_2(\text{L-met})]^{2+}$	31.55	285	29.76	+2.53	78
			38.46	+0.24	
dl-$[Coen_2(\text{L-phe})]^{2+}$	20.50	104	18.87	−0.09	118
			20.36	+0.03	
			22.2	−0.26	
d_{546}-$[Coen_2(\text{L-phe})]^{2+}$	20.50	104	19.84	+2.33	118
			22.96	−0.17	
l_{546}-$[Coen_2(\text{L-phe})]^{2+}$	20.50	104	19.72	−2.32	118
l-$[Coen_2(\text{L-pro})]^{2+}$	20.40	100	19.20	−1.86	79
			22.20	−0.26[c]	

[a] Not given.
[b] Taken from figure.
[c] Shoulder.

not likely to be greater than 200 cm^{-1}, the A_2 transition remains at lowest energy even when the E level is split. Because of the relative size of the two observed CD bands, it could be assumed that the first two components are both positive and are dominating the negative band. If this is the case, the *d* isomer has the D configuration.

For *d*-$[Coen_2Cl_2]^+$ the A_2 transition is the lower in energy, and in the absorption spectrum it is situated at about 16.6 kK [92]. Two CD bands are observed at 16.26 and 18.59 kK in the cobalt(III) complex and at 16.94 and 19.23 kK in the corresponding chromium(III) complex [124]. Bosnich has studied the CD of *cis*-α-dichloro(R-*N*,*N*,′-bis(2′-picolyl)-1-methyl-1,2-diaminoethane)cobalt(III) and chromium(III) with the D configuration enforced by the stereospecific nature of the ligand [17]. The features of the CD spectrum of the cobalt(III) complex are very similar to those of the bis-(ethylenediamine) complex. The CD spectrum for the chromium complex, however, shows a +, −, + pattern under the T_{1g} band. This suggests that for the corresponding cobalt complex the low-energy positive band has been canceled completely and that in *d*-$[Coen_2Cl_2]^+$ the lowest energy transition (A_2) has a positive Cotton effect and the configuration is D. This conclusion and that for the *cis*-diamine complex are consistent with Sargeson's proposals based on reaction mechanism studies [128]. Nevertheless, more work is required on these complexes in an attempt to make the assignment of the transitions and thus the configurations more definite.

CD data for a number of complexes with the general formula *cis*-[$Coen_2$-XY]$^{n+}$, which have a rhombic chromophore, have been included in Table 5-6. Unfortunately, again it is very difficult to interpret the spectra with the amount of information that is currently available.

[$Coen_2aa$]$^{n+}$. Absorption and CD data for complexes of the general type [$Coen_2$aa]$^{n+}$, where aa is a bidentate, are also given in Table 5-6. The series with aa amino acid was treated separately above. As found for that series, the CD spectra are dominated by one band under the T_{1g} envelope. McCaffery, Mason, and Norman proposed that if this band is positive, the complex has a D configuration, and if it is negative, the complex is L [124].

Consider the CD spectrum for *d*-[$Coen_2CO_3$]$^+$. The A_2 component is at the lower energy and is calculated to be at approximately 18.5 kK. The CD shows one positive Cotton effect at 18.87 kK. It has been proposed that the position of this band and its dominance are due to the lower-energy component being of the same sign as the A_2 [92]. The CD spectra for some of the other complexes—[$Coen_2$ox]$^+$, [Co(R-pn)$_2$ox]$^+$, and [$Coen_2$acac]$^{2+}$, for example—could be rationalized in the same way.

Although salicylate is an unsymmetrical bidentate, its complexes [$Coen_2$-sal]$^+$ and [$Coen_2$salH]$^{2+}$, in which the hydroxyl is deprotonated and protonated, respectively, have CD spectra that are comparable with those of the kind being discussed here. The levo isomer of the protonated form shows only one negative Cotton effect under the T_{1g} band. In the deprotonated complex with the same absolute configuration, however, a −, +, − pattern is observed [64].

[Co(aa)$_2$en]$^{n+}$. In this group are included complexes of the type [$Coox_2$en]$^-$, [$Comal_2$en]$^-$, and [$Coox_2$(R-pn)]$^-$. The CD of l_{546}-[$Comal_2$en]$^-$ shows the +, −, + pattern expected for a complex with the D-configuration [50]. Solubility data suggest that it has the enantiomeric configuration to d_{546}-[$Coox_2$en]$^-$, and this is supported by the CD, because two negative Cotton effects are found for the latter compound.

trans-[Co(am)$_2$ox]$^-$. Complexes with this general formula have been studied for the amino acids glycine, L-alanine, and β-alanine [94]. The d_{546} isomers all have CD showing two negative Cotton effects. An inflection in the curve for the β-alanine complex suggests that the middle energy component is positive. This is the same sign pattern as found for the vicinal effect imposed by L-alanine. The $A_2(D_{4h})$ transition is lowest in energy here, in contrast to the [$Coox_2$en]$^-$ complexes, where it is highest in energy.

trans-[*Co*(*am*)$_3$]. The central metal ion chromophore of *trans*-tris(α aminoacetato)cobalt(III) complexes has a rhombic symmetry, which give rise to three transitions under the first ligand field band. The transition tha transforms as the axis of strongest field lies lowest in energy, and the transition corresponding to the axis of weakest field is at highest energy. However, very little structure is observed in the CD spectra for a whole range of amino acids other than that due to the vicinal effect of the optically active ligands [42,51 109,180,181,182]. The so-called α-[Co(L-glu)$_3$], as opposed to α′, is an exception, because it shows the three components (−,+,−) [109]. In genera the *l* complexes have a dominant negative Cotton effect and the *d* a positive Cotton effect. In the *l*-tris(prolinato) [42] and *l*-tris(valinato) [181] complexes there is evidence of two negative Cotton effects.

The absolute configurations can be determined by relating the CD spectra to those for *l*-tris(L-prolinato) and *d*-tris(L-alaninato)cobalt(III), whose configurations are unambiguously known. In its *trans*-tris complex, L-prolinate is stereospecific, forming the L complex; the D complex can be seen from Dreiding models to be very unstable because of severe nonbonded interactions. The absolute configuration of the alanine complex has been determined by X-ray studies to be D [52].

The results for this series of compounds again support the empirical rule above: the L configuration gives rise to two transitions with negative Cotton effects.

[*CoEDTA*]$^-$ *and related complexes.* The absolute configuration of l_{546}-[CoEDTA]$^-$ has been determined by the comparison of its ORD [128] and CD spectra [50,67] with those of l_{546}-[Co(*d*-PDTA)]$^-$, which is known to be D (positive octant sign; see structure in Fig. 5-47) because of the stereospecific complex formation by *d*-PDTA. The CD spectra for these complexes show a large positive Cotton effect followed by a smaller negative band. Unfortunately, we are unable to assign the spectra objectively. The D configuration,

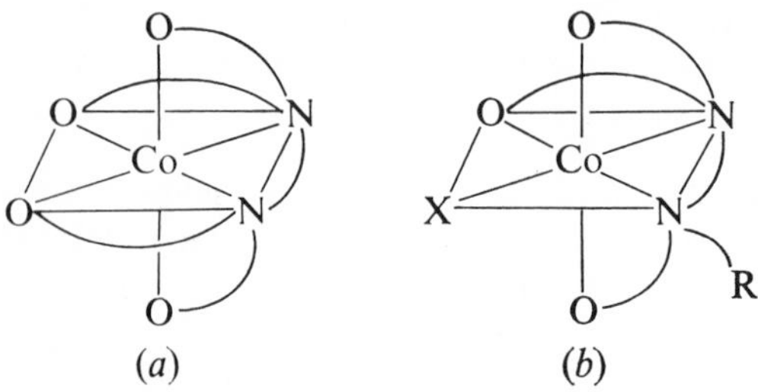

Fig. 5-47 Structure of (*a*) D-[CoEDTA]$^-$, and (*b*) the related complex with the quinquidentate ligand.

however, is thought to impose a positive Cotton effect on the $A_2(D_{4h})$ transition, and if this is correct, the observed Cotton effects derive from the $E(D_{4h})$ band.

Douglas and his coworkers have studied a series of EDTA and related complexes in which one of the equatorial chelate rings has been unattached and replaced at its coordination site by a unidentate [76,187]. The CD data are summarized in Table 5-7. These complexes have no octant sign and no chirality according to the ring-pair method, but their distributions of chelates are not achiral. This emphasizes one of the weaknesses of the present methods for determining chirality. The central metal ion chromophore for the quinquidentate complexes has a rhombic symmetry. The relative order of the three

Table 5-7 Absorption and CD data for complexes of EDTA and related ligands

	Absorption		CD		
Complex	$\bar{\nu}$, kK	ε	$\bar{\nu}$, kK	$\Delta\varepsilon$	Ref.
l_{546}-[CoEDTA]$^-$	18.63	324	17.01	+1.7	67
			19.42	−0.9	
l_{546}-[Co(*d*-PDTA)]$^-$	18.65	380	17.04	+1.7	67
			19.42	−0.9	
d_{546}-[Co(EDTA)Cl]$^{2-}$	17.10	250	15.45	+0.05	76
			17.70	−1.48	
			22.35	+0.07	
l_{546}-[Co(MEDTA)Cl]$^-$	17.30	210	17.50	+1.50	187
			22.40	−0.04	
l_{546}-[Co(YOH)Cl]$^-$	17.10	278	17.60	+1.45	187
			22.50	−0.03	
d_{546}-[Co(EDTA)Br]$^{2-}$	17.05	293	17.50	−1.44	76
			22.10	+0.08	
l_{546}-[Co(MEDTA)Br]$^-$	17.10	215	17.42	+1.38	187
			22.30	−0.04	
l_{546}-[Co(YOH)Br]$^-$	17.00	241	17.35	+1.41	187
			22.20	−0.05	
l_{546}-[Co(EDTA)NO_2]$^{2-}$	17.10	116[a]	17.15	+1.01	76
	20.05	236	19.45	−1.13	
			21.75	−0.42[a]	
d_{546}-[Co(MEDTA)NO_2]$^-$	17.10	90[a]	17.40	−0.33	187
	20.30	182	19.30	+0.40	
			21.80	+0.32	
d_{546}-[Co(YOH)NO_2]$^-$	17.10	110[a]	17.30	−0.69	187
	20.20	232	19.20	+0.82	
			21.80	+0.35[a]	

[a] Shoulder.

transitions depends on the ligand-field splitting parameter for X, although the transition that transforms as the O—Co—O axis (Γ_{OO}) is the highest energy component for the complexes considered. When X is a halide, the transition corresponding to the N—Co—O axis (Γ_{NO}) is at lowest energy, but when X is NO_2^-, the transition that transforms as the N—Co—X axis (Γ_{NX}) is at lowest energy.

The d_{546} isomer of chloro(ethylenediaminetetraacetato)cobalt(III) is formed from D(l_{546})-[CoEDTA]$^-$ with complete retention of configuration [76]. This complex shows a CD spectrum with a +, −, + pattern. Thus for the stereochemistry shown in Fig. 5-47 Γ_{OO} and Γ_{NO} are positive and Γ_{NX} is negative and dominant. A similar pattern is shown by the d_{546} bromo complex, and therefore it has the same configuration. The l_{546}-nitro complex has a large positive band at lowest energy followed by two negative Cotton effects. Haines and Douglas assigned this isomer the same configuration as the d_{546} halo compounds [76]. However, it is surely the enantiomer because the positive band is Γ_{NX} and the two negative bands are Γ_{NO} and Γ_{OO}. This conclusion is supported by the solubility data of Hidaka, Shimura, and Tsuchida [95].

Van Saun and Douglas have studied the related complexes with *N*-2-hydroxyethylethylenediaminetriacetate (YOH) and *N*-methylethylenediaminetriacetate (MEDTA) [187]. The circular-dichroism spectra for these complexes are very similar to those for the quinquidentate EDTA complexes enabling the absolute configurations to be assigned. They are included in Table 5-7.

CD spectra have been published for some *cis*-α complexes of *N*,*N′*-ethylenediaminediacetate, in which the two coordinated carboxylate groups are trans to one another [113]. The fifth and sixth coordination sites are occupied by R-propylenediamine and L-alanine. The metal ion chromophore for the R-propylenediamine complex has a tetragonal symmetry, with the $A_2(D_{4h})$ component at higher energy than $E(D_{4h})$. The *d* and *l* isomers have almost enantiomeric CD spectra, with the *d* having a large positive Cotton effect followed by a smaller negative band. These two components are thought to arise from the E level and are comparable with the two components observed for [CoEDTA]$^-$. The absolute configuration of the *d* isomer has been found to be D by comparing its CD with that for the equivalent ethylenediamine-*N*,*N′*-di-L-α-propionate complex, whose absolute configuration is known from nmr studies [179].

cis-[CotrienX$_2$]$^{n+}$ and related complexes. The CD spectra of *cis*-α-triethylenetetraminecobalt(III) complexes are very similar to the equivalent *cis*-bis(ethylenediamine) complexes [174]. By comparison with the proposal above for *d*-*cis*-[Coen$_2$Cl$_2$]$^+$ it is expected that a positive Cotton effect lies at lowest energy but is hidden by a large negative CD band, and the complex has the

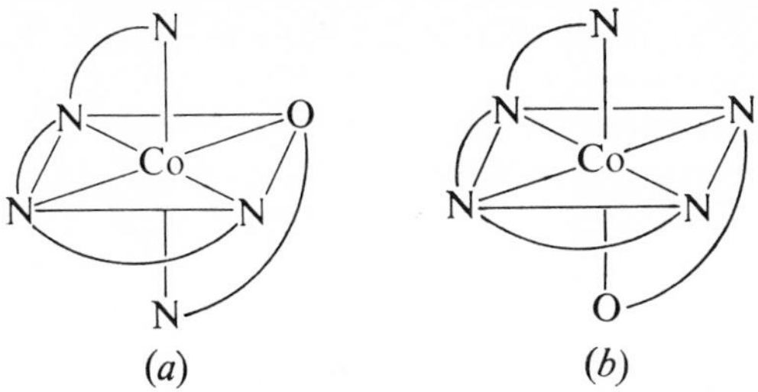

Fig. 5-48 Structures of (*a*) L-β_1- and (*b*) L-β_2-[Cotrien(am)]$^{2+}$.

D configuration. The absolute configurations of the other trien complexes have also been assigned on the basis of the similarity between the trien and en complexes.

The *cis*-β isomers have much lower resultant rotational strengths than the equivalent *cis*-α complexes. In the cases studied—$X_2 = Cl_2$; $Cl(H_2O)$; $(H_2O)_2$; CO_3, and $(NO_2)_2$—the *d* isomers have a dominant positive Cotton effect and have been assigned the D-configuration [174].

Marzilli and Buckingham [134] and Lin and Douglas [116] have studied some complexes of the type β[Cotrien(am)]$^{2+}$, for which there are two geometrical isomers (see Fig. 5-48). The CD spectra are again very similar to the corresponding [Coen$_2$am]$^{2+}$ complexes. With glycine only one positive Cotton effect is found for the *d*-β_2 isomer, but two negative bands are observed for the *l*-β_1 compound. With sarcosine, the *l*-β_2 isomer shows only one negative CD band. By comparison with the bis(ethylenediamine) complexes, *l*-β_2 sarcosinate and *l*-β_1 glycinate are assigned the L configuration and *d*-β_2 glycinate the D configuration [134].

Complexes of tetraethylenepentamine have recently been prepared, and their CD spectra are presently being studied [29,96]. House and Garner have published the CD spectrum of a chloro compound that they claim has the structure in Fig. 5-49. The spectrum shows a positive Cotton effect followed

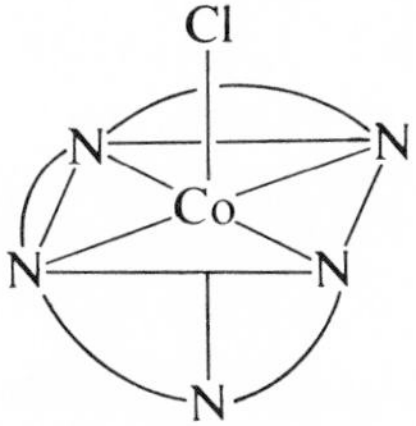

Fig. 5-49 One isomer of chlorotetraethylenepentaminecobalt(III).

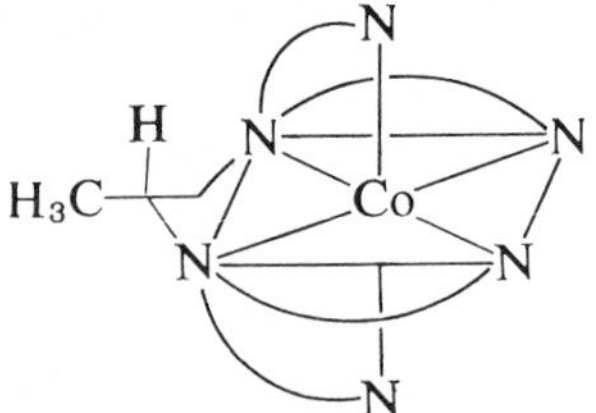

Fig. 5-50 L-[Co(R-mepenten)]$^{3+}$.

by a large negative band. This negative band could originate from both the higher energy E component and the $A_2(D_{4h})$ transition, and this could suggest the configuration shown in Fig. 5-49, which has a negative octant sign. Without additional information, however, an objective assignment of the configuration based simply on the CD is impossible.

R-*N*,*N*,*N′*,*N′*-tetrakis(2′-aminoethyl)-1,2-diaminopropane (R-mepenten) coordinates stereospecifically to cobalt(III) as a sexadentate with the methyl group equatorial and the chelate rings distributed with an L chirality [69,70, 149] (see Fig. 5-50). The CD of *l*-[Copenten]$^{3+}$ closely resembles that of the above complex (*l*-[Co(R-mepenten)]$^{3+}$) and, therefore, has the same configuration. This conclusion has been supported by the X-ray studies of Saito and his coworkers [184] but is in conflict with an earlier analysis of the CD data by Mason and Norman [138].

Distribution of Unidentates

Some complexes are dissymmetric solely because of the way in which their unidentate ligands are distributed about the central metal ion. Little information is available about the optical activity that is derived from this source. Russian workers [34] have studied some platinum(IV) complexes with three cis pairs of unidentates, and ORD results have been published, but, to date, no CD spectrum has been published for a complex containing only unidentate ligands whose distribution is responsible for the activity. The complex *d*-*cis*-*cis*-[Coen$(NH_3)_2Cl_2$]$^+$ has been obtained in an optically pure state and its CD published [91]. Although it has a chelate ring, any preference for one conformation would be extremely small, and so its activity is in fact due to the distribution of the unidentates. To a first approximation it has a tetragonal symmetry with a slight rhombic perturbation, and its absorption spectrum has similar features to that of *cis*-[Coen$_2Cl_2$]$^+$. As in the case of *d*-*cis*-[Coen$_2Cl_2$]$^+$ its CD has a low-energy negative Cotton effect followed by a positive band, but here the higher-energy Cotton effect is by far the smaller in contrast to the bis(ethylenediamine) complex, and, further, the position of the positive suggests that it does not derive from the highest energy transition. If the positive Cotton effect corresponded to the highest energy component of the E, any cancelation by the other transition with E parentage would push the observed maximum to higher energy than the actual absorption maximum. The fact that it is found at an energy equal to or less than that of the E absorption band indicates that there is a higher energy negative component. Therefore the $A_2(D_{4h})$ transition, which has the lowest energy, can be assigned to the observed large negative Cotton effect. In *d-cis*-$[Coen_2Cl_2]^+$ the lowest-energy observed negative band has been assigned to the first E component.

The absolute configuration of *d-cis-cis*-$[Coen(NH_3)_2Cl_2]^+$ is known to be R because it is derived with retention of configuration from D(*d*)-*cis*-$[Coen(NH_3)_2CO_3]^+$, whose configuration has in turn been determined by comparison of its CD with that of D(*d*)-*cis*-$[Coen_2CO_3]^+$ [91,92]. It must be remembered when comparing the CD of R-*cis-cis*-$[Coen(NH_3)_2Cl_2]^+$ with that of D-$[Coen_2Cl_2]^+$ that the two attain their dissymmetry in different ways. There is no a priori reason for suggesting that the two should have the same sign of Cotton effect for the comparable transitions. Much more work needs to be done on compounds like this in order that empirical rules can be derived to determine the distribution of unidentate ligands.

NONEMPIRICAL METHODS

As stated above, it is possible for two or more electric-dipole-allowed transitions with polarizations dissymmetrically arrayed to couple to yield component transitions that are intrinsically optically active and that have Cotton effects whose signs are determined by the phase relationships of the individual dipoles. Because of the unambiguous way in which the signs of the Cotton effects are determined, when the component transitions are correctly identified, the absolute configuration of the molecule can be determined directly from the observed circular dichroism.

Following the early work of Kuhn [106,107], Mason and his coworkers applied the coupled-oscillator method to a number of metal-complex systems by studying the circular dichroism of transitions localized principally on the ligands [110,120,122,135,136,139,140,141]. More recently, other research groups have extended the method's applications [18,19,20,21,25,58]. It has become apparent, however, that the identification of the component transitions is not a simple matter and that it is important to understand fully the nature of the ligand absorption bands. Because the discussion in this section is mainly restricted to complexes of 1,10-phenanthroline and 2,2′-bipyridine, the absorption spectra of these two chelates are briefly reviewed below.

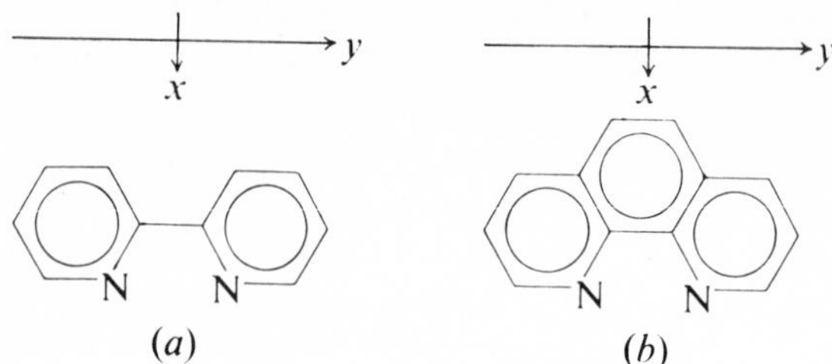

Fig. 5-51 Short axis (x) and long axis (y) polarizations in (a) 2,2′-bipyridine and (b) 1,10-phenanthroline.

Ultraviolet Absorption Spectra of 2,2′-Bipyridine and its Monoprotonated Form

The UV absorption spectrum of 2,2′-bipyridine, whose structure is given in Fig. 5-51, is presented in Fig. 5-52 along with the spectrum of the monoprotonated form. There are two main absorptions in both forms, and they can be assigned on the basis of the theoretical calculations of Gondo [72]. For the monoprotonated form, which has the cis configuration in solution [117], he predicted that the lowest state should have $^1B_1$ symmetry (37.1 kK) with a $^1A_1$ state at higher energy (38.2 kK). These were estimated to be separated from a similar group at 43.3 and 44.6 kK. The calculations indicated that the transition to the $^1B_1$ state is long-axis-polarized and should

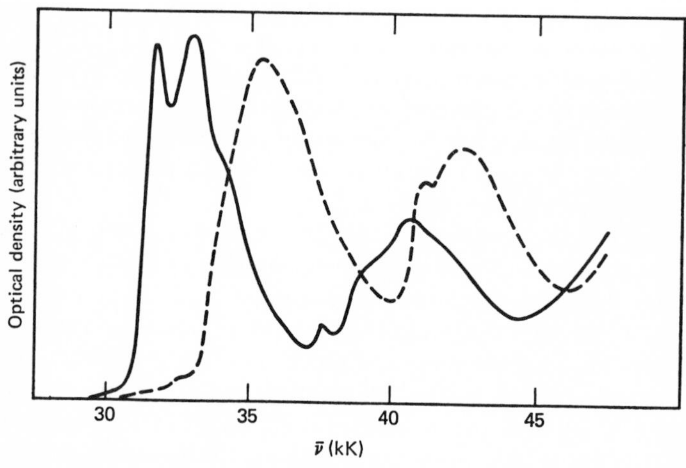

Fig. 5-52 Ultraviolet absorption spectra at 77°K of 2,2′-bipyridine (– – –) and its monoprotonated form (——) in ethanol-methanol (4:1) glass [25].

be more intense by more than an order of magnitude than the transition to the $^1A_1$ state. The low-energy absorption band in Fig. 5-52 at 31.5 to 33.5 kK is thus assigned to a transition that has long-axis polarization, and the observed structure of the band at 77° is vibrational in origin. The structure corresponds to the excitation of a progression of a 1400 cm^{-1} vibration that occurs in the electronic absorption spectra of many aromatic molecules. This is supported by the spectra of the complexes discussed below.

Ultraviolet Absorption Spectra of 1,10-Phenanthroline and its Monoprotonated Form

Following the very recent Pariser-Parr-Pople treatment of 1,10-phenanthroline and its iron(II) tris complex [98], the α, p, β', and β in-plane $\pi \rightarrow \pi^*$ transitions can be assigned as presented in Table 5-8. The structure in the α band in the 30 to 33 kK region of the unprotonated spectrum (see Fig. 5-53)

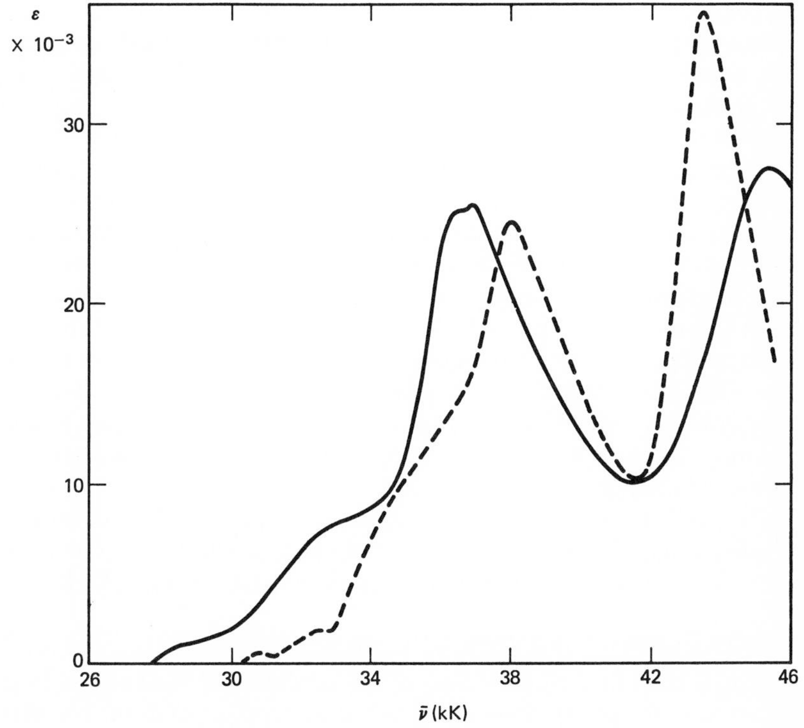

Fig. 5-53 Room temperature ultraviolet absorption spectra of 1,10-phenanthroline (– – –) and its mono-protonated form (——) in ethanol-methanol (4:1) [25].

Table 5-8 Calculated and observed transition energies for the spectrum of 1,10-phenanthroline

	Calculated for phen [98]			Observed for phen [25]		Observed for phenH$^+$ [25]		
Symmetry	$\bar{\nu}$, kK	f	Band	$\bar{\nu}$, kK	ε	$\bar{\nu}$, kK	ε	Polarization
A_1	30.90	0.0008	α	30.9	800	28.5	1,200	x
				32.6	2,200			
B_1	32.59	0.10	p	35.5[a]	12,200	33.0	8,000	y
B_1	37.19	0.73	β'	38.0	25,000	37.0	26,000	y
A_1	40.74	0.15						
B_1	43.00	0.30						
A_1	44.94	0.36	β	43.4	37,000	45.3	28,200	x
B_1	47.28	0.90						
A_1	49.94	0.11						

[a] Shoulder.

arises from a progression of a vibration of about 1500 cm^{-1}, which is probably of the same kind as the 1378 cm^{-1} vibration observed by McClure for the α band of phenanthrene [127].

Prior to the work of Ito and his coworkers [98], the spectrum of 1,10-phenanthroline was discussed in relation to the parent hydrocarbon, phenanthrene, although the energy levels of the latter were incompletely understood, because some controversy existed regarding the relative position of the β and β' states. According to the most recent analysis [48], these states are nearly degenerate, and together are responsible for the very intense ($\varepsilon \sim$ 70,000) absorption band near 40.0 kK. There is no real uncertainty about the position of the p band ($\sim$ 35.0 kK), which has medium intensity ($\varepsilon \sim$ 15,000), or the α band ($\sim$ 29.0 kK), which is very weak ($\varepsilon \sim$ 300). The combined intensity of the 38.0 and 43.4 kK bands in 1,10-phenanthroline is nearly equal to the intensity of the 40.0 kK band of phenanthrene, which is consistent with the interpretation above.

Mason assigned the 30 to 40 kK region of 1,10-phenanthroline to the α and p bands and assigned the band at 43.4 kK to β with the β' state at still higher energy [136]. He seemingly ignored the band at about 35.5 kK.

Excitation Resonance Interaction in Bis and Tris Complexes

The crucial issue in the application of the nonempirical method lies in the assignment of the components of the ligand absorption bands for which the circular dichroism has been measured. For the tris complexes electrostatic interactions among the three ligands, as well as interactions involving the metal, remove the threefold degeneracy associated with electronic excitation energy on one of the ligands. As a result, there are two excited states, an A_2 state and a twofold degenerate E state, within the D_3 symmetry of the complex. For the bis complexes, the corresponding states are A and B for the C_2 symmetry of the molecule.

It has been customary to estimate the splitting due to excitation resonance interaction, sometimes called the *exciton* splitting,* by neglecting any contribution from the metal and considering only the effect of the electrostatic interaction among the ligands by a Coulombic term V.

For a tris complex with three identical ligands A, B, C the total Hamiltonian is

$$H = H_A + H_B + H_C + V_{AB} + V_{BC} + V_{CA} \tag{5-49}$$

where H_A, H_B, and H_C represent the ligand Hamiltonian in the absence of the interactions. If the following functions define the ground and excited states:

$$\phi = \psi_A\psi_B\psi_C \qquad \text{ground state} \tag{5-50}$$

$$\phi_A = \psi'_A\psi_B\psi_C \qquad \phi_B = \psi_A\psi'_B\psi_C \qquad \phi_C = \psi_A\psi_B\psi'_C \tag{5-51}$$

where the primed functions denote electronically excited molecules, then the excitation energies are given by

$$\Delta\mathscr{E}(A_2) = w' - w + 2(V' - V) + 2\Sigma \tag{5-52}$$

$$\Delta\mathscr{E}(E) = w' - w + 2(V' - V) - \Sigma \tag{5-53}$$

In these equations, w and w' refer to electronic energies of the ground and excited states, respectively, and the remaining terms are defined by

$$V' = \psi'_A\psi'_A V_{AB}\psi_B\psi_B \, d\tau \tag{5-54}$$

$$V = \psi_A\psi_A V_{AB}\psi_B\psi_B \, d\tau \tag{5-55}$$

$$\Sigma = \psi'_A\psi_A V_{AB}\psi'_B\psi_B \, d\tau \tag{5-56}$$

The A_2 and E states of the tris complex are therefore separated by an energy equal to 3Σ, where Σ is the excitation resonance energy. Usually this is calculated by assuming an approximate form of V_{AB}. Under certain conditions—see, for example, Ref. 38—V_{AB} can be expanded as a point multipole series, of which only the first term is retained in order to evaluate Σ. This is the dipole–dipole approximation that leads to

$$\Sigma = \frac{-e^2[3(\mathbf{p}_A\cdot\mathbf{R})(\mathbf{p}_B\cdot\mathbf{R}) + R^2(\mathbf{p}_A\cdot\mathbf{p}_B)]}{R^5} \tag{5-57}$$

* Although the term "exciton" has been used for intramolecular interactions of this kind recently—for example, see [146]—it was originally defined for interactions in crystals in which there was translational symmetry [59]. When a local excitation in a crystal spreads outward through it, the process can be likened to the movement of a quasi-particle, the exciton, through the crystal.

where $\mathbf{p}_A$ and $\mathbf{p}_B$ are transition dipoles located at the molecular centers, and R is the separation between them. For long-axis-polarized transitions on phenanthroline or bipyridine, this reduces to

$$\Sigma = \frac{3p^2}{4R^3} \tag{5-58}$$

which is positive. Thus, the dipole–dipole approximation predicts that the A_2 component must lie at higher energy than the E.

The integrals V' and V are usually neglected, because they produce the same energy shift in both A_2 and E excitation energies. Later we shall relate the levels of tris complexes to the energy of the mono. There are good reasons for neglecting the $V' - V$ term explicitly when doing this. The term measures the change in the dispersion energy of interaction between two ligands after one has been electronically excited and, therefore, is of the same kind that gives rise to a decrease of excitation energy on going from the vapor to the condensed phase. It is just one of a large number of similar terms if a more general Hamiltonian, which includes Coulombic interactions between the ligands and solvent molecules, is considered. When comparing the energy levels of the solvated "trimer" to those of the solvated "monomer," $V' - V$ can be absorbed into $w' - w$. This procedure is substantiated by analyses of the spectra of dimers of aromatic hydrocarbons [32,33,57].

For a bis complex, MX_2Y, in which there is no coupling between the X and Y ligands, the energy levels are

$$\begin{aligned} \Delta\mathscr{E}(\mathrm{B}) &= w' - w + (V' - V) + \Sigma \\ \Delta\mathscr{E}(\mathrm{A}) &= w' - w + (V' - V) - \Sigma \end{aligned} \tag{5-59}$$

with V', V, and Σ defined in the same way as for the tris complex. If Σ is approximated by the dipole–dipole term, the B state is found to be at higher energy than the A. This has been assumed by Bosnich [19].

Recently Mason has attempted to calculate the excitation resonance energy by way of a Hückel molecular-orbital calculation [136]. He found values of $+970$ and $703\ \mathrm{cm}^{-1}$ for Σ for 2,2′-bipyridine and 1,10-phenanthroline, respectively, using a Coulomb integral increment appropriate for neutral nitrogen of $\Delta\alpha_N = 0.5\beta$. As pointed out by Murrell and Tanaka [147], however, this kind of calculation overestimates by at least a factor of 5 the resonance energies. Nevertheless, Mason has viewed this as a lower limit and dismisses values of 190 and $420\ \mathrm{cm}^{-1}$ calculated by the dipole–dipole approximation as being too small.

A fundamental objection to these approaches is the neglect of bonding interactions, both σ and π, between the metal and the ligand, as well as ligand–ligand, overlap. Originally Mason estimated that the metal interactions would reverse the order of the A_2 and E transitions for complexes of the type M-(bipy)$_3$, seemingly independently of M [122]. His most recent estimations of the various contributions to the A_2 and E splitting have shown, however, that, for phenanthroline, both the interligand and metal–ligand π bonding tend to oppose the excitation resonance energy, but, for bipyridine, the interligand and metal–ligand interaction contributions have opposite signs, and, for both, the A_2 transition lies at higher energy. Nevertheless, as will be seen later, the calculated splittings are far too large for some complexes, and, in others, are of the wrong sign.

In an attempt to determine whether the magnitude and sign of the excitation resonance energy could be determined experimentally, the absorption spectra of some divalent metal ion complexes have been studied in detail [25]. It was found necessary to use vibronic-coupling theory to analyze the spectra.

Vibronic-coupling Theory for Tris Complexes

Perrin and Gouterman [158] have presented a theory of vibronic coupling in trimers that provides a suitable basis for the discussion of the absorption spectra of the tris complexes. Because the complexes of 2,2′-bipyridine display considerable fine structure in the long-axis-polarized band, this ligand has been taken as an example.

According to the model each molecule is assumed to have only one (totally symmetric) vibration. A 1400 cm^{-1} vibration dominates the spectrum of 2,2′-bipyridine, and, although the assumption above is not strictly correct, it allows the spectral features to be described reasonably well. Following Perrin and Gouterman [158], the absorption spectra of M(bipy)$_3$ can be computed in terms of two parameters Σ and λ, where λ is a measure of the displacement of the excited-state vibrational oscillator. The latter parameter is fixed to give a vibrational intensity distribution that is close to the observed distribution (at 77°K) of the monoprotonated 2,2′-bipyridine, that is, with Σ set equal to zero. The value chosen was 0.9 [25]. The spectrum of the monoprotonated species was chosen because it resembles that of the complexes in energy and shape more closely than that of the free base.

Allowing for the fact that the intensity of the A_2 component absorption is twice that of the E (see below), it is possible to compute the energy levels and absorption and emission intensities for various values of Σ in units of energy equal to the vibrational frequency (1400 cm^{-1}). A spectrum can be synthesized for each value of Σ allowing the peaks to be Gaussian in shape. Computed spectra [25] are shown in Figs. 5-54 and 5-55 for various positive and negative values of Σ between 0 and 1. The emission spectrum was computed to arise from the lowest level.

From Figs. 5-54 and 5-55 it can be seen that for small values of Σ, the effect of vibronic coupling is simply a change in intensity distribution away

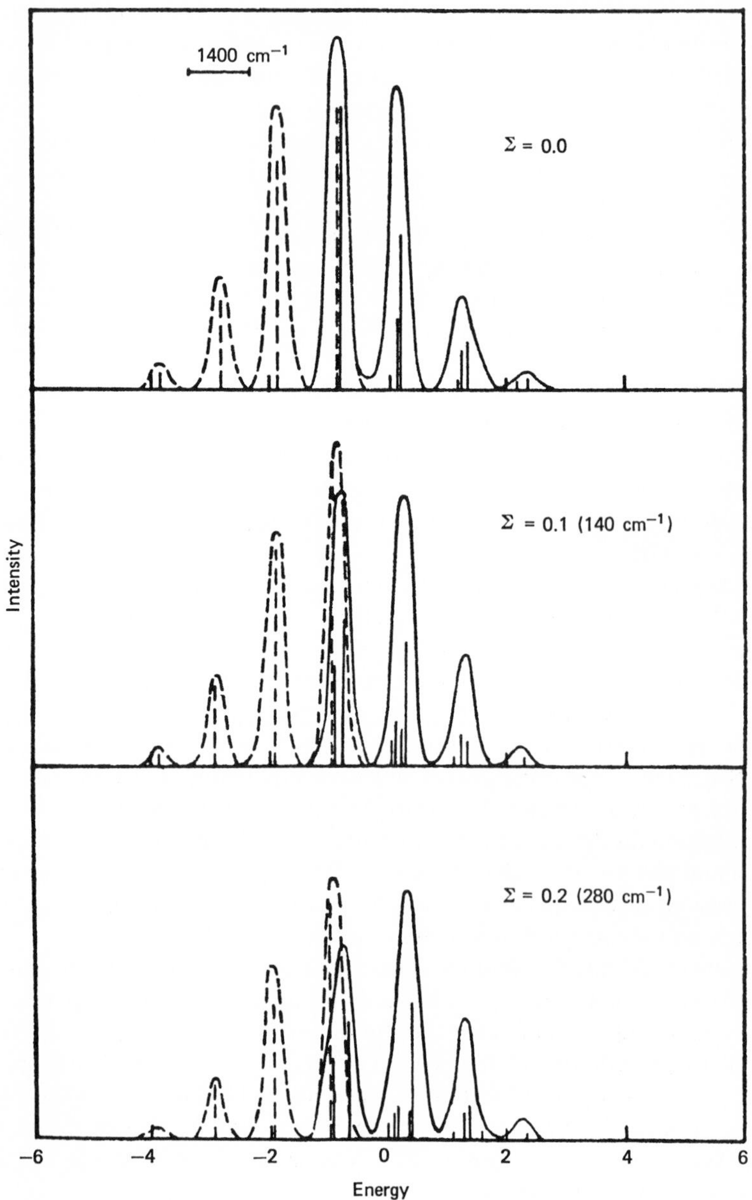
1400 cm^{-1}
Σ = 0.0
Σ = 0.1 (140 cm^{-1})
Σ = 0.2 (280 cm^{-1})
Intensity
−6
−4
−2
0
2
4
6
Energy

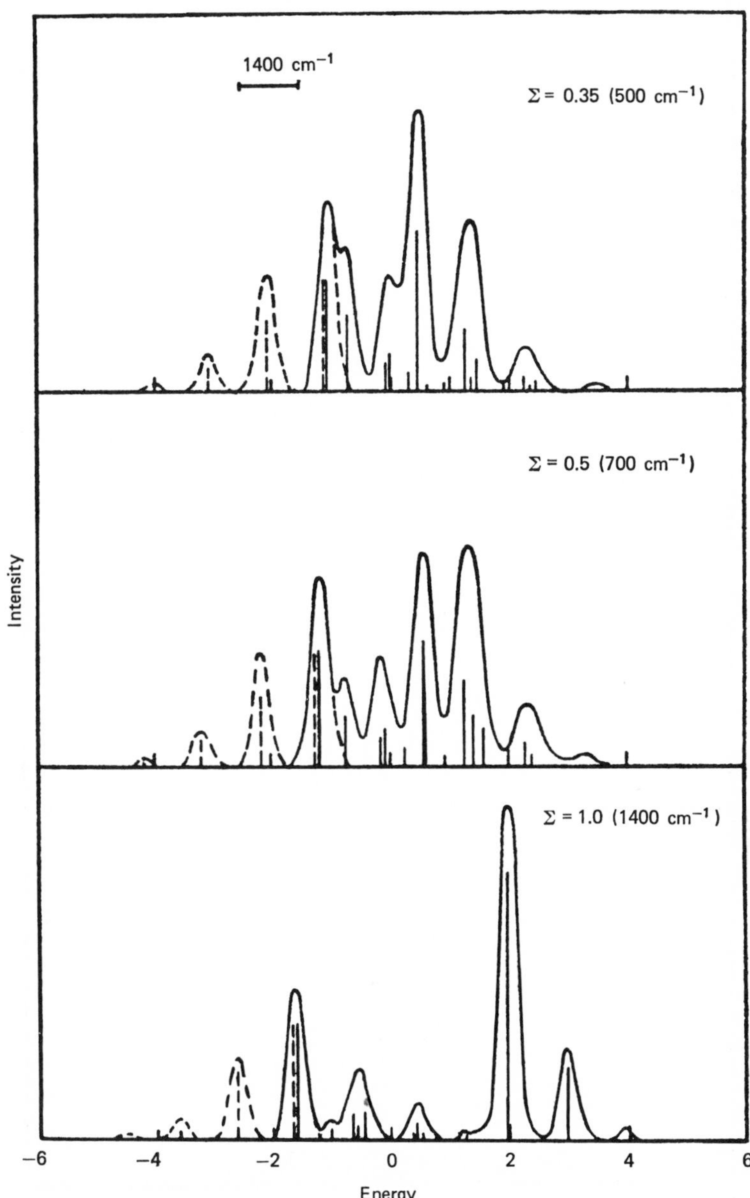

Fig. 5-54 Computed absorption and emission spectra for positive values of Σ. The case $\Sigma = 0.0$ corresponds to the monomer. All spectra were computed with $\lambda = 0.9$ and a line width $\alpha = 0.2$ [25].

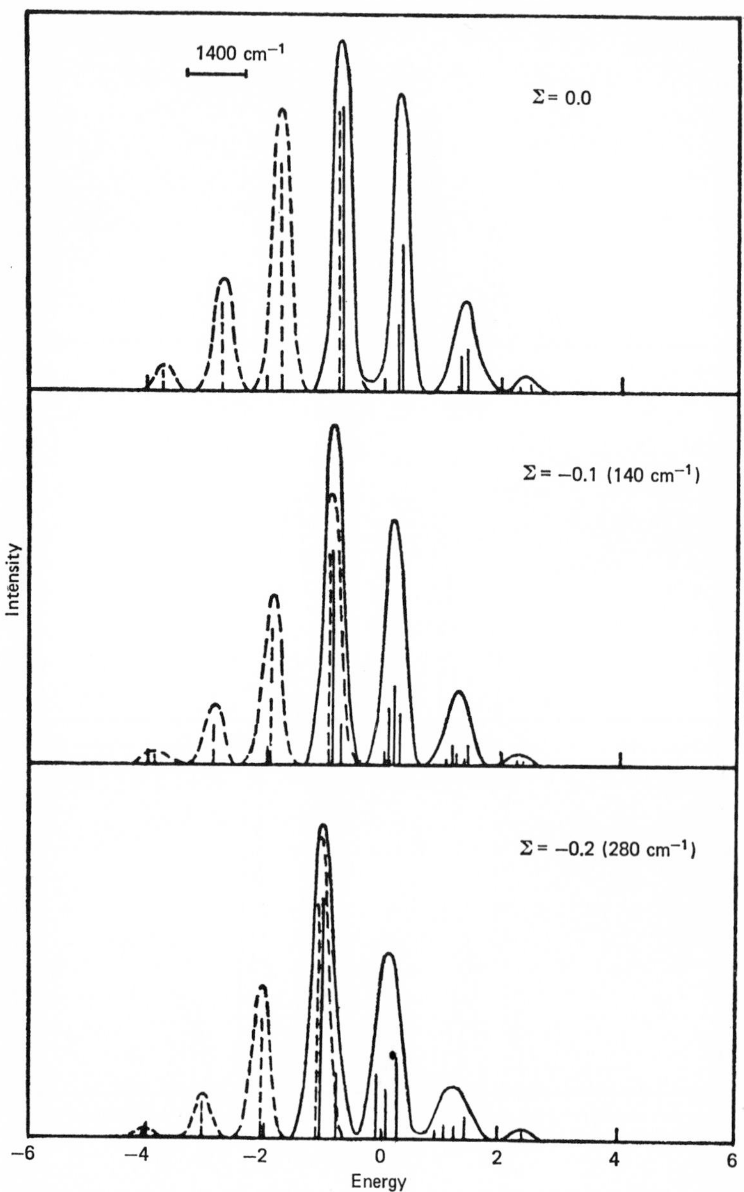
1400 cm^{-1}
Σ = 0.0
Σ = −0.1 (140 cm^{-1})
Σ = −0.2 (280 cm^{-1})
Intensity
−6
−4
−2
0
2
4
6
Energy

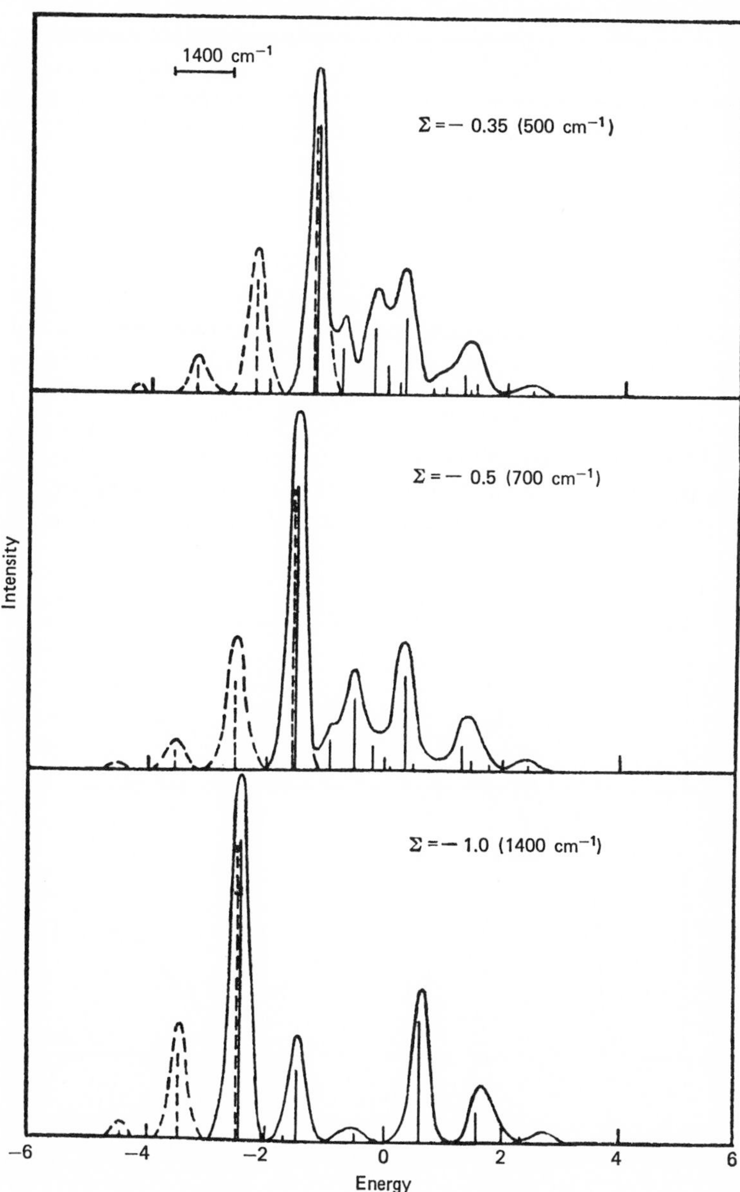

Fig. 5-55 Computed absorption and emission spectra for negative values of Σ [25].

from that of the monomer, and it is only for Σ near ± 1, that is, 1,400 cm^{-1}, that the A_2 and E systems can be clearly distinguished. Thus from a comparison of an observed spectrum with the computed spectra it should be possible to estimate the sign and approximate size of Σ. Further, if it is possible to observe fluorescence, a comparison between absorption and fluorescence spectra will help fix Σ, because the mirror-image relationship is quickly lost as Σ increases from zero. Because the fluorescence spectrum of $[Zn(bipy)_3]^{2+}$ is well resolved, this system is dealt with first.

Absorption and Fluorescence Spectra of $[Zn(bipy)_3]^{2+}$

The absorption and fluorescence spectra of $[Zn(bipy)_3]^{2+}$ are shown in Fig. 5-56 [25]. Note that the frequency scales run in opposite directions and that the two curves have been normalized for the peak that corresponds to one quantum of the 1400 cm^{-1} vibration, rather than the pure electronic peak, because the intensity of the latter is always reduced because of reabsorption effects. Allowing for the fact that the fluorescence spectrum is slightly sharper because of being measured at a lower temperature, it can be seen that a mirror-image relationship exists between the two spectra. This strongly sug-

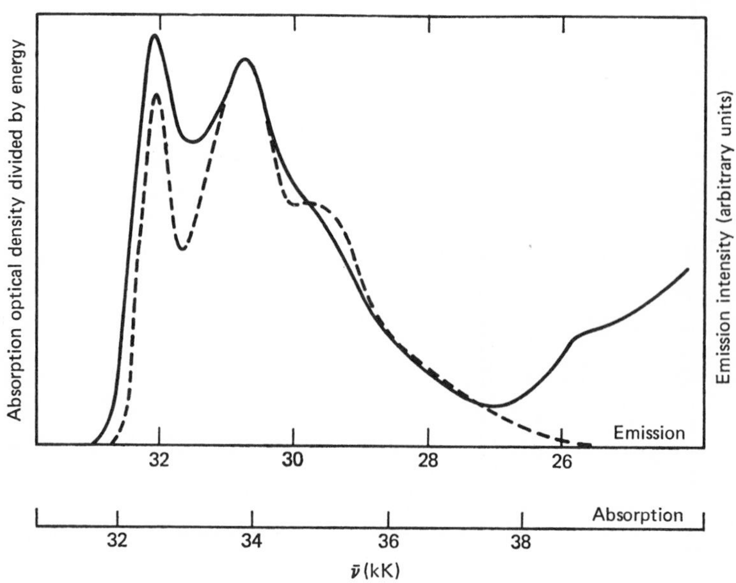

Fig. 5-56 Fluorescence spectrum (– – –) at 77°K of $[Zn(bipy)_3]^{2+}$ superposed on the absorption spectrum (——) at about 150°K, with intensities that are directly comparable [25].

gests that the resonance splitting energy Σ is small. This conclusion is supported by the similarity of the intensity distribution in the absorption spectra of monoprotonated 2,2′-bipyridine (see Fig. 5-52) and $[Zn(bipy)_3]^{2+}$. The value of Σ was estimated to be less than about 100 cm^{-1} on the basis of the computed spectra (see Figs. 5-54 and 5-55), but it was not possible to determine its sign [25].

Absorption Spectra of $[Nien_2bipy]^{2+}$ and $[Ni(bipy)_3]^{2+}$

At 77°K marked vibrational structure is observed in the spectra of these complexes (see Fig. 5-57). A comparison of the spectra of the mono and tris complexes again showed no evidence for a measurable resonance splitting, and an upper limit of $\pm 100\ cm^{-1}$ has been placed on Σ [25]. These findings show that Mason's calculated lower limit of Σ is out by over an order of magnitude.

Absorption Spectra of $[Nien_2phen]^{2+}$ and $[Ni(phen)_3]^{2+}$

The spectra of $[Nien_2phen]^{2+}$ and $[Ni(phen)_3]^{2+}$ (see Fig. 5-58) lack vibrational structure at 77°K [25], preventing a detailed comparison of the spectra. The broadening of the main absorption band near 37.0 kK in the spectrum of the tris complex, however, is consistent with a small positive value of Σ. In these spectra, the weak α system is found between 29.0 and 33.0 kK, overlapping the p band, which runs from approximately 33.0 to about 36.0 kK.

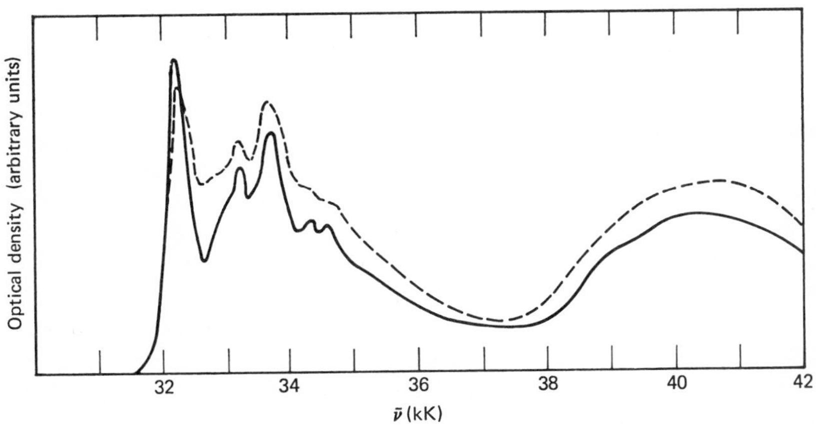

Fig. 5-57 Absorption spectra of $[Nien_2bipy]^{2+}$ (– – –) and $[Ni(bipy)_3]^{2+}$ (——) at 77°K in ethanol-methanol (4:1) glass [25].

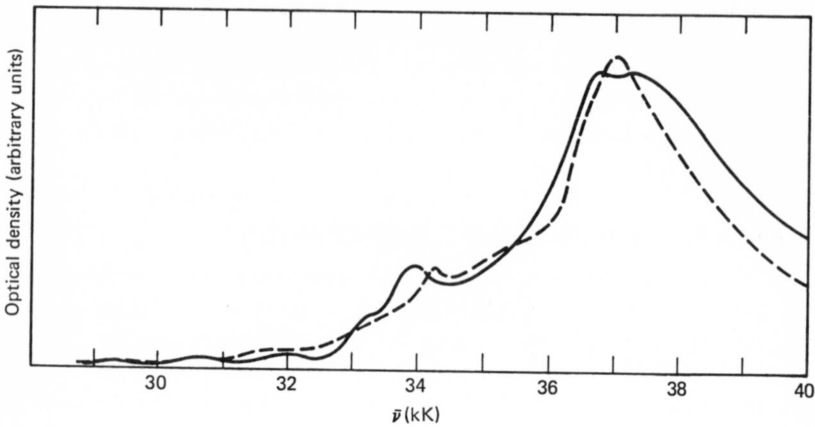

Fig. 5-58 Absorption spectra of $[Nien_2phen]^{2+}$ (– – –) and $[Ni(phen)_3]^{2+}$ (——) at 77°K in ethanol-methanol (4:1) glass [25].

Absorption Spectra of Mixed Phenanthroline: Bipyridine Complexes

In heterochelated complexes of phenanthroline and bipyridine, excitation resonance interactions are possible not only between the two identical ligands but also between the phenanthroline and bipyridine. The 77°K spectra of two such complexes, $[Ni(phen)_2bipy]^{2+}$ and $[Ni(bipy)_2phen]^{2+}$, are given in Fig. 5-59. The analysis of the spectra is complicated by the overlap of the intermediate system (*p* band) of the phenanthroline and the lowest absorption band of bipyridine, both of which are of comparable intensity. From a comparison of the spectra of $[Ni(bipy)_2phen]^{2+}$ and $[Nien_2bipy]^{2+}$ it can be seen that the bipyridine-bipyridine and bipyridine-phenanthroline interactions do not affect the absorption by the bipyridine because the positions of the vibrational peaks are essentially unchanged in the two spectra. The main absorption band of the phenanthroline near 37.0 kK is, however, broader in the spectrum of $[Ni(phen)_2bipy]^{2+}$, and the center of gravity of the intensity is at higher energy than in the spectrum of $[Ni(bipy)_2phen]^{2+}$. This is consistent with a small positive value of Σ for the phenanthroline resonance splitting energy. The overlap between the 37.0 kK phenanthroline and the 40.0 kK bipyridine band, however, made it impossible to obtain a quantitative estimate of Σ [25].

These heterochelated nickel(II) complexes [136] and the equivalent osmium(II) and (III) complexes [136,140] have been studied by Mason and the equivalent ruthenium(II) complexes by Bosnich [20]. Bosnich presented the theory of the interactions in some detail but did not make any quantitative estimates of the energy levels. These estimates have been made by Mason (see Table 5-9) based on Hückel molecular orbitals. He found values of 970 and

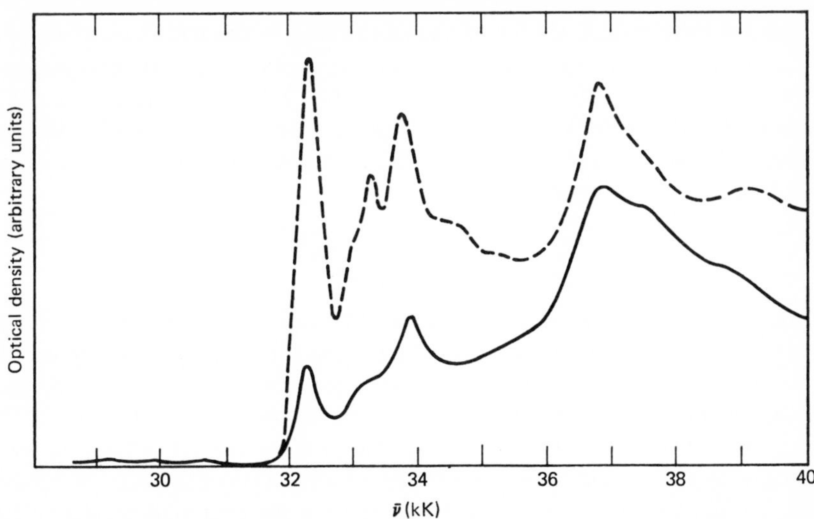

Fig. 5-59 Absorption spectra of $[Ni(phen)_2bipy]^{2+}$ (——) and $[Ni(bipy)_2phen]^{2+}$ (– – –) at 77°K in ethanol-methanol (4:1) glass [25].

703 cm^{-1} for Σ for phenanthroline-phenanthroline and bipyridine-bipyridine interactions, respectively, and 850 cm^{-1} for the phenanthroline-bipyridine interaction energy [136]. Mason used the energies of the free bases to represent the energies of the noninteracting ligands, although the energies of the monoprotonated species are more closely representative. Thus he estimated the bipyridine band to be at 35.0 kK. It should be pointed out here that the experimental energies that Mason has attributed to bipyridine belong to the

Table 5-9 Observed and calculated energies of the levels of $[Os(phen)_2bipy]^{2+}$, $[Os(bipy)_2phen]^{2+}$, and $[Ni(phen)_2bipy]^{2+}$ from the work of Mason and Norman [140] and Mason [136]

Complex		Energy, kK					
		Phen			Bipy		
$[Os(phen)_2bipy]^{2+}$	(Theory)	37.1		38.9		35.2	
	(Exp.)	37.2		38.8		34.5	
$[Os(bipy)_2phen]^{2+}$	(Theory)		38.6		34.5		35.8
	(Exp.)		37.9		34.1		35.5
$[Ni(phen)_2bipy]^{2+}$	(Theory)	37.1		38.9		35.2	
	(Exp.)	36.5		38.2		33.8	

band maxima whose intensities are derived mainly from the long-axis-polarized intermediate system of phenanthroline, which has its main absorption peak near 34.0 kK. Therefore, although Mason has shown excellent agreement between theoretical and calculated values for the energies, his results should be completely ignored. Further, from the experimental evidence presented here, the above values of the excitation resonance energies calculated by Mason are far too high.

Method for Assigning the Absorption Spectra

Because the theoretical estimations of the order of the component transitions are misleading, an alternative method has been sought [58]. It can be readily shown, with the aid of three-dimensional geometry, that the $A_1 \rightarrow A_2$ (in-phase) component of the tris complexes should be twice as intense as the $A_1 \rightarrow E$ (out-of-phase). Unfortunately, the splittings are not sufficiently great to allow an assignment to be made by simple inspection of the data. Nevertheless, an assignment can be made by comparing the center of gravity of the $\pi \rightarrow \pi^*$ ligand transition in the tris to that of a corresponding mono complex —M en_2phen^{n+} or M en_2bipy^{2+}, for example. The A_2 and E components should be displaced to either side of the transition for the "monomer" due to the various interactions—see Eqs. (5-52) and (5-53). Because the A_2 is twice as intense as the E, the center of gravity of the band for the tris should be displaced toward the A_2 component. If the center of gravity of the tris spectrum lies above that of the mono, it follows that the A_2 state lies above the E state. If it is below, the A_2 state lies lower in energy.

Similar arguments can be applied to the case of the bis complex. Here the transition to the B state corresponds to the in-phase addition of transition moments, and its intensity is three times that of the transition to the A state. The relative order of the two states is then decided by comparison of the centers of gravity of the intensities of the spectra of the mono and the bis complexes.

Mason has criticized this approach, claiming that the composite character of the isotropic absorption makes the method unreliable [136]. However, the relatively low intensity of the short-axis-polarized transitions that Mason suggests will interfere, do not seriously affect the analysis.

The centers of gravity can be estimated by using a planimeter, drawing a smooth continuous background under each band connecting the low- and high-energy sides of the band. Any contribution from a short-axis-polarized transition or from a charge-transfer band should, if possible, be subtracted. The centers of gravity determined in this way for a series of cobalt(III) and chromium(III) complexes are indicated in Figs. 5-60 and 5-61 by vertical lines at the bottom of each group proportional to the number of ligands.

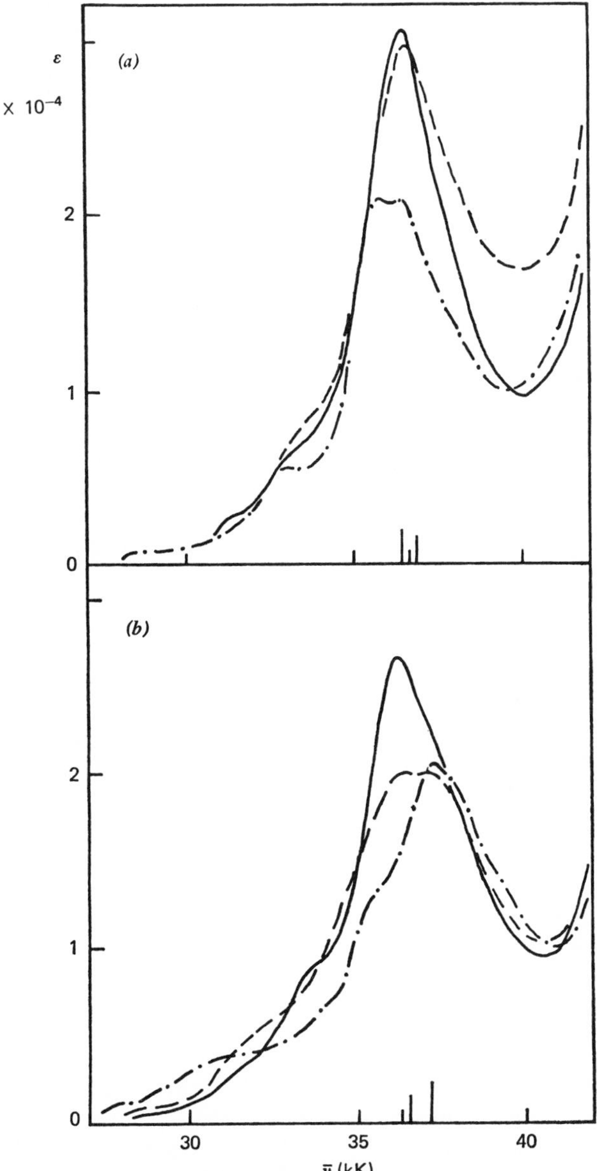

Fig. 5-60 Ultraviolet absorption spectra per mole of 1,10-phenanthroline for (*a*) *dl*-$[Coen_2phen]^{3+}$ (——), *l*-$[Co(phen)_2ox]^{+}$ (– – –), *d*-$Co(phen)_3]^{3+}$ (–·–·), and (*b*) *dl*-$[Crox_2phen]^{-}$ (——), *d*-$[Cr(phen)_2ox]^{+}$ (– – –), *d*-$[Cr(phen)_3]^{3+}$ (–·–·) in aqueous solution [58].

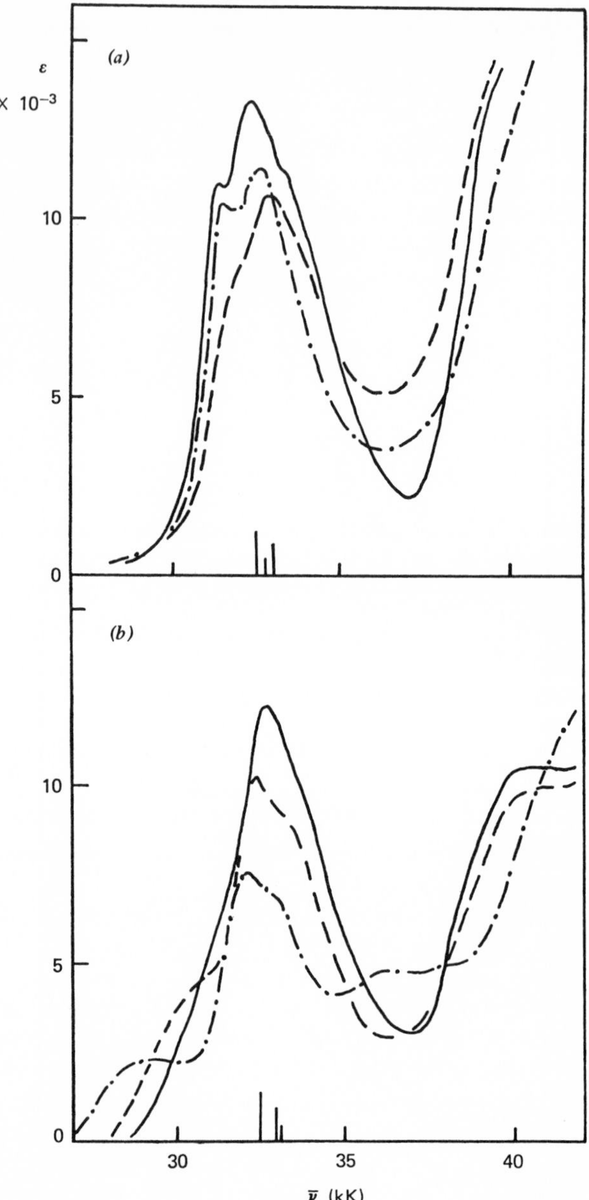

Fig. 5-61 Ultraviolet absorption spectra per mole of 2,2′-bipyridine for (*a*) *dl*-[Coen$_2$-bipy]$^{3+}$ (——), *l*-[Co(bipy)$_2$ox]$^+$ (– – –), *l*-[Co(bipy)$_3$]$^{3+}$ (–·–·), and (b) *dl*-[Crox$_2$bipy]$^-$ (——), *l*-[Cr(bipy)$_2$ox]$^+$ (– – –), *dl*-[Cr(bipy)$_3$]$^{3+}$ (–·–·) in aqueous solution [58].

Circular Dichroism of the Tris Complexes

The signs of the Cotton effects of the two components of the long-axis-polarized absorption bands can be determined simply by the approach outlined above. For the tris complexes with the D absolute configuration, the total magnetic dipole moment of the in-phase (A_2) component is antiparallel with its electric dipole moment, and the rotational strength is therefore negative; whereas the two dipole moments are parallel for the out-of-phase (E) component, and its rotational strength is positive. (The phase relationships are shown in Fig. 5-17.) For the bis complexes, the excitations corresponding to the in-phase (B) and out-of-phase (A) oscillations of the component transition moments have negative and positive rotational strengths, respectively, for the D configuration. For the L configurations the signs are, of course, reversed. For both the bis and tris complexes, the Cotton effects of the two components are calculated to have rotational strengths that are equal in size but opposite in sign. The short-axis-polarized transitions couple to give components that are optically inactive.

In the observed CD spectra the positive and negative components significantly cancel one another, and the observed differences in their maxima do not reflect the actual size of the splitting of the two components. Further, the vibronic coupling modifies the intensity distribution for weak and intermediate coupling systems. Theory shows that the higher-energy vibronic system has an altered intensity distribution in which intensity moves out of the lowest (pure electronic) band into the vibrational bands, whereas the lower-energy system has a normal distribution of intensity. Moffitt and Moscowitz [145] have shown that the rotational strength associated with an electronic transition has the same spectral distribution as the vibronic system, and, therefore, it is essential to apply the vibronic-coupling theory if we are to understand the observed CD curve.

This has been done for $[Ni(bipy)_3]^{2+}$, using values for λ and Σ equal to 0.9 and 0.05 (70 cm^{-1}), respectively, to calculate the energy levels of the complex [27].* In order to approximate the room-temperature conditions, Gaussian curves with a line width $\alpha = 0.5$ were superimposed to obtain a curve for each band system that is proportional to the vibronic rotational strength. These are shown in Fig. 5-62, along with the difference between the two vibronic curves, which is proportional to the circular dichroism. The computed curve is in excellent agreement with the observed CD of l-$[Ni(bipy)_3]^{2+}$. In Fig. 5-63 the computed curve has been drawn so that the computed and experimental curves coincide at 32.5 kK.

* The sign of Σ was arbitrarily chosen as positive [25]. Because CD studies have shown, however, that the levo isomer has the L configuration [58], Σ must, in fact, be positive in order to account for the observed sign pattern of the Cotton effects.

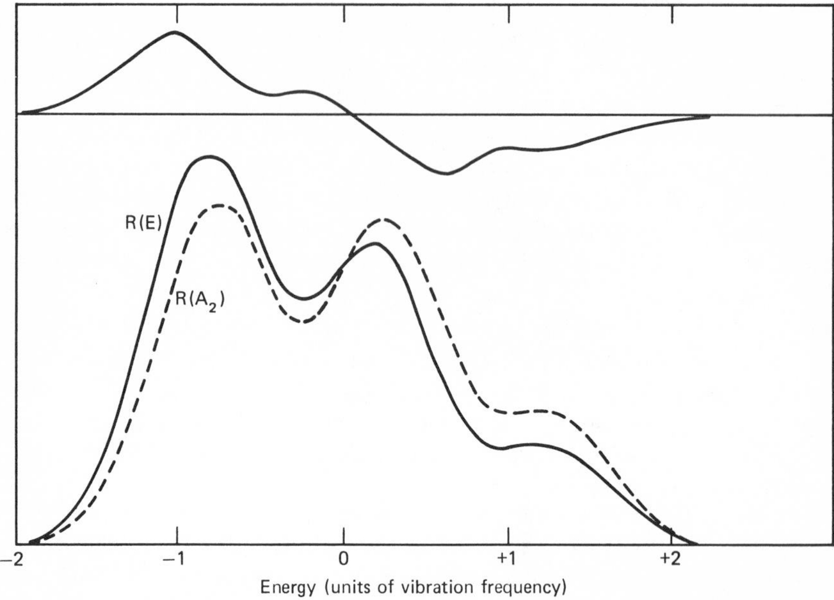

Fig. 5-62 Calculated vibronic envelopes of the rotational strengths (arbitrary units) for the E and A_2 systems of $[Ni(bipy)_3]^{2+}$ ($\lambda = 0.9$, $\Sigma = 0.05$) in the lower section. The upper curve gives $R(E) - R(A_2)$ which is proportional to the circular dichroism [25].

Because of the lack of vibrational structure in the long-axis-polarized phenanthroline band, the room-temperature CD spectra of $[Ni(phen)_3]^{2+}$, $[Ni(phen)_2bipy]^{2+}$, and $[Ni(bipy)_2phen]^{2+}$ (see Fig. 5-64) cannot be analyzed in such detail. It is possible, however, to assign the observed Cotton effects to the various absorptions [25]. In the spectrum of *l*-$[Ni(phen_3)]^{2+}$ the main positive and negative bands at 38.8 and 36.6 kK derive from the β' band, whereas the negative Cotton effect at about 34 kK corresponds to the *p* band, and the two weak CD bands at about 30 kK come from the α system. Similarly, in the spectrum of *l*-$[Ni(phen)_2bipy]^{2+}$ the two β' phenanthroline components are found at 38.5 kK (positive) and 36.6 kK (negative). The next most important Cotton effect is the negative band at 33.9 kK, which also appears for $[Ni(phen)_3]^{2+}$ and is due, in the main, to the phenanthroline *p* band. The phenanthroline α system is again present at about 30 kK. A slight shoulder on the low-energy side of the 33.9-kK CD band could be assigned to the long-axis-polarized absorption band of bipyridine, which occurs in this complex at 32.5 kK. This assignment differs from that of Mason and Norman [140], Mason [136], and Bosnich [20] in relation to the Cotton effect near 34 kK. The complex structure of the CD of *d*-$[Ni(bipy)_2phen]^{2+}$ in the region of 32

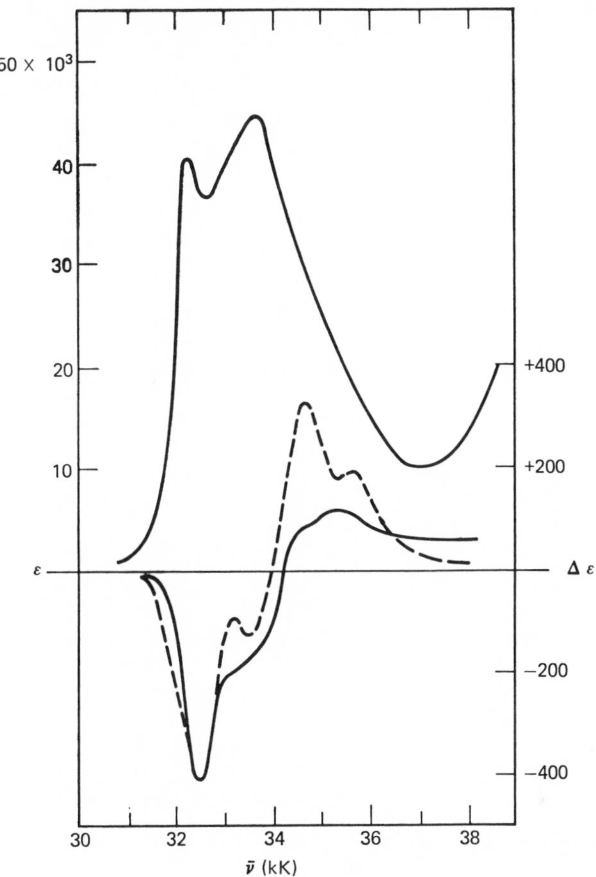

Fig. 5-63 Absorption and CD spectra of *l*-$[Ni(bipy)_3]^{2+}$ at room temperature. The broken curve gives the calculated CD spectrum [25].

to 35 kK probably arises from interaction between the phenanthroline *p* band and the intense low-energy system of bipyridine. A complete assignment of this spectrum should await the measurement of the CD spectrum at low temperatures.

For the above nickel(II) complexes and $[Zn(bipy)_3]^{2+}$, the excitation resonance splittings are too small to allow the center-of-gravity method to be used to determine the relative order of the component transitions. However, the splittings for some bis and tris chromium(III) and cobalt(III) complexes were sufficiently large to allow this method to be applied. The circular dichroism spectra for these complexes are presented in Figs. 5-65 and 5-66. The order of the component transitions was determined from Figs. 5-60

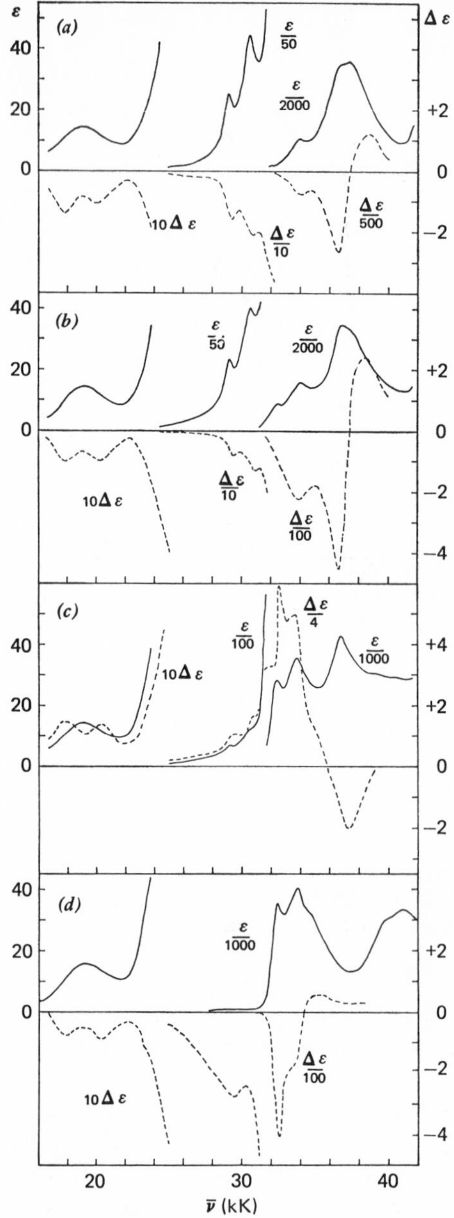

Fig. 5-64 Absorption (——) and CD (– – –) spectra of (*a*) *l*-$[Ni(phen)_3]^{2+}$; (*b*) *l*-$[Ni(phen)_2bipy]^{2+}$, (*c*) *d*-$[Ni(bipy)_2phen]^{2+}$, and (*d*) *l*-$[Ni(bipy)_3]^{2+}$ in alcoholic solution at about 0° [58].

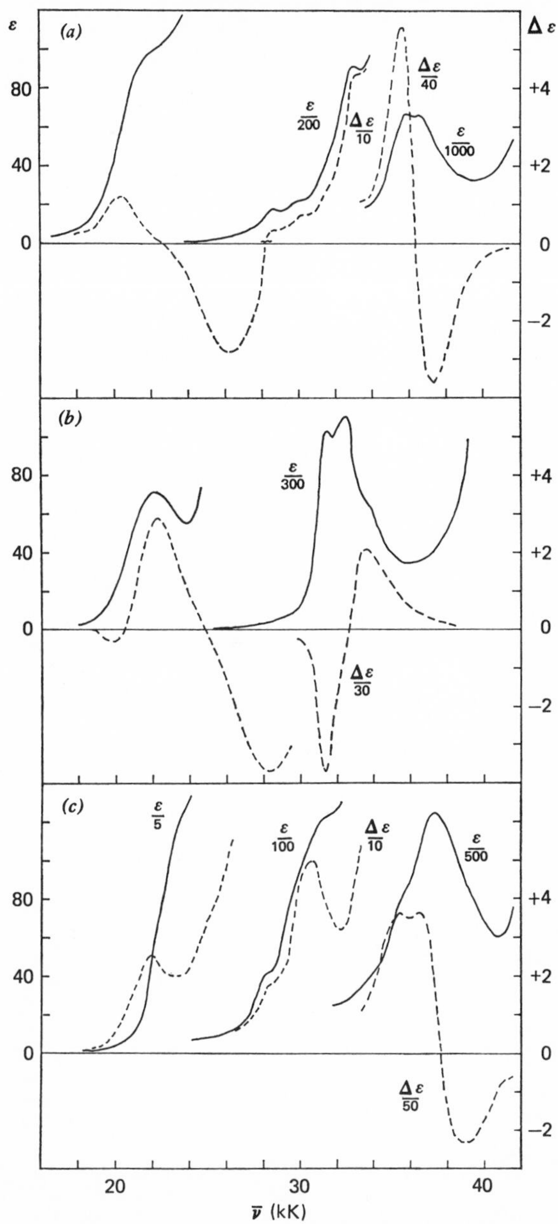

Fig. 5-65 Absorption (——) and CD (– – –) spectra of (*a*) *d*-$[Co(phen)_3]^{3+}$, (*b*) *l*-$[Co(bipy)_3]^{3+}$, and (*c*) *d*-$[Cr(phen)_3]^{3+}$ in aqueous solution [58].

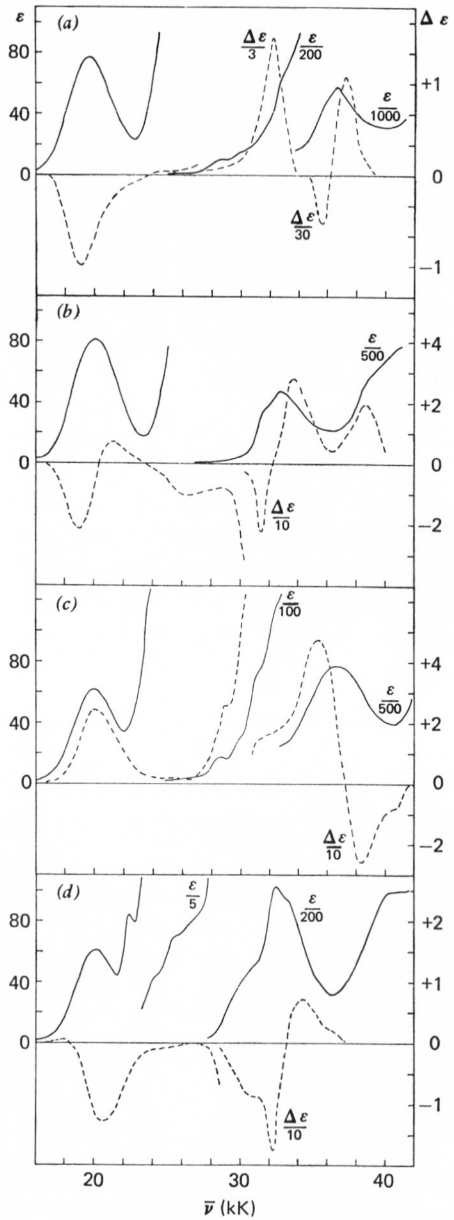

Fig. 5-66 Absorption (——) and CD (- - -) spectra of (*a*) *l*-$[Co(phen)_2ox]^+$, (*b*) *l*-$[Co(bipy)_2ox]^+$, (*c*) *d*-$[Cr(phen)_2ox]^+$, and (*d*) *l*-$[Cr(bipy)_2ox]^+$ in aqueous solution [58].

and 5-61, and the D absolute configuration was assigned to the following: *l*-$[Co(phen)_3]^{3+}$, *d*-$[Cr(phen)_3]^{3+}$, *l*-$[Co(bipy)_3]^{3+}$, *d*-$[Co(phen)_2ox]^{2+}$, *d*-$[Cr(phen)_2ox]^{+}$, *d*-$[Co(bipy)_2ox]^{+}$, and *l*-$[Cr(bipy)_2ox]^{+}$ [58]. Mason and Norman [139] had previously assigned the D configuration to *d*-$[Co(phen)_3]^{3+}$, based on the dipole–dipole approximation for the "exciton splitting" and supported this assignment by applying the normal empirical method to the *d–d* transitions. However, the region of the T_{1g} band is overlapped by a large negative CD band from an adjacent absorption, and, therefore, the dominant band directly under the T_{1g} band need not be the $^1A_1 \rightarrow {}^1E$ transition, as stipulated by Mason and Norman.

The simple rule proposed by Mason [122] in 1964 that the A_2 component of the intense long-axis-polarized $\pi \rightarrow \pi^*$ transition lies lower in energy than the E for tris(bipyridine) and higher for tris(phenanthroline) complexes has no justification, and the absolute configurations assigned on this basis should be disregarded. Each system has to be individually investigated because the amount of metal-ligand σ and π bonding varies from complex to complex. The series of cobalt and chromium(III) complexes above has examples in conflict with the rule, and the levo isomers of $[Ni(phen)_3]^{2+}$ and $[Ni(bipy)_3]^{2+}$ have the same configuration, and not enantiomeric configurations, as predicted by the rule.

The nonempirical method has been applied to complexes of ligands other than 2,2′-bipyridine and 1,10-phenanthroline—acetylacetonate [110], catecholate [135], *N,N′*-bis(salicylidene)-*l*-propylenediamine [18], and to a series of related sexadentates [21], for example. These will not be discussed in detail here. It must be reemphasized, however, that great care has to be taken in the correct assignment of the component transitions, because in most cases the interactions leading to the excitation resonance splittings cannot be simplified to dipole–dipole interactions.

GENERAL READING

Absorption Spectroscopy

1. Ferguson, J., Spectroscopy of 3d Complexes, *Prog. Inorg. Chem.*, **12** (1970), in press.
2. Sutton, D., "Electronic Spectra of Transition Metal Complexes," McGraw-Hill, London, 1968.
3. Hare, C. R., Visible and Ultraviolet Spectroscopy, in K. Nakamoto and P. J. McCarthy (eds.), "Spectroscopy and Structure of Metal Chelate Compounds," Wiley, New York, 1968.
4. Jørgensen, C. K., "Absorption Spectra and Chemical Bonding in Complexes," Pergamon, Oxford, 1962.

Optical Activity

5. Lowry, T. M., "Optical Rotatory Power," Dover, New York, 1964.
6. Jaeger, F. M., "Optical Activity and High Temperature Measurements," McGraw-Hill, New York, 1930.

Circular Dichroism of Metal Complexes

7. Shimura, Y., Optical Dispersion and Circular Dichroism, in K. Nakamoto and P. J. McCarthy (eds.), "Spectroscopy and Structure of Metal Chelate Compounds," Wiley, New York, 1968.
8. Woldbye, F., Technique of Optical Rotatory Dispersion and Circular Dichroism, in H. B. Jonassen and A. Weissberger (eds.), "Technique of Inorganic Chemistry," vol. IV, p. 249, Interscience, New York, 1965.

Circular Dichroism Instrumentation

9. Velluz, L., M. Legrand, and M. Grosjean, "Optical Circular Dichroism," Verlag Chemie, Weinheim, 1965.

REFERENCES

10. Ballhausen, C. J., and W. Moffitt, *J. Inorg. Nucl. Chem.*, **3**:178 (1956).
11. Barclay, G. A., E. Goldschmied, N. C. Stephenson, and A. M. Sargeson, *Chem. Commun.*, **1966**:540.
12. Beddoe, P. G., and S. F. Mason, *Inorg. Nucl. Chem. Letters*, **4**:433 (1968).
13. Biot, J. B., *Mém. Inst.*, **1**:1 (1812).
14. Blount, J. F., H. C. Freeman, A. M. Sargeson and K. R. Turnbull, *Chem. Commun.*, **1967**:324.
15. Born, M., *Physik. Z.*, **16**:251 (1915).
16. Born, M., *Ann. Physik.*, **55**:177 (1918).
17. Bosnich, B., *Proc. Roy. Soc. London, Ser. A*, **297**:88 (1967).
18. Bosnich, B., *J. Am. Chem. Soc.*, **90**:627 (1968).
19. Bosnich, B., *Inorg. Chem.*, **7**:178 (1968)
20. Bosnich, B., *Inorg. Chem.*, **7**:2379 (1968).
21. Bosnich, B., and A. T. Phillip., *J. Am. Chem. Soc.*, **90**:6352 (1968).
22. Boys, S. F., *Proc. Roy. Soc. London, Ser. A*, **144**:655 (1934).
23. Bragg, W. H., *Proc. Roy. Soc. London, Ser. A*, **89**:575 (1914).
24. Bragg, W. H., and R. E. Gibbs, *Proc. Roy. Soc. London, Ser. A*, **109**:405 (1925).
25. Bray, R. G., J. Ferguson, and C. J. Hawkins, *Aust. J. Chem.*, **22**:2091 (1969).
26. Buckingham, D. A., L. G. Marzilli, and A. M. Sargeson, *J. Am. Chem. Soc.*, **89**:825 (1967).
27. Buckingham, D. A., L. G. Marzilli, and A. M. Sargeson, *Inorg. Chem.*, **7**:915 (1968).
28. Buckingham, D. A., P. A. Marzilli, and A. M. Sargeson, *Inorg. Chem.*, **6**:1032 (1967).
29. Buckingham, D. A., P. A. Marzilli, and A. M. Sargeson, private communication, 1969.
30. Buckingham, D. A., P. A. Marzilli, A. M. Sargeson, S. F. Mason, and P. G. Beddoe, *Chem. Commun.*, **1967**:433.
31. Buckingham, D. A., S. F. Mason, A. M. Sargeson, and K. R. Turnbull, *Inorg. Chem.*, **5**:1649 (1966).

32. Chandross, E. A., and J. Ferguson, *J. Chem. Phys.*, **45**:397, 3554 (1966).
33. Chandross, E. A., J. Ferguson, and E. G. McRae, *J. Chem. Phys.*, **45**:3546 (1966).
34. Chernyaev, I. I., L. S. Korablina, and G. S. Muraveiskaya, *Russ. J. Inorg. Chem.*, **20**:567 (1965).
35. Condon, E. U., *Rev. Mod. Phys.*, **9**:432 (1937).
36. Condon, E. U., W. Altar, and H. Eyring, *J. Chem. Phys.*, **5**:753 (1937).
37. Cotton, A., *Ann. Chim. Phys.*, **8**:347 (1896).
38. Craig, D. P., and T. Thirunamachandran, *Proc. Phys. Soc. London*, **84**:781 (1964).
39. Crum Brown, A., *Proc. Roy. Soc. Edinburgh*, **17**:181 (1890).
40. Curradini, P., G. Paiaro, A. Panunzi, S. F. Mason, and G. H. Searle, *J. Am. Chem. Soc.*, **88**:2863 (1966).
41. Denning, R. G., *Chem. Commun.*, **1967**:120.
42. Denning, R. G., and T. S. Piper, *Inorg. Chem.*, **5**:1056 (1966).
43. Dingle, R., *Chem. Commun.*, **1965**:304.
44. Dingle, R., *J. Chem. Phys.*, **46**:1 (1967).
45. Dingle, R., and C. J. Ballhausen, *Kgl. Danske Videnskab. Selskab, Mat. Fys. Medd.*, **35**(12):1 (1967).
46. Dingle, R., and R. A. Palmer, *Theoretica Chim. Acta*, **6**:249 (1966).
47. Djerassi, C., and L. E. Geller, *J. Am. Chem. Soc.*, **81**:2789 (1959).
48. Dörr, F., G. Hohlneicher, and S. Schneider, *Ber. Bunsenges. Phys. Chem.*, **70**:803, (1966).
49. Douglas, B. E., *Inorg. Chem.*, **4**:1813 (1965).
50. Douglas, B. E., R. A. Haines, and J. G. Brushmiller, *Inorg. Chem.*, **2**:1194 (1963).
51. Douglas, B. E., and S. Yamada, *Inorg. Chem.*, **4**:1561 (1965).
52. Drew, M. G. B., J. H. Dunlop, R. D. Gillard, and D. Rogers, *Chem. Commun.*, **1966**:42.
53. Drouard, E., and J.-P. Mathieu, *Compt. rend.*, **236**:2395 (1953).
54. Drude, P., "Lehrbuch der Optik," Hirzel, Leipzig, 1900. Eng. trans., Dover, New York, 1959.
55. Dunlop, J. H., and R. D. Gillard, *Tetrahedron*, **23**:349 (1967).
56. Dunlop, J. H., R. D. Gillard, N. C. Payne, and G. B. Robertson, *Chem. Commun.*, **1966**:874.
57. Ferguson, J., *J. Chem. Phys.*, **44**:2677 (1966).
58. Ferguson, J., C. J. Hawkins, N. A. P. Kane-Maguire, and H. Lip, *Inorg. Chem.*, **8**:771 (1969).
59. Frenkel, J., *Phys. Rev.*, **37**:17, 1276 (1931).
60. Fresnel, A., *Oeuvres*, **1**(XXVIII):748–749 (1822).
61. Fresnel, A., *Ann. Chim. Phys.*, **28**:147 (1825).
62. Fujita, J., T. Yasui, and Y. Shimura, *Bull. Chem. Soc. Japan*, **38**:654 (1965).
63. Garbett, K., and R. D. Gillard, *J. Chem. Soc.*, **1965**:6084.
64. Garbett, K., and R. D. Gillard, *J. Chem. Soc., A*, **1968**:979.
65. Gibbs, R. E., *Proc. Roy. Soc. London, Ser. A*, **110**:443 (1926).
66. Gillard, R. D., *J. Inorg. Nucl. Chem.*, **26**:1455 (1964).
67. Gillard, R. D., *Spectrochim. Acta*, **20**:1431 (1964).
68. Gillard, R. D., and R. Wooton, *J. Chem. Soc., B*, **1967**:921.
69. Gollogly, J. R., and C. J. Hawkins, *Chem. Commun.*, **1966**:873.
70. Gollogly, J. R., and C. J. Hawkins, *Aust. J. Chem.*, **20**:2395 (1967).
71. Gollogly, J. R., and C. J. Hawkins, *Chem. Commun.*, **1968**:689.
72. Gondo, Y., *J. Chem. Phys.*, **41**.3928 (1964).

73. Gray, F., *Phys. Rev.*, **7**:472 (1916).
74. Guye, P. A., *Compt. rend.*, **110**:714 (1890).
75. Haidinger, W., *Ann. Phys.*, **70**:531 (1847).
76. Haines, R. A., and B. E. Douglas, *Inorg. Chem.*, **4**:452 (1965).
77. Hall, S. K., and B. E. Douglas, *Inorg. Chem.*, **7**:533 (1968).
78. Hall, S. K., and B. E. Douglas, *Inorg. Chem.*, **7**:530 (1968).
79. Hall, S. K., and B. E. Douglas, *Inorg. Chem.*, **8**:372 (1969).
80. Halpern, B., A. M. Sargeson, and K. R. Turnbull, *J. Am. Chem. Soc.*, **88**:4630 (1966).
81. Hamer, N. K., *Mol. Phys.*, **5**:339 (1962).
82. Hatfield, W. E., *Inorg. Chem.*, **3**:605 (1964).
83. Hawkins, C. J., *Acta. Chem. Scand.*, **18**:1564 (1964).
84. Hawkins, C. J., *Chem. Commun.*, **1969**:777.
85. Hawkins, C. J., and E. Larsen, *Acta. Chem. Scand.*, **19**:185 (1965).
86. Hawkins, C. J., and E. Larsen, *Acta. Chem. Scand.*, **19**:1969 (1965).
87. Hawkins, C. J., E. Larsen, and I. Olsen, *Acta. Chem. Scand.*, **19**:1915 (1965).
88. Hawkins, C. J., and P. J. Lawson, *Inorg. Chem.*, **9**:6 (1970).
89. Hawkins, C. J., and P. J. Lawson, *Aust. J. Chem.*, **23**:1735 (1970).
90. Hawkins, C. J., and P. J. Lawson, Unpublished results, 1969.
91. Hawkins, C. J., J. Niethe, and C. L. Wong, *Chem. Commun.*, **1968**:427.
92. Hawkins, C. J., J. Niethe, and C. L. Wong, *Aust. J. Chem.*, Submitted for publication.
93. Hawkins, C. J., and C. L. Wong, *Aust. J. Chem.*, **23** (1970), in press.
94. Hidaka, J., and Shimura, Y., *Bull. Chem. Soc. Japan*, **40**:2312 (1967).
95. Hidaka, J., Y. Shimura, and R. Tsuchida, *Bull. Chem. Soc. Japan*, **33**:847 (1960),
96. House, D. A., and C. S. Garner, *Inorg. Chem.*, **6**:272 (1967).
97. Ito, H., J. Fujita, and K. Saito, *Bull. Chem. Soc. Japan*, **40**:2584 (1967).
98. Ito, T., N. Tanaka, I. Hanazaki, and S. Nagakura, *Bull. Chem. Soc. Japan*, **42**:702 (1969).
99. James, D. W., and M. J. Nolan, *Prog. Inorg. Chem.*, **9**:195 (1968).
100. Jensen, K. A., S. Burmester, G. Cederberg, R. B. Jensen, C. T. Pedersen, and E. Larsen, *Acta. Chem. Scand.*, **19**:1239 (1965).
101. Karipides, A., and T. S. Piper, *J. Chem. Phys.*, **40**:674 (1964).
102. Kirkwood, J. G., *J. Chem. Phys.*, **5**:479 (1937).
103. Kuhn, W., *Z. Physik. Chem. (Leipzig), B*, **4**:14 (1929).
104. Kuhn, W., *Trans. Faraday Soc.*, **26**:293 (1930).
105. Kuhn, W., *Z. Physik. Chem. (Leipzig), B*, **20**:325 (1933).
106. Kuhn, W., *Naturwissenschaften*, **19**:289 (1938).
107. Kuhn, W., and K. Bein, *Z. Physik. Chem. (Leipzig), B*, **24**:335 (1934).
108. Kuhn, W., and K. Freudenberg, "Hand- und Jahrbuch der chemischen Physik," vol. 8, part III, p. 59, Akadem. Verlagsgesellschaft, Leipzig, 1932.
109. Larsen, E., and S. F. Mason, *J. Chem. Soc., A*, **1966**:313.
110. Larsen, E., S. F. Mason, and G. H. Searle, *Acta Chem. Scand.*, **20**:191 (1966).
111. Larsen, E., and I. Olsen, *Acta Chem. Scand.*, **18**:1025 (1964).
112. Larsen, S., K. J. Watson, A. M. Sargeson, and K. R. Turnbull, *Chem. Commun.*, **1968**:847.
113. Legg, J. I., D. W. Cooke, and B. E. Douglas, *Inorg. Chem.*, **6**:700 (1967).
114. Legg, J. I., and B. E. Douglas, *J. Am. Chem. Soc.*, **88**:2697 (1966).
115. Liehr, A. D., *J. Phys. Chem.*, **68**:665, 3629 (1964).
116. Lin, C., and B. E. Douglas, *Inorg. Nucl. Chem. Letters*, **4**:15 (1968).

117. Linnell, R. H., and A. Kaczmarczyk, *J. Phys. Chem.*, **65**:1196 (1961).
118. Liu, C. T., and B. E. Douglas, *Inorg. Chem.*, **3**:1356 (1964).
119. McCaffery, A. J., and S. F. Mason, *Mol. Phys.*, **6**:359 (1963).
120. McCaffery, A. J., and S. F. Mason, *Proc. Chem. Soc.*, **1963**:211.
121. McCaffery, A. J., S. F. Mason, and R. E. Ballard, *J. Chem. Soc.* **1965**:2883.
122. McCaffery, A. J., S. F. Mason, and B. J. Norman, *Proc. Chem. Soc.*, **1964**:259.
123. McCaffery, A. J., S. F. Mason, and B. J. Norman, *Chem. Commun.*, **1965**:49.
124. McCaffery, A. J., S. F. Mason, and B. J. Norman, *J. Chem. Soc.*, **1965**:5094.
125. McCaffery, A. J., S. F. Mason, and B. J. Norman, *Chem. Commun.*, **1966**:661.
126. McCaffery, A. J., S. F. Mason, B. J. Norman, and A. M. Sargeson, *J. Chem. Soc.*, *A*, **1968**:1304.
127. McClure, D. S., *J. Chem. Phys.*, **25**:481 (1956).
128. MacDermott, T. E., and A. M. Sargeson, *Aust. J. Chem.*, **16**:334 (1963).
129. McLauchlan, A. D., and M. A. Ball, *Mol. Phys.*, **8**:581 (1964).
130. Malleman, R. de, *Compt. rend.*, **181**:298 (1925).
131. Malleman, R. de, *Rev. Gen. Sci.*, **38**:453 (1927).
132. Malleman, R. de, *Trans. Faraday Soc.*, **26**:281 (1930).
133. Martin, R. B., Tsangaris, J. M., and J. W. Chang, *J. Am. Chem. Soc.*, **90**:821 (1968).
134. Marzilli, L. G., and D. A. Buckingham, *Inorg. Chem.*, **6**:1042 (1967).
135. Mason, J., and S. F. Mason, *Tetrahedron*, **23**:1919 (1967).
136. Mason, S. F., *Inorg. Chim. Acta Revs.*, **2**:89 (1968).
137. Mason, S. F., and B. J. Norman, *Proc. Chem. Soc.*, **1964**:339.
138. Mason, S. F., and B. J. Norman, *Chem. Commun.*, **1965**:73.
139. Mason, S. F., and B. J. Norman, *Inorg. Nucl. Chem. Letters*, **3**:285 (1967).
140. Mason, S. F., and B. J. Norman, *Chem. Phys. Letters*, **2**:22 (1968).
141. Mason, S. F., and J. W. Wood, *Chem. Commun.*, **1968**:1512.
142. Matsumoto, K., Y. Kushi, S. Ooi, and H. Kuroya, *Bull. Chem. Soc. Japan*, **40**:2988 (1967).
143. Michelsen, K., *Acta Chem. Scand.*, **19**:1175 (1965).
144. Moffitt, W., *J. Chem. Phys.*, **25**:1189 (1956).
145. Moffitt, W., and A. Moscowitz, *J. Chem. Phys.*, **30**:648 (1959).
146. Murrell, J. N., "The Theory of the Electronic Spectra of Organic Molecules," Methuen, London, 1963.
147. Murrell, J. N., and J. Tanaka, *Mol. Phys.*, **7**:363 (1964).
148. Muto, A., F. Marumo, and Y. Saito, *Inorg. Nucl. Chem. Letters*, **5**:85 (1969).
149. Muto, A., F. Marumo, and Y. Saito, To be published.
150. Nakahara, A., Y. Saito, and H. Kuroya, *Bull. Chem. Soc. Japan*, **25**:331 (1952).
151. Ogino, K., K. Murano, and J. Fujita, *Inorg. Nucl. Chem. Letters*, **4**:351 (1968).
152. Ohkawa, K., J. Hidaka, and Y. Shimura, *Bull. Chem. Soc. Japan*, **40**:2830 (1967).
153. Orgel, L. E., *J. Chem. Phys.*, **23**:1004 (1955).
154. Orgel, L. E., *J. Chem. Soc.*, **1961**:3683.
155. Oseen, C. W., *Ann. Physik.*, **48**:1 (1915).
156. Pariser, R., and R. G. Parr, *J. Chem. Phys.*, **21**:466, 767 (1953).
157. Pasteur, L., *Ann. Chim. Phys.*, **24**:442 (1848).
158. Perrin, M. H., and M. Gouterman, *J. Chem. Phys.*, **46**:1019 (1967).
159. Pfeiffer, P., and W. Christeleit, *Z. Physiol. Chem.*, **245**:197 (1937).
160. Piper, T. S., *J. Am. Chem. Soc.*, **83**:3908 (1961).
161. Piper, T. S., and A. G. Karipides, *Mol. Phys.*, **5**:475 (1962).
162. Piper, T. S., and A. G. Karipides, *J. Am. Chem. Soc.*, **86**:5039 (1964).

163. Polder, D., *Physica*, **9**:709 (1942).
164. Pople, J. A., *Trans. Faraday Soc.*, **49**:1375 (1953).
165. Robinson, W. T., D. A. Buckingham, G. Chandler, L. G. Marzilli, and A. M. Sargeson, *Chem. Commun.*, **1969**:539.
166. Roos, B., *Acta Chem. Scand.*, **20**:1673 (1966).
167. Roothaan, C. C. J., *Rev. Mod. Phys.*, **32**:179 (1960).
168. Rosenfeld, L., *Z. Physik.*, **52**:161 (1928).
169. Saburi, M., M. Homma, and S. Yoshikawa, *Inorg. Chem.*, **8**:367 (1969).
170. Saburi, M., Y. Tsujito, and S. Yoshikawa, *Inorg. Nucl. Chem. Letters*, **5**:203 (1969).
171. Saburi, M., and S. Yoshikawa, *Inorg. Chem.*, **7**:1890 (1968).
172. Saito, Y., and H. Iwasaki, *Bull. Chem. Soc. Japan*, **35**:1131 (1962).
173. Sargeson, A. M., private communication, 1969.
174. Sargeson, A. M., and G. H. Searle, *Inorg. Chem.*, **4**:45 (1965).
175. Schäffer, C. E., Abstract of paper presented at Symposium on the Structure and Properties of Coordination Compounds, Bratislava, 1964.
176. Schäffer, C. E., *Proceedings of 8th I.C.C.C.*, p. 77, Vienna, 1964.
177. Schäffer, C. E., *Proc. Roy. Soc. London, Ser. A*, **297**:96 (1967).
178. Schäffer, C. E., and C. K. Jørgensen, *Kgl. Danske Videnskab. Selskab., Mat. Fys. Medd.*, **34**(13):1 (1965).
179. Schoenberg, L. N., D. W. Cooke, and C. F. Liu, *Inorg. Chem.*, **7**:2386 (1968).
180. Shibata, M., H. Nishikawa, and K. Hosaka, *Bull. Chem. Soc. Japan*, **40**:236 (1967).
181. Shibata, M., H. Nishikawa, and Y. Nishida, *Bull. Chem. Soc. Japan*, **39**:2310 (1966).
182. Shibata, M., H. Nishikawa, and Y. Nishida, *Inorg. Chem.*, **7**:9 (1968).
183. Shinada, M., *J. Phys. Soc. Japan*, **19**:1607 (1964).
184. Smith, H. L., and B. E. Douglas, *Inorg. Chem.*, **5**:784 (1966).
185. Sugano, S., *J. Chem. Phys.*, **33**:1883 (1960).
186. Treptow, R. S., *Inorg. Chem.*, **7**:1229 (1968).
187. Van Saun, C. W., and B. E. Douglas, *Inorg. Chem.*, **7**:1393 (1968).
188. Walden, P., *Z. Physik. Chem. (Leipzig)*, **17**:245 (1895).
189. Wellman, K. M., S. Bogdansky, W. Mungall, T. G. Mecca, and C. R. Hare, *Tetrahedron Letters*, **37**:3607 (1967).
190. Wellman, K. M., T. G. Mecca, W. Mungall, and C. R. Hare, *J. Am. Chem. Soc.*, **89**:3646 (1967).
191. Wellman, K. M., W. Mungall, T. G. Mecca, and C. R. Hare, *J. Am. Chem. Soc.*, **89**:3647 (1967).
192. Wellman, K. M., T. G. Mecca, W. Mungall, and C. R. Hare, *J. Am. Chem. Soc.*, **90**:805 (1968).
193. Wentworth, R. A. D., *Chem. Commun.*, **1965**:532.
194. Wentworth, R. A. D., *Inorg. Chem.*, **5**:496 (1966).
195. Wentworth, R. A. D., and T. S. Piper, *Inorg. Chem.*, **4**:202 (1965).
196. Yamada, S., A. Nakahara, Y. Shimura, and R. Tsuchida, *Bull. Chem. Soc., Japan*, **28**:222 (1955).
197. Yamatera, H., *Bull. Chem. Soc. Japan* **31**:95 (1958).
198. Yasui, T., *Bull. Chem. Soc. Japan*, **38**:1746 (1965).
199. Yasui, T., J. Hidaka, and Y. Shimura, *J. Am. Chem. Soc.*, **87**:2762 (1965).
200. Yasui, T., J. Hidaka, and Y. Shimura, *Bull. Chem. Soc. Japan*, **39**:2417 (1966).
201. Yoshikawa, S., T. Sekihara, and M. Goto, *Inorg. Chem.*, **6**:169 (1967).

6

Nuclear Magnetic Resonance

Although nuclear magnetic resonance (nmr) has been extensively applied to stereochemical problems involving purely organic molecules, its application to the stereochemistry of coordination compounds has been surprisingly limited. Nevertheless, its potential value in this field is tremendous. A detailed coverage of the theory of nmr is outside the scope of the present work. Many suitable texts are available at elementary and advanced levels (see, for example, the reading list), and, if necessary, they should be consulted prior to reading the present chapter. Certain aspects of the theory, however, which are important for the subsequent discussion, are dealt with briefly first.

THEORY

The nuclei of particular isotopes of elements have an intrinsic mechanical spin, which, because the nuclei are charged, gives rise to a magnetic moment

$$\mu_N = g_N \frac{M_H}{M_N} \beta_N[I(I + 1)]^{1/2} \tag{6-1}$$

where g_N is the nuclear equivalent of the Landé splitting factor, M_H and M_N are the masses of the hydrogen nucleus and the nucleus in question, β_N is the nuclear magneton, and I is the nuclear spin, which has the values 0, $\frac{1}{2}$, 1, $1\frac{1}{2}$, 2, ... Nmr is concerned with transitions between the nuclear energy levels corresponding to the quantized values of the nuclear magnetic moment.

When the nucleus is placed in a static uniform magnetic field H_0, there are $(2I + 1)$ orientations available to it. The separation between the highest and lowest of these levels is $2\mu H_0$, and the energy difference between adjacent levels is $\mu H_0/I$, where μ is the maximum measurable value of the nuclear magnetic moment. Of course, when I is equal to $\frac{1}{2}$, as is the case for the proton, the energy gap between the two states corresponding to the two orientations of the nuclear magnet with respect to the field (parallel or antiparallel) is given by

$$\mathscr{E} = h\nu = 2\mu H_0 \qquad (6\text{-}2)$$

If the resonance frequency of a particular nucleus, say, the proton, depended solely on μ and H_0, a measurement of ν would not provide us with stereochemical information. It turns out that the resonance frequency depends to a small degree on the nucleus' environment. This comes about because the extranuclear electrons screen the nucleus and, in so doing, vary the field felt by the nucleus according to the expression

$$H_X = H_0(1 - \sigma) \qquad (6\text{-}3)$$

where σ, the screening or shielding constant, is a nondimensional constant independent of H_0 but dependent on the chemical environment. Thus, for a particular irradiating frequency, fields different from H_0 and from each other are necessary to cause the resonance of nuclei in different structural environments. The separation of these fields from an arbitrarily chosen reference H_R is termed the *chemical shift*

$$\delta = (H_A - H_R)H_0^{-1} \times 10^6 \text{ ppm} \qquad (6\text{-}4)$$

The usual references for protons are tetramethylsilane (TMS) and sodium trimethylsilylpropanesulfonate (NaTMS), 0.04 ppm downfield from TMS, and the H_A values are normally downfield from these, giving rise to negative values of δ. In the present work, however, the negative sign is omitted.

Factors Governing the Chemical Shift

It is profitable to divide the shielding constant into contributions from electrons localized on atoms and in chemical bonds. An expression for σ_A is given in Eq. (6-5):

$$\sigma_A = \sigma_{AA}{}^{\text{dia}} + \sigma_{AA}{}^{\text{para}} + \sum_{B \neq A} \sigma_{AB} + \sigma_A{}^{\text{deloc}} \qquad (6\text{-}5)$$

The diamagnetic shielding constant associated with electrons localized on A, $\sigma_{AA}{}^{\text{dia}}$, can be pictured as arising in the following way. The uniform magnetic field causes the electrons to rotate about their nucleus in such a direction as to produce a secondary magnetic field, which, in the region of the nucleus, is opposed to the applied field and proportional to it (see Fig. 6-1). These local diamagnetic currents provide a source of positive shielding; they reduce the total field experienced by the nucleus, and, therefore, a higher applied field is necessary to cause resonance. The molecular environment affects this shielding

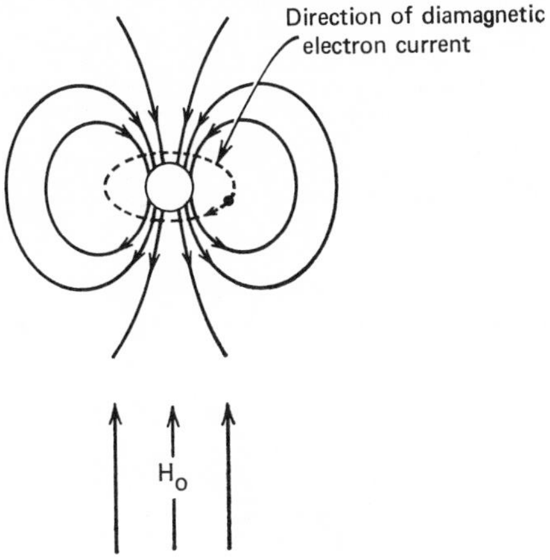

Fig. 6-1 Diamagnetic shielding.

by changing the electron density around the nucleus. For example, in a compound of the type CH_3X, the electronegativity of the group X influences the chemical shift of the protons.

It has been found that the protons of ammonia groups coordinated trans to an electronegative ligand X appear at higher field than the protons in ammonia ligands cis to X [17,35]. This difference in the diamagnetic shielding at the two different kinds of protons arises, at least in part, from polarization effects that are more effectively transmitted to groups trans to the electronegative ligand. The difference is quite marked and is very important in distinguishing geometrical isomers.

The second term in Eq. (6-5), σ_{AA}^{para}, derives from the induced paramagnetic currents on A, which arise as a consequence of the mixing of ground and excited states by the applied magnetic field. This is zero when the electrons localized on A are in pure S states. When A is hydrogen, σ_{AA}^{para} is small compared with σ_{AA}^{dia}. However, when A is ^{19}F, σ_{AA}^{para} is the dominant term.

The term σ_{AB} is known as the *magnetic anisotropic shielding parameter* and arises from induced currents localized on atoms other than A or in bonds distant from A. If the electronic distribution on B is spherically symmetrical, the induced field is independent of the direction of the applied field. The relative orientation of A and B with respect to the applied field, however, determines the component of the induced field at A (see Fig. 6-2). If the molecule is in the liquid or gaseous state, the consequent random motion leads to the field at A because of the induced currents in B averaging to zero. If B

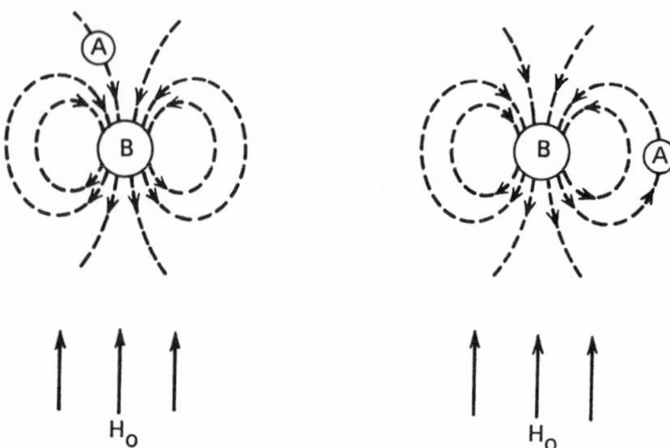

Fig. 6-2 The dependence of the shielding of a nucleus A by B on the relative orientation of A and B with respect to the applied field H_0.

is not spherically symmetrical, the induced field depends on the direction of the applied field, and the field at A due to B averages to a finite value. This long-range shielding depends directly on the magnetic anisotropy of B and is particularly important for bonds [46,63]. If B is axially symmetric, the shielding is given by

$$\sigma_{AB} = \frac{\Delta\chi_B r_{AB}^{-3}(1 - 3\cos^2\theta)}{3N_0} \tag{6-6}$$

where $\Delta\chi_B$ is the difference between the magnetic susceptibility parallel and perpendicular to the symmetry axis, r_{AB} is the distance between A and the electrical center of gravity of B, θ is the acute angle that a line joining these two points makes with the symmetry axis, and N_0 is Avogadro's number.

A graph of $(1 - 3\cos^2\theta)$ against θ is shown in Fig. 6-3. It is seen from this that the shielding changes sign at $\theta = 55°44'$. If $\Delta\chi_{XY}$ is positive—$\Delta\chi_{C-C}$, $\Delta\chi_{C-H}$, for example—a nucleus situated directly above a bond XY would be shielded by that bond, whereas a nucleus situated along the axis of the bond would be deshielded (see Fig. 6-4). It must be remembered that σ_{AB} depends on the distance of A from B, and because this is an r^{-3} dependence, σ_{AB} quickly tends to zero at large distances.

The difference in chemical shift for axial and equatorial hydrogen nuclei in cyclohexane of 0.46 ppm has been explained in terms of the diamagnetic anisotropy of the C—C and C—H bonds [53]. Consider Fig. 6-5. H_a and H_e are identically oriented with respect to C_1—C_6, C_1—C_2, and the equatorial C—H bonds at C_2 and C_6, and thus these bonds do not contribute to δ_{ae}, the difference in the chemical shifts of H_a and H_e. The principal contributions to the difference in the magnetic anisotropic shielding of H_a and H_e come from

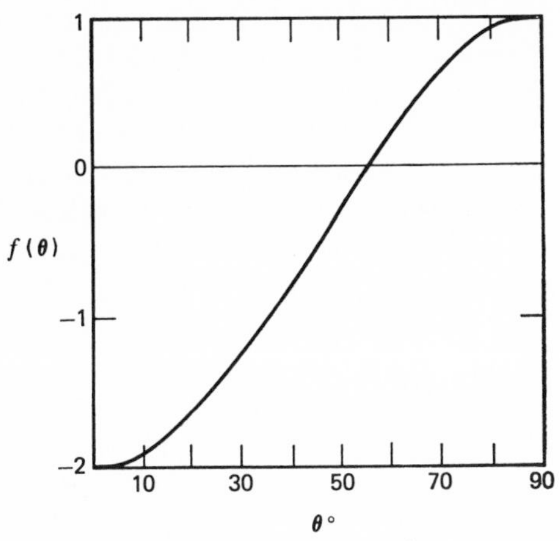

Fig. 6-3 Graph of $f(\theta) = 1 - 3\cos^2\theta$.

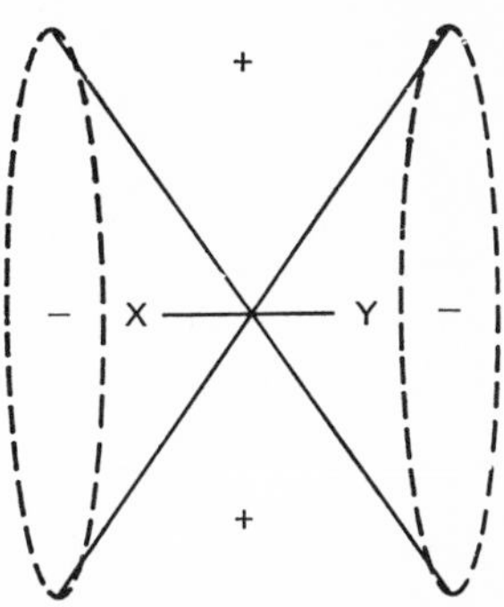

Fig. 6-4 Long-range magnetic anisotropic shielding effect of the bond XY.

C_2—C_3 and C_5—C_6 and the axial C—H bonds at C_2 and C_6. From Fig. 6-5 and Eq. (6-6) it can be seen that C_2—C_3 and C_5—C_6 deshield H_e and shield H_a, whereas the axial protons on C_2 and C_6 shield H_e and deshield H_a. Moritz and Sheppard [53] calculated the difference in the shielding to be given by

$$\Delta\sigma_{ae} = 0.1073(\Delta\chi_{C-C} - \Delta\chi_{C-H}) \tag{6-7}$$

The last term in Eq. (6-5), σ_A^{deloc}, arises from induced currents involving electrons that are not localized on any one atom or bond in the molecule.

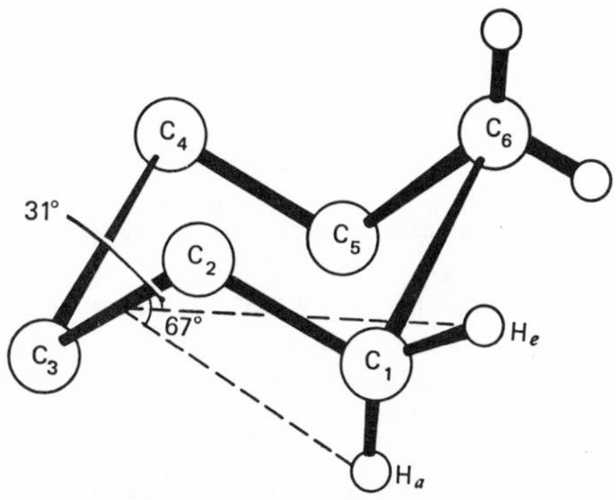

Fig. 6-5 Long-range shielding in cyclohexane.

It is particularly important for cyclic compounds, such as aromatic hydrocarbons, in which electron delocalization is especially marked, but it is also thought to be important for saturated ring systems. The long-range shielding by a benzene ring can be understood in terms of Fig. 6-6. The electron currents around the benzene ring give rise to a field that enhances the applied field at the aromatic protons but opposes the applied field at A, thus deshielding H but shielding A. The marked magnetic anistropy of the benzene ring results in the shielding arrangement shown in Fig. 6-6*b*.

Spin-Spin Coupling

The nuclear resonances often show fine structure, which derives from the fact that the field experienced by a nucleus is influenced by the spins of nuclei in neighboring groups. Take, for example, a molecule AX_n, where A and X are nonequivalent nuclei resonating at different field strengths*; the spectrum

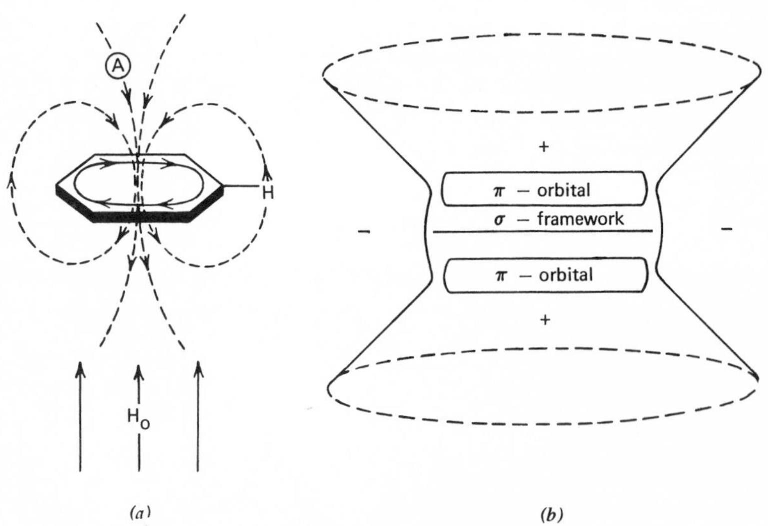

Fig. 6-6 Long-range shielding of the benzene ring.

* Nonequivalent nuclei of the same species that resonate at fields separated by shifts of similar magnitude to the coupling constants involved are denoted by the letters A, B, C, etc., M, P, Q, etc., or X, Y, Z, etc., depending on the chemical shift. A, B, C and X, Y, Z are used when the two sets of nuclei resonate at fields separated by large chemical shifts, whereas M, P, Q are reserved for nuclei of intermediate field. Nuclei which have the same chemical shift but which are not magnetically equivalent, in that they couple differently with other nuclei, are distinguished by primes (e.g., A and A′, B and B′).

for nucleus A shows $(2nI_X + 1)$ lines, where I_X is the spin quantum number of X. The relative intensities of the lines are given by the *n*th binomial coefficients. The lines are equally spaced, and the magnitude of the splitting is known as the coupling or spin-spin interaction constant and is given by the symbol J_{AX}. The above holds as long as the A and X resonances are well separated. If this is not the case, the relative intensities of lines differ from that proposed, and depending on the system, more or fewer lines appear than predicted by the simple first-order analysis. As an example, let us consider the molecule CH_3—$CH(ND_3^+)$—COO^-. The methyl group is free to rotate; and thus the three methyl protons are equivalent, and the system is of the AX_3 type. The methyl signal appears as a doublet at $\delta = 1.5$ ppm from TMS, with $J_{HH'} = 7$ Hz, because of the coupling with the —CH— proton, which resonates at lower field ($\delta = 3.8$ ppm) as a 1:3:3:1 quartet (see Fig. 6-7).

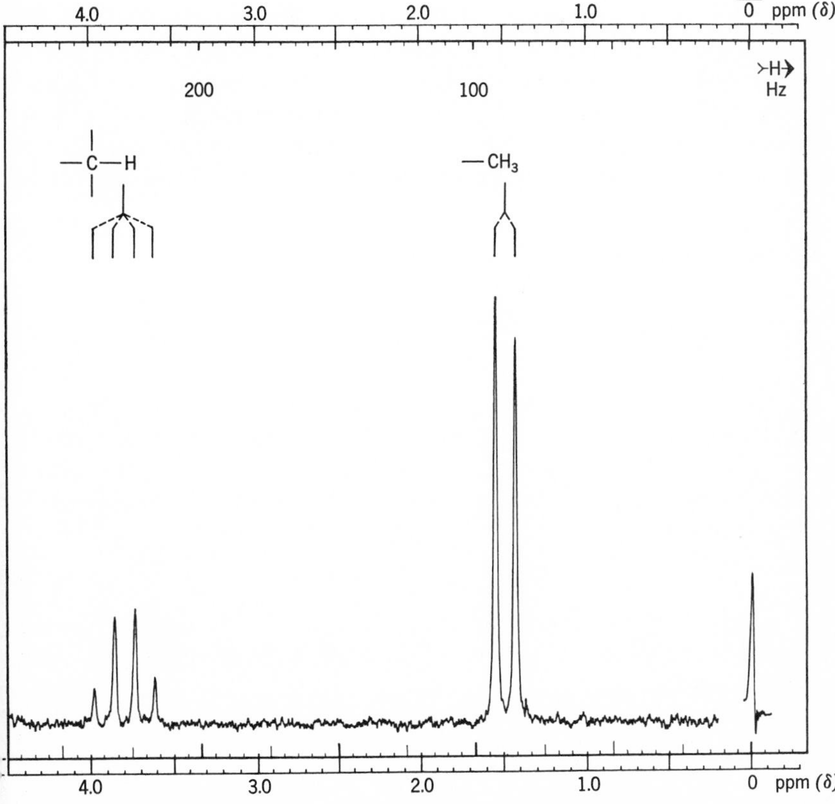

Fig. 6-7 Sixty-megahertz spectrum of alanine in D_2O with TMS as an external reference.

It is found that (a) the coupling constant is independent of the applied field in contrast to the chemical shift, which increases with increasing field, and (b) the nuclei of a magnetically equivalent group—the three protons of the methyl group of alanine, for example—do not couple to yield observable multiplicity. Further, it is observed that, in general, the magnitude of the coupling constant decreases as the number of bonds separating the interacting nuclei increases. The coupling constants have absolute signs, which can be determined experimentally for some system [39,42]. The observations above have been explained by a coupling mechanism involving the bonding electrons [57,58]. For ^{19}F—^{19}F interactions a *through-space* mechanism has also been suggested [55].

Replacement of specific protons by deuterium ($I = 1$) is commonly used to assist an analysis of high-resolution spectra. ^2H—^1H coupling constants are considerably less than $J_{^1H^1H}(J_{^2H^1H} = 0.154J_{^1H^1H})$, and for the complexes being considered in the present chapter, they are so small that the coupling is not observable and is not taken into account in the analyses.

The coupling constants are capable of providing considerable stereochemical information, because the magnitude of J depends on certain geometrical relationships involving the interacting nuclei. In most cases, however, if the maximum information is to be extracted from the spectra, complex calculations have to be carried out preferably with the aid of a computer. Emsley, Feeney, and Sutcliffe discuss these calculations in their treatise "High Resolution Nuclear Magnetic Resonance" [3].

Vicinal coupling constants. Consider the molecular fragment

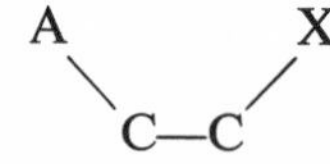

It has been found that the coupling between A and X depends on the dihedral angle between the A—C and X—C bonds. Karplus [37] used a valence-bond approach to calculate $J_{HH'}$ for a H—C—C—H fragment as a function of the dihedral angle. The calculated $J_{HH'}$ values were found to fit approximately a $\cos^2 \phi$ function

$$J_{HH'} = k_1 \cos^2 \phi - k_2 \qquad (6\text{-}8)$$

that has maxima at 180° and 0° and a minimum at 90°. The constants k_1 and k_2 depend on the nature of the system [4,7,45,65]. A more refined calculation by Karplus [40] led to

$$J_{HH'} = A + B \cos \phi + C \cos 2\phi \qquad (6\text{-}9)$$

where A, B, and C are constants that depend on the system. Because the coupling constant also depends on the H—C—C angles—as these angles increase, $J_{HH'}$ decreases—and may be affected by substituents of greater electronegativity than hydrogen and by bond length changes, one must be careful in attempting structural analysis based solely on the dihedral-angle dependence of the vicinal coupling constant [40]. Nevertheless, it is generally found, for example, that $J_{HH'}{}^{trans}$ is larger than $J_{HH'}{}^{gauche}$ and that such a generalization can be profitably applied to stereochemical problems.

Whereas $J_{HH'}$ values are usually small, spin-spin interactions involving the ^{19}F nucleus are much larger and, because of this, are a more sensitive probe for stereochemistry. Gutowsky and his coworkers [28] found that J^{trans} is also larger than J^{gauche} for H—F and F—F interactions and may sometimes be of opposite sign. More recently, from a study of a series of rigid molecules, it has been shown that $J_{HF}{}^{vic}$ has the same dependence on ϕ as $J_{HH'}{}^{vic}$, and an equation identical to Eq. (6-9) has been proposed to account for this [66]. The coupling constants for ^{19}F also depend on the C—C—F angles, on the electronegativity of other substituents, and on the bond lengths [66]. Further, for $J_{FF'}$, some dependence on the through-space separation introduces an additional reason for caution in a quantitative application to stereochemistry.

Geminal coupling constant. Gutowsky, Karplus, and Grant [30] have applied a valence-bond approach to the estimation of the angular dependence of the geminal ^1H interactions in HCH fragments. Their estimated constants, however, have the same sign as $J_{HH'}{}^{vic}$ whereas from experiment it is known that they have opposite signs [29]. Although the theoretical variation of $J_{HH'}{}^{gem}$ with $\angle$HCH is incorrect, it is probable that $J_{HH'}{}^{gem}$ and other geminal coupling constants have an angular dependence.

Coupling constants between nuclei separated by four σ bonds. Karplus [38] has suggested that the magnitude of the H—H spin coupling through four σ bonds should be $\leqslant$ 0.5 Hz, although in practice the observed values usually lie in the range of 0 to 1.5 Hz, with larger values sometimes being found. The H—C—C—C—H coupling is known to depend on the dihedral angle, but the exact form of the angular dependence is open to debate. Anet [5] found for camphane-2,3-diols that when $\phi = 180°$, the 1,3 constants were in the range of 1 to 5 Hz but otherwise were zero. Other evidence supports this view [52]. Bothner-By and Naar-Colin [8], however, found for 2,3-disubstituted butanes that the magnitude of J for the CH_3—C—C—H fragment is almost zero when $\phi = 180°$ and approximately 0.30 Hz when $\phi = 60°$.

Spin Decoupling

In order that the spin-spin coupling may be observed, the interacting nuclei should spend a time in a given spin-state arrangement that is long in comparison with the reciprocal of the frequency separation of the multiplet components [31]. If the coupling constant, in hertz, is less than the frequency at which the spin arrangements of the group are changing, only a single absorption line results for the other nucleus, because it experiences a field that is the average of the various spin arrangements. This spin decoupling can be promoted by irradiating one of the coupled nuclei with its appropriate resonance frequency. The technique, also known as *double resonance*, is most useful in the analysis of complex nmr spectra.

Partial or complete decoupling can occur naturally if one of the nuclei has an electric quadrupole moment, because the lifetimes of such nuclei in any particular spin state are limited by rapid spin-lattice relaxation. This can lead to extensive broadening of resonances, which depends on the interaction of the quadrupole moment of the nucleus with the electric field gradients within the molecule. If the symmetry of the molecule, and thus the electric field, is high, the interaction is small, and the broadening is minimized. However, in the compounds we are interested in the symmetry is low, and quadrupole relaxation usually obliterates fine structure that could be most informative.

An important example of a nucleus with a quadrupole moment is ^{14}N with $I = 1$. The resonance of protons bound to nitrogen are usually broad because of the interaction above. They can be made to sharpen by irradiating the sample with the nitrogen resonance frequency, decoupling the ^{14}N and ^{1}H nuclei. An example of the effect of this decoupling is shown in Fig. 6-8 for formamide. The technique has a great potential in the application of nmr to metal-complex stereochemical problems, but as yet it has been little exploited.

Time-Averaging Processes

Resonances of two protons or other nuclei in nonequivalent environments appear as two discrete spectral lines or multiplet sets if environmental exchange processes take place at rates that are small compared with $2\pi\delta_{AA'}$, where $\delta_{AA'}$, in hertz, is the chemical-shift difference between the two bands. If the rate of exchange is faster, a single resonance is observed, corresponding to the time-averaged environment, and any spin-spin coupling between the two nuclei would not be observed. In some cases it is possible to slow down the rate of environmental exchange by lowering the temperature. For example, at room temperature, the axial and equatorial protons of cyclohexane yield one resonance because of conformational interchange. Chair-to-chair interconversion puts an equatorial hydrogen into the axial position, and vice versa. On cooling a solution of cyclohexane in carbon disulfide, however, the resonance shows a progressive broadening until at $-70°$ two distinct bands appear [34]. At lower temperatures the band separation remains constant.

Nmr of Metal Complexes

The application of nmr to the stereochemistry of metal complexes is sometimes affected by factors that, although not peculiar to this class of compound,

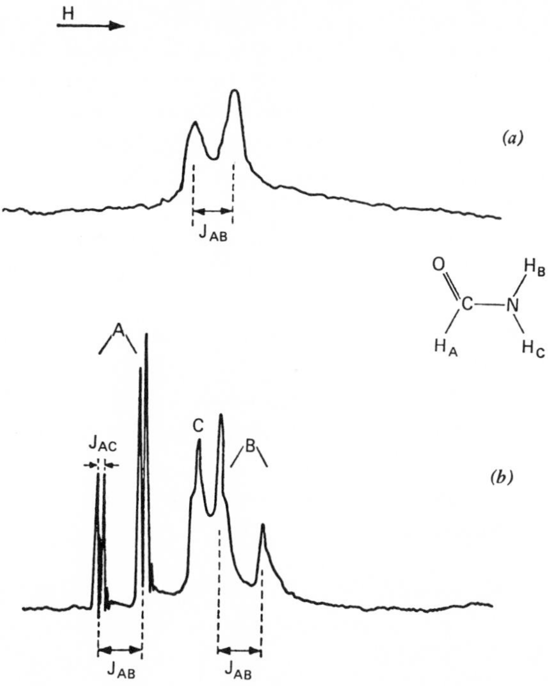

Fig. 6-8 The hydrogen resonance spectrum of formamide at 40 MHz (*a*) without spin decoupling, (*b*) with simultaneous irradiation of the ^{14}N nuclei by means of an applied radiofrequency of 2.8904 MHz. (By courtesy of Varian Associates. Figure taken with permission from [3].)

are often associated with the central metal ion and the coordination sphere. Most of the transition metals have isotopes with nuclear spin, and a high percentage of them have spin greater than $I = \frac{1}{2}$. Those with $I > \frac{1}{2}$ have a quadrupole moment and are capable of destroying the fine structure of the ligand nuclei resonances but in practice this quadrupole broadening does not seem to be significant; for example, sharp resonances are observed for complexes of cobalt(III) (^{59}Co, $I = \frac{7}{2}$; 100 percent abundance) and palladium(II) and (IV) (^{105}Pd, $I = \frac{5}{2}$; 22 percent abundance). On the other hand, the coupling between some metal ions—^{195}Pt ($I = \frac{1}{2}$; 34 percent abundance), for example—and ligand protons is a good source of structural information because of the large size of the coupling constants.

If a molecule is paramagnetic, the nmr signals are often very broad, yielding a minimum of information. In fact, the effects of electronic paramagnetism are so large compared with those of the nuclear moments that nuclear resonance is sometimes undetectable in paramagnetic samples. This limits considerably the number of metal ions that are suitable for these studies. The effect arises, because the paramagnetic metal ion produces a very intense fluctuating magnetic field leading to greatly reduced spin-lattice relaxation times. Nevertheless, under certain circumstances the presence of paramagnetic ions leads to large shifts in nmr spectra resulting from nucleus-electron isotropic hyperfine contact interactions [47] and pseudocontact interaction [48]. This can be used to advantage in determining the coordination sites in multidentate ligands [49], in separating signals from diastereoisomers, in studying planar-diamagnetic $\rightleftarrows$ tetrahedral-paramagnetic equilbria [54], and in conformational studies.*

In order that time-averaging processes do not interfere with the studies, the compounds must not be labile. This further limits the number of complexes that can be studied effectively. In fact, most of the work in this field has been with the inert diamagnetic complexes of cobalt(III), platinum and palladium-(II) and (IV), and rhodium(III).

GEOMETRICAL ISOMERS

The nmr spectra of a series of pentamminecobalt(III) complexes

$$[Co(NH_3)_5X]^{n+}$$

show two NH_3 resonances with an intensity ratio of 12:3, the intense band being at lower field [13,17,35] (see Fig. 6-9). The 12 equivalent protons belong to the NH_3 groups cis to the ligand X, and the 3 to the trans NH_3 group.

Clifton and Pratt [17] suggested that the cis protons may have been shifted to lower field than the trans by:

1. Intramolecular hydrogen bonding between the cis ammonia groups and the ligand X
2. Magnetic anisotropic effects
3. The reduced polarity of the trans ammine group due to polarization of the metal ion by the ligand X

If the magnetic anistropy of the M—X bond is the important factor, because of the relative orientations of the trans and cis NH_3 groups with respect to the M—X and M—N bonds, $\Delta\chi_{M-X}$ would have to be less positive than $\Delta\chi_{M-N}$: the θ values for the trans NH_3 groups are approximately 0° for M—X and 80° for four M—N bonds; on the other hand, the θ values for

* Note added in proof: In two important recent papers Ho and Reilley have used nucleus-electron isotropic hyperfine contact shift data for a series of nickel(II)-diamine complexes to determine the geometry of the preferred conformations and the energy differences between various structures [F. F.-L. Ho and C. N. Reilley, *Anal. Chem.*, **41**:1835 (1969); *idem. ibid.*, **42**: 600 (1970)].

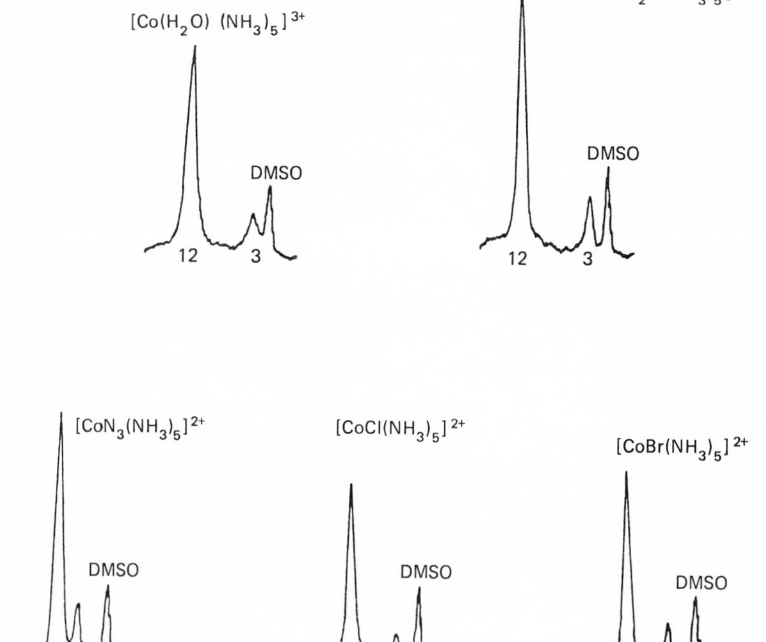

Fig. 6-9 Sixty-megahertz spectra of some pentamminecobalt(III) complexes in dimethylsulfoxide-d_6 [13].

each cis NH_3 group are approximately 80° for M—X and three M—N bonds and 0° for one M—N bond—see Eq. (6-6). Since the chemical-shift difference has been observed for a large range of substituents with widely differing hydrogen-bonding capabilities, it seems unlikely that intramolecular hydrogen bonding is of major importance. The polarization effects would most certainly contribute to the observed separation, but the extent of this contribution is difficult to estimate.

Similar differences in shielding have been found for ethylenediamine and other bidentate and multidentate amines. For example, protons attached to nitrogens trans to the ligand X in the complexes *cis*-$[Coen_2X_2]^{n+}$ [10,17] and *cis*-$[CotrienX_2]^{n+}$ [51,60] resonate at higher field than the cis N—H protons. This separation is quite marked even for *cis*-$[Coen_2(NH_3)_2]^{3+}$, in which any polarization contribution to the splitting should be very small (see Fig. 6-10).

This phenomenon provides an excellent means of distinguishing cis and trans isomers. Take as an example the *cis*- and *trans*-$[Coen_2X_2]^{n+}$ complexes.

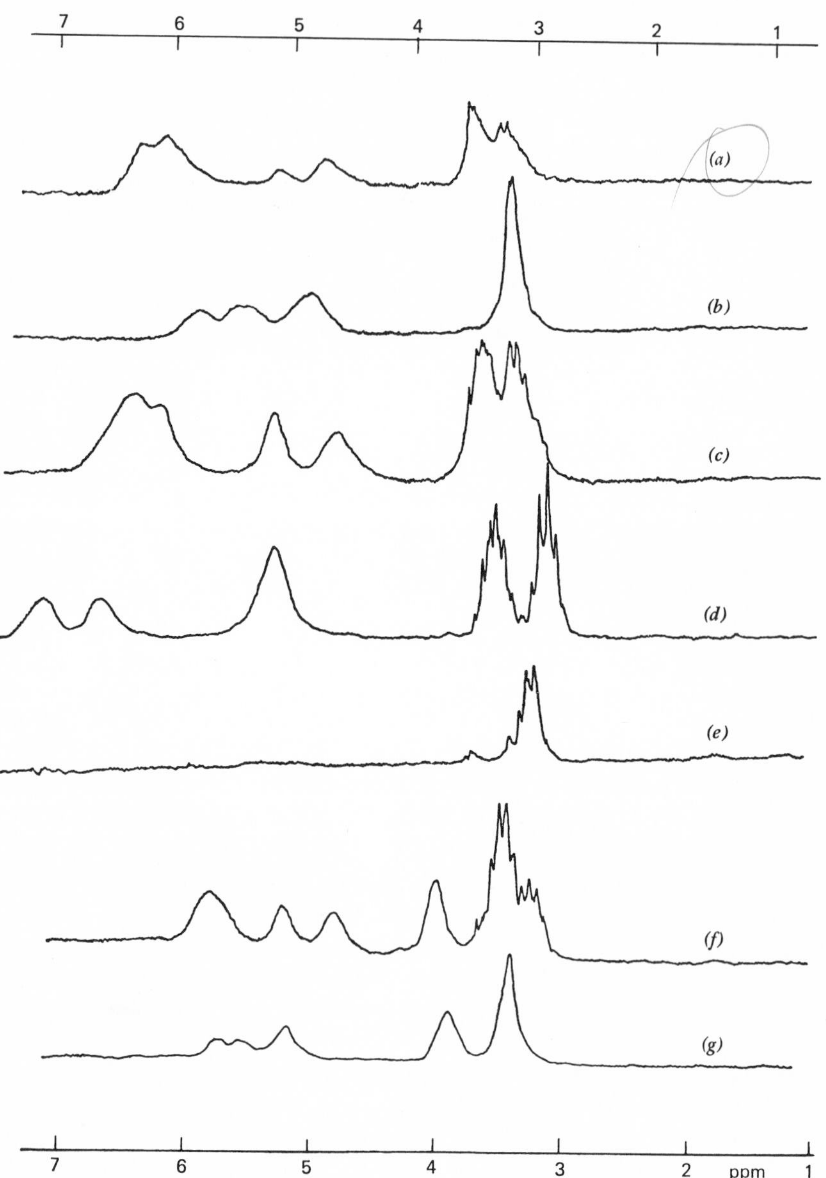

Fig. 6-10 One-hundred-megahertz spectra of *cis*-$[Coen_2X_2]^{n+}$ complexes in 1.8 *M* D_2SO_4, where X_2 is (*a*) Cl_2, (*b*) $(NO_2)_2$, (*c*) $Cl(H_2O)$, (*d*) $(H_2O)_2$, (*e*) $(N_3)_2$ in D_2O, (*f*) $Cl(NH_3)$, and (*g*) $(NH_3)_2$ [10].

There is no marked splitting of the NH_2 resonances in the latter, because each NH_2 group is trans to another NH_2, in contrast to the cis isomers, in which, as we have seen, at least two different NH resonances are observed in the nmr spectra (see Figs. 6-10 and 6-11). Sargeson and Searle [60] have similarly distinguished between the cis α- and β-[Cotrien $X_2]^{n+}$ complexes: the β isomer has five NH resonances with the relative intensities 1:1:1:1:2, whereas the α has three peaks with the intensity ratio 2:2:2 (see Fig. 6-12). These observations are consistent with the different site symmetries of the N protons. Marzilli and Buckingham [51] used the fact that proton chemical shifts for coordinated amines are in the order $NHR_1R_2 > NH_2R > NH_3$ and that the NH protons trans to an electronegative group absorb at higher fields than those that are cis in order to distinguish the β_1 and β_2 isomers of [Cotrien(gly)]$^{2+}$ (see Fig. 5-43). The β_1 isomer shows two broad NH resonances at 7.0 and 6.4 ppm from TMS, corresponding to the NH group trans to NH_2 and COO^-, respectively. Both NH protons resonate at 6.9 ppm for the β_2 because both are trans to $-NH_2$, and the NH_2 group trans to $-COO^-$ is found at highest field.

Nmr spectroscopy has also been successfully applied to the differentiation of the cis and trans isomers of $[M(AB)_3]^{n+}$ (14,15,20,24,25,36]. In the trans isomer the three ligands are symmetrically distinct, and a particular magnetic nucleus should, in principle, have different chemical shifts for the three ligands. In contrast, the three ligands in the cis isomer are equivalent because of their distribution about the C_3 axis, and a single resonance should be observed for the above nucleus. In practice the trans "triplets" do not always appear as separate peaks because two of the resonances are often relatively close together as a result of the near equivalent environments of two of the ligands.

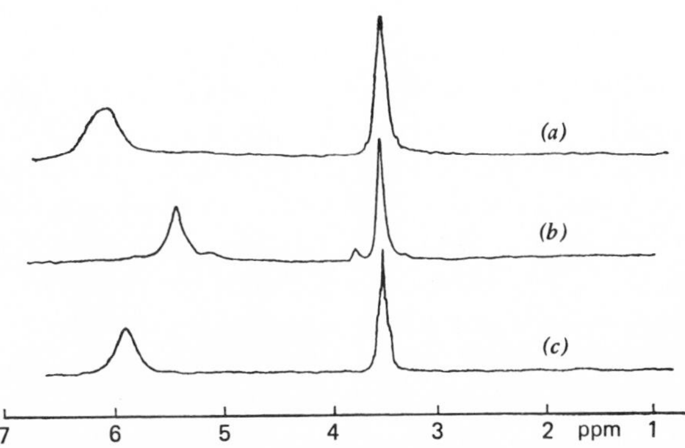

Fig. 6-11 One-hundred-megahertz spectra of *trans*-$[Coen_2X_2]^{n+}$ complexes in D_2O/D_2SO_4 where X_2 is (*a*) $Cl(H_2O)$, (*b*) $(NH_3)_2$, (*c*) Cl_2. [10].

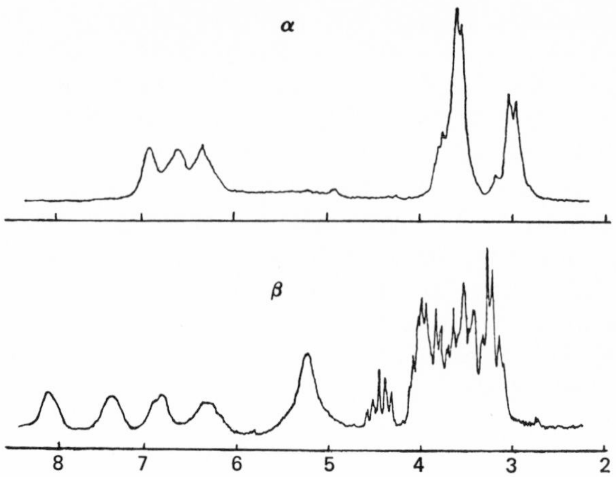

Fig. 6-12 One-hundred-megahertz spectra of α- and β-[Cotrien$(OH_2)_2$]$(C10_4)_3$ in 1.8 M D_2SO_4 relative to TMS (external) [60].

We shall consider two examples. The nmr spectrum of tris(N-methyl-2-hydroxyacetophenimine)cobalt(III) (**LI**) is given in Fig. 6-13:

O
Co/3
C=N
CH_3 CH_3

LI

Three CH_3—C=N signals appear with two of the resonances closely spaced at low field. This allows the complex to be unambiguously assigned the trans configuration [15]. Only two N—CH_3 resonances are observed, and they have the intensity ratio of 1:2, the less intense line appearing at lower field. From a consideration of structures **XXIV** and **XXV**, it can be seen that it is possible to convert the trans isomer to the cis by turning one ligand by 180° while keeping the rest of the molecule fixed. In other words, two of the ligands are symmetrically placed about a pseudo-C_3 axis and probably give rise to the closely spaced resonances.

Tris(L-alaninato)cobalt(III) has been chosen as the second example. The nmr of the cis and trans isomers have been published [20], and the CH_3 resonances are presented in Fig. 6-14. One isomer has a simple doublet with $J = 7$ Hz for the CH_3 resonance, whereas the other shows five peaks with the

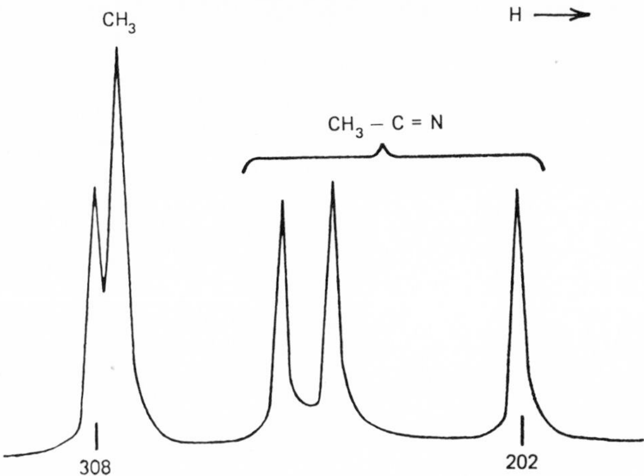

Fig. 6-13 Nuclear resonance spectrum of tris(*N*-methyl-2-hydroxyacetophenimine)-cobalt(III) in $CDCl_3$. Numerical figures refer to chemical shifts in hertz from TMS at 100 MHz [15].

intensity ratio of 1:1:1:2:1. The latter pattern arises from three sets of doublets with $J = 7$ Hz. Again the two C—CH_3 signals at low field have similar chemical shifts. An unambiguous assignment of configuration can be made, based solely on this evidence. It is supported by the CH resonances, which appear for the first isomer (cis) as a simple 1:3:3:1 quartet with $J = 7$ Hz,

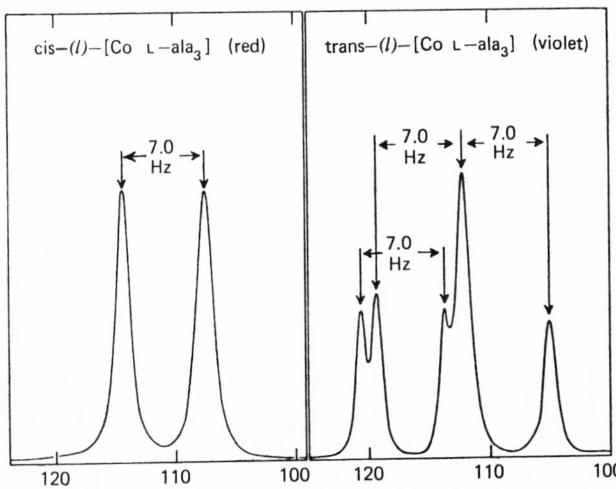

Fig. 6-14 Methyl proton resonances with the chemical shift in hertz downfield from TMS at 60 MHz [20].

and for the second (trans), a complex spectrum, in which three sets of quartets with $J = 7$ Hz can be traced.

Cooke and his coworkers have used nmr to identify the geometrical isomers of a number of iminodicarboxylate complexes. As an example, let us consider the three possible geometrical isomers of (iminodiacetato)diethylenetriamine-cobalt(III), **LII**, **LIII**, and **LIV**, which have been studied by Legg and Cooke [44]:

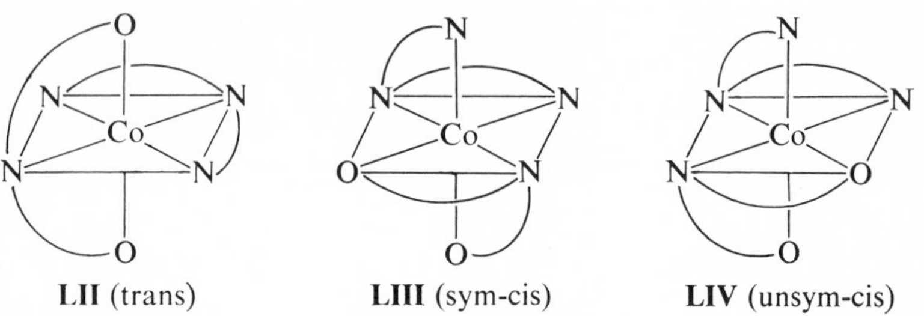

LII (trans) **LIII** (sym-cis) **LIV** (unsym-cis)

The nmr spectra of the three isomers are presented in Figs. 6-15 to 6-17. The spectra are composed of broad NH and quite sharp CH resonances. The sharpness in the CH region is due to the rigidity of the chelate rings. The methylene protons of the acetate rings appear at lower field than the ethylene protons of diethylenetriamine because of the combined deshielding effects on the acetate protons from the —NH_2 and —COO^- groups. The iminodiacetate (IDA) proton orientations in the cis and trans isomers are shown in Fig. 6-18. The H_a and H_b protons are in environmentally different situations and are capable of coupling to produce AB quartets. The number of AB patterns that are found allows an unambiguous assignment of configuration. Structure **LIII** would give rise to one AB quartet because both acetate rings are in the same

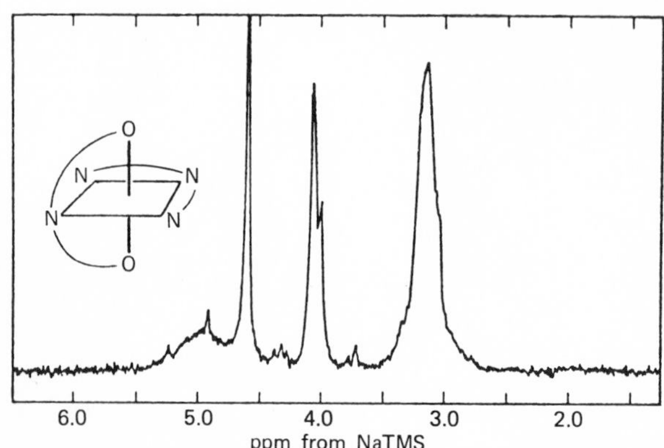

Fig. 6-15 Sixty-megahertz spectrum of *trans*-[Co(dien) (IDA)]$^+$ [44].

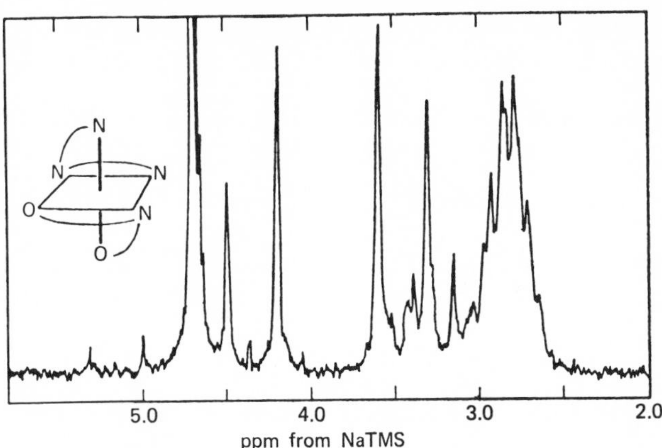

Fig. 6-16 Sixty-megahertz spectrum of sym-*cis*-$[Co(dien)(IDA)]^+$ (trace of HCl present) [44].

environment. On the other hand, structures **LII** and **LIV** would exhibit two AB quartets, perhaps overlapping, because the two acetate rings are in slightly different environments. A single AB pattern is found in Fig. 6-16, and thus this isomer must be the symmetrical cis isomer. Molecular models show that the magnetic anisotropy of the C—N bond in one acetate ring would lead to H_b in the other ring being shielded and H_a deshielded. Therefore the high field doublet at about 3.42 ppm from NaTMS has been assigned to the H_b protons and the low field doublet at about 4.32 ppm to the H_a protons.

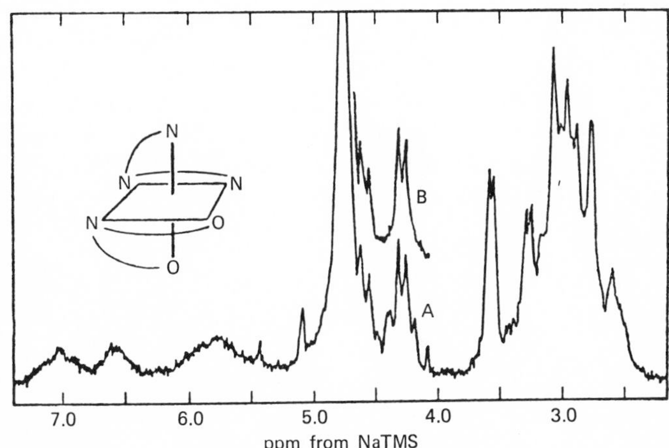

Fig. 6-17 Sixty-megahertz spectra of unsym-*cis*-$[Co(dien)(IDA)]^+$ (trace of HCl present) (A) before amine proton exchange and (B) after amine proton exchange [44].

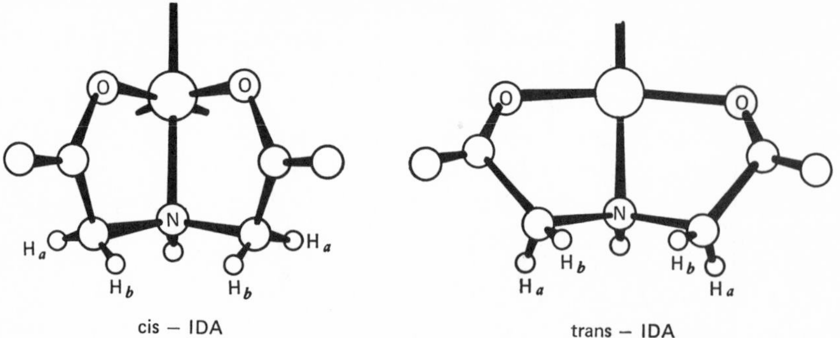

Fig. 6-18 Orientation of IDA in cis and trans configuration.

In the unsymmetrical cis isomer the two acetate methylene groups are in slightly different environments with respect to the diethylenetriamine. Thus in Fig. 6-17 the H_b protons can be seen to give rise to two overlapping doublets with δ values similar to that found for the symmetrical cis compound. The H_a resonances also appear as two overlapping doublets, but at slightly lower field than for the other cis isomer. The NH proton of IDA couples with the low-field acetate methylene protons with $J = 7.8$ Hz, but no coupling was observed with the high-field acetate methylene protons. This supports the prior assignment of the H_a and H_b resonances, because molecular models show that the dihedral angle between C—H_a and N—H is small, whereas the corresponding angle for H_b is approximately 90°, and from the Karplus equation (6-9) it is known that $J_{HH'}$ values are sizable for small dihedral angles but very small for $\phi \simeq 90°$.

The remaining isomer, the trans, has a relatively simple nmr spectrum. This is due to the fact that the H_a and H_b protons are similarly oriented with respect to the C—N bond in the opposite acetate ring and also with respect to other bonds that are likely to contribute significantly to a chemical-shift difference, and thus their doublets are almost superimposed (see Fig. 6-15).

Cooke [18] has also studied the isomers of $[Co(IDA)_2]^-$ (structures **LV**, **LVI**, and **LVII**):

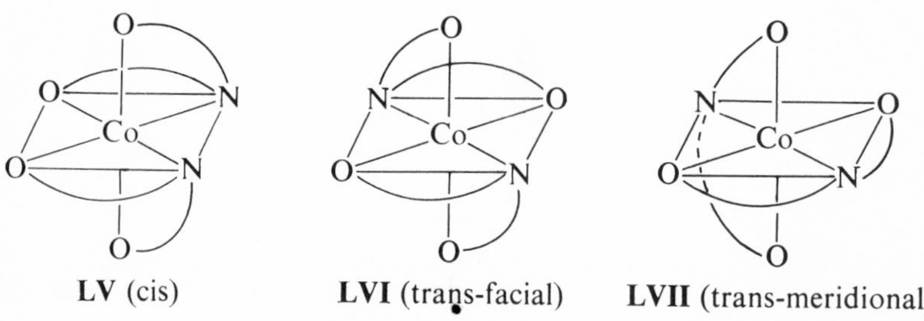

LV (cis) **LVI** (trans-facial) **LVII** (trans-meridional)

In the trans-facial isomer, the plane of symmetry through the two nitrogens and bisecting the ligands must give rise to a single AB quartet. For both the other isomers, the two acetate methylene groups are in slightly different fields and would be expected to show two overlapping AB patterns. The nmr of one of the trans isomers is shown in Fig. 6-19. When the N—H protons are replaced by deuterium, an AB quartet is observed. This must be the trans-facial isomer with H_b and H_a at high and low field, respectively. The undeuterated compound exhibits fine structure from the H—N—C—H coupling. Again J_{NH-CH_b} is found to be small (= 1 Hz) as $\phi \simeq 90°$ and J_{NH-CH_a} to be large (= 7.5 Hz), because ϕ is small. The cis isomer of $[Co(IDA)_2]^-$ shows the expected pattern arising from two overlapping AB patterns (see Fig. 6-20). The undeuterated sample gives rise to a complex fine structure due to H—N—C—H coupling. Cooke [18] has assigned the various peaks, and his assignment is included in Fig. 6-20. The third geometrical isomer (**LVII**) has not been prepared.

Similar reasoning to the above has been used to distinguish the cis and trans isomers of (β-aminoethyliminodiacetato)ethylenediaminecobalt(III) [16]. In the trans isomer the two acetate rings are equivalent and should yield one AB quartet, whereas in the cis, the two are different and two overlapping quartets are expected. The nmr spectra of the two isomers were consistent with this analysis.

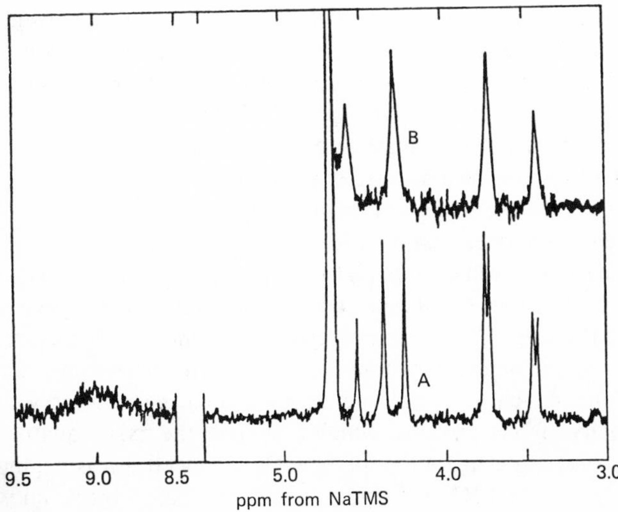

Fig. 6-19 Sixty-megahertz spectra of *trans*-$H[Co(IDA)_2]$, (A) before and (B) after amine proton exchange [18].

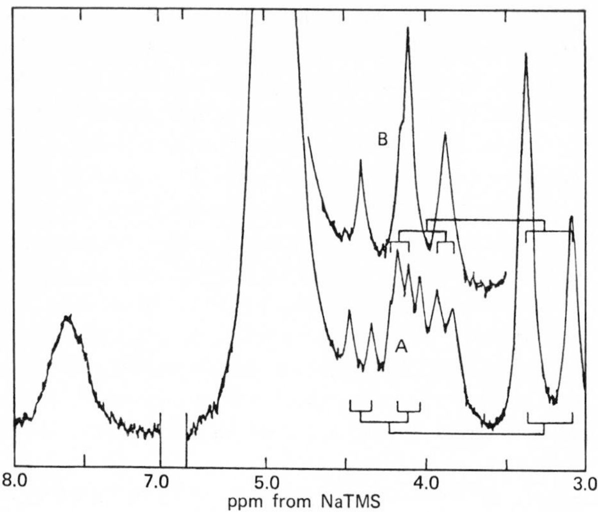

Fig. 6-20 Sixty-megahertz spectra of *cis*-H[Co(IDA)$_2$], (A) before and (B) after amine proton exchange [18].

DIAMINE COMPLEXES

Although chelated diamines have been extensively studied with nmr, the results have been very disappointing as far as stereochemical information is concerned. In most cases it has not even been possible to determine unambiguously whether the spectrum derives from a time-averaged conformation or from conformations that are slow to invert on the nmr time scale.

Consider the complex $[Co(NH_3)_4en]^{3+}$. From conformational-analysis studies it appears that each chirality of the puckered ethylenediamine ring has a variety of conformations available to it with little or no energy barrier between them. They are separated from an enantiomeric range of conformations by an energy barrier that is thought to be of the order of 6 kcal mol^{-1}. Even if the barrier is sufficiently high for the conformation inversion to be slow, the nmr spectrum is still essentially a time-averaged one because of the flexibility of the ring. This time-averaged conformation, however, is most probably of the symmetric skew type, in which there are two equivalent pairs of CH protons, which are axial and equatorial to the ring, as well as two equivalent pairs of NH protons, which also have the axial-equatorial characteristics. Assuming that the coupling with the metal ion is negligible, the system is of the AA′BB′XX′YY′ type. A high-resolution spectrum would be expected to show an extremely complex multiplet system, but if the lines are close enough, they coalesce into a broad envelope showing little or no fine structure. When the nitrogens are deuterated, the system should simplify to an AA'BB' type, but this is still capable of giving rise to 24 lines. Alternatively, if the lifetime of a ring with a particular chirality is short, the CH and NH protons separately become equivalent, and the spectrum should be of the $A_2A'_2X_2X'_2$ type, which would collapse to a single sharp resonance on deuteration of the NH_2 group.

The nmr spectrum of $[Co(NH_3)_4en]^{3+}$ is shown in Fig. 6-21. In 0.1 *M* D_2SO_4 the NH_2 and NH_3 resonances appear as broad bands at 5.25 and 3.5 ppm from NaTMS. Two of the ammonia group are cis to —NH_2 and two are trans. Although the NH_3 resonance is unsymmetrical, no separation of the resonances is obvious for the two kinds of NH_3 groups.* The CH signal, which is found at about 2.8 ppm, shows some fine structure but not sufficient to allow an analysis of the system. On deuteration in D_2O, the spectrum collapses to a single absorption with a half-width of 7 Hz. These results are consistent with rapid conformation inversion if it is assumed that the unexpected width of the CH signal in the deuterated complex is due to coupling with the central cobalt atom ($I = \frac{7}{2}$), the deuterium, or the ^{14}N nuclei ($I = 1$). In other complexes, however, such as $[Coox_2en]^-$ (see Fig. 6-22), the equivalent resonance is much sharper (3 Hz), and this suggests that the coupling with ^{59}Co, ^2H, and ^{14}N is not significant.

When ethylenediamine is replaced by R-propylenediamine, the spectrum retains its overall features, but the CH and NH_3 resonances are more strongly overlapping (see Fig. 6-23). The CH_3 resonance occurs as a sharp doublet at 1.5 ppm with $J_{CH_3—CH} = 6$ Hz. The —CH_2—CH—CH_3 grouping should give

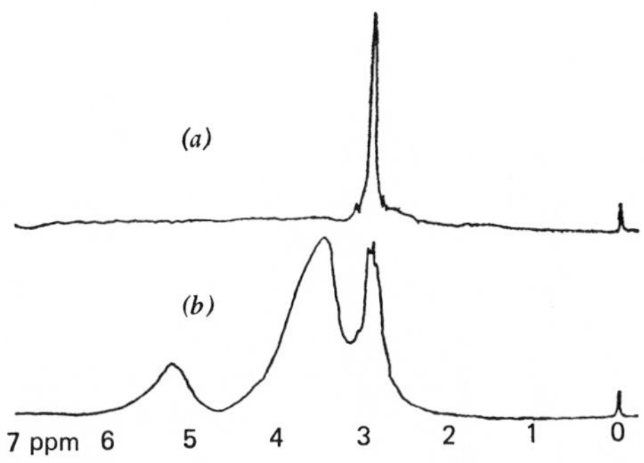

Fig. 6-21 Sixty-megahertz spectra of $[Co(NH_3)_4en]^{3+}$ in (*a*) D_2O and (*b*) 0.1 *M* D_2SO_4 [10].

* Note added in proof: A more recent 60 MHz study has found a separation of 5 Hz between the resonances for the two kinds of NH_3 groups. (C. J. Hawkins and M. M. McKenna, Unpublished results, 1970).

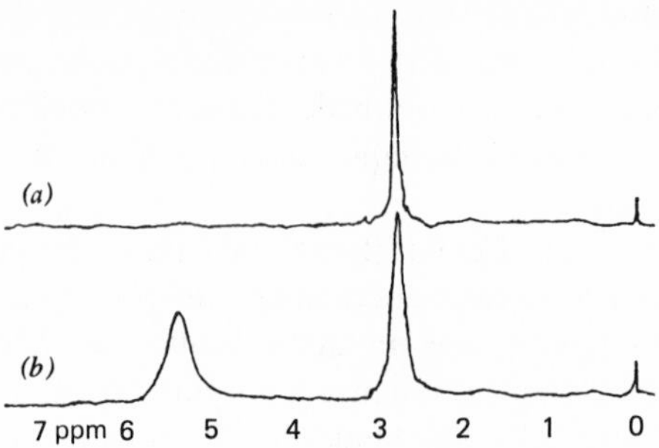

Fig. 6-22 Sixty-megahertz spectra of *K*[Co(ox)$_2$en] in (*a*) D_2O, and (*b*) 0.1 *M* D_2SO_4 [10].

rise to an $ABMX_3$ spectrum, irrespectively of the rate of conformational interchange. The nmr spectrum of the *N*-deuterated complex shows extensive fine structure in the CH region.

In the related complex with the methyl group substituted on the nitrogen atom, $[Co(NH_3)_4Meen]^{3+}$ (see Fig. 6-24), the NH, NH_2, and NH_3 protons

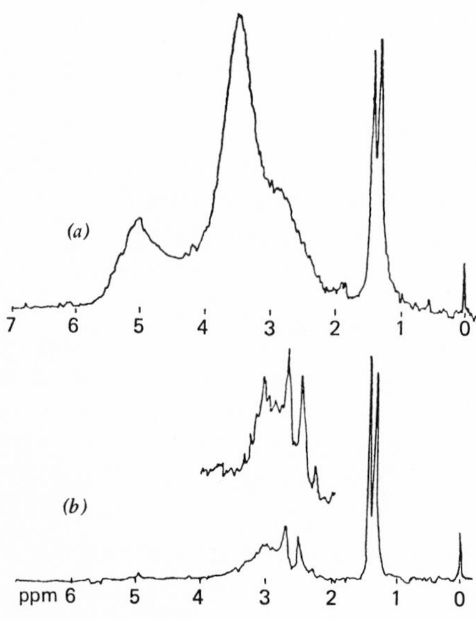

Fig. 6-23 Sixty-megahertz spectra of $[Co(NH_3)_4(l\text{-pn})]^{3+}$ in (*a*) 0.1 *M* D_2SO_4 and (*b*) D_2O [10].

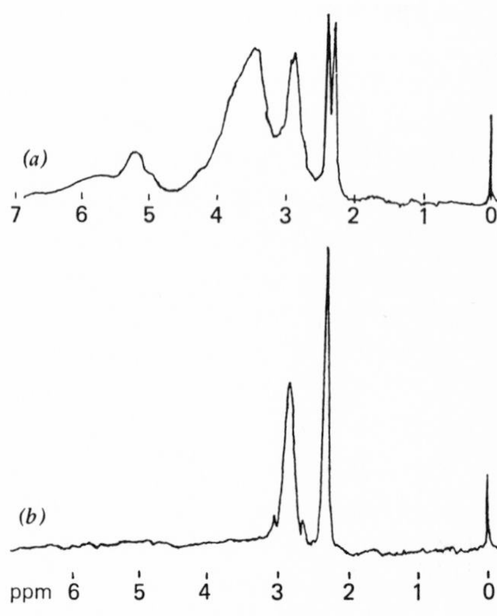

Fig. 6-24 Sixty-megahertz spectra of $[Co(NH_3)_4(Meen)]^{3+}$ in (*a*) 0.1 *M* D_2SO_4 and (*b*) D_2O [10].

resonate at 5.8, 5.2, and 3.5 ppm from NaTMS. The CH_3 signal is found at 2.35 ppm as a doublet with $J_{CH_3-NH} = 6$ Hz. On deuteration the methyl signal collapses to a single line. The —CH_2—CH_2— grouping is of the ABCD type, irrespectively of the rate of ring inversion. The deuterated spectrum shows only a relatively broad central band flanked by two satellite peaks.

The methyl resonance of propylenediamine has been used to assign absolute configurations to the isomers of $[Coox_2(l\text{-pn})]^-$ [9]. The nmr spectra of the racemic and the resolved species are shown in Fig. 6-25. Froebe [26] found from a study of platinum(IV) complexes containing *l*- and *d*-propylenediamine that δ_{CH_3} for a methyl group attached to a C—C bond axis that is oblique to a pseudo-C_3 axis of the complex ion is greater than δ_{CH_3} when the C—C axis is parallel to the pseudo-C_3 axis. Because δ_{CH_3} for l-$[Coox_2(l\text{-pn})]^-$ is less than δ_{CH_3} for the *d* isomer, according to this proposal, the *l* isomer must have the L configuration because *l*-propylenediamine has the λ conformation.

Bis(ethylenediamine) complexes have been extensively studied [10,17,27, 33,41]. First, let us consider *trans*-$[Coen_2X_2]^{n+}$. Rapid conformational interchange should lead to the CH and NH protons becoming equivalent just as found for $[Co(NH_3)_4en]^{3+}$. The CH resonance for *trans*-$[Coen_2Cl_2]^+$, however, shows a quintet structure that arises from an $A_2A'_2X_2X'_2$ coupling between the CH and NH protons. It is possible to observe this even if conformational interchange is rapid [27]. As stated above, the NH protons show only one broad band.

The 60-MHz spectrum of $[Pten_2]Cl_2$ in D_2O exhibits a sharp (2 Hz) CH band with satellite peaks because of ^{195}Pt—CH coupling ($J_{Pt-CH} = 41$ Hz). The band is, in fact, sharper than that for $[Pt(ND_2-CH_3)_4]Cl_2$ (2.5 Hz) (see Fig. 6-26) [6]. The 100-MHz spectrum of the ethylenediamine complex also shows a sharp resonance that does not change over the temperature range of 2.5 to 92° [21]. In D_2SO_4/D_2O a broad NH_2 resonance with ^{195}Pt satellites ($J_{Pt-NH} = 55$ Hz) is observed at 60 MHz, and the CH_2 band is found to be a quintuplet with the fine structure reproduced in the ^{195}Pt satellites [6]. This evidence strongly supports the suggestion that the conformations invert rapidly. Eidson and Liu [21] have reported, however, that they were able to resolve $[Pten_2]^{2+}$, and the inversion had a half-life of approximately 1 hr. If this is indeed the case, then the nmr observations above can be rationalized only if the axial and equatorial protons resonate at the same field.

The cis complexes show a variety of spectra for the —CH_2—CH_2— region. The two ethylenediamine rings are related by a C_2 axis of symmetry. In each ring one CH_2 group is bonded to a NH_2 that is trans to X, and the other is next to a NH_2 group trans to —NH_2. Further, under no circumstances do the two protons of each CH_2 group become environmentally equivalent, although the signals could accidentally coincide. The complex with X = D_2O shows the most interesting fine structure. According to Henney and coworkers [33]

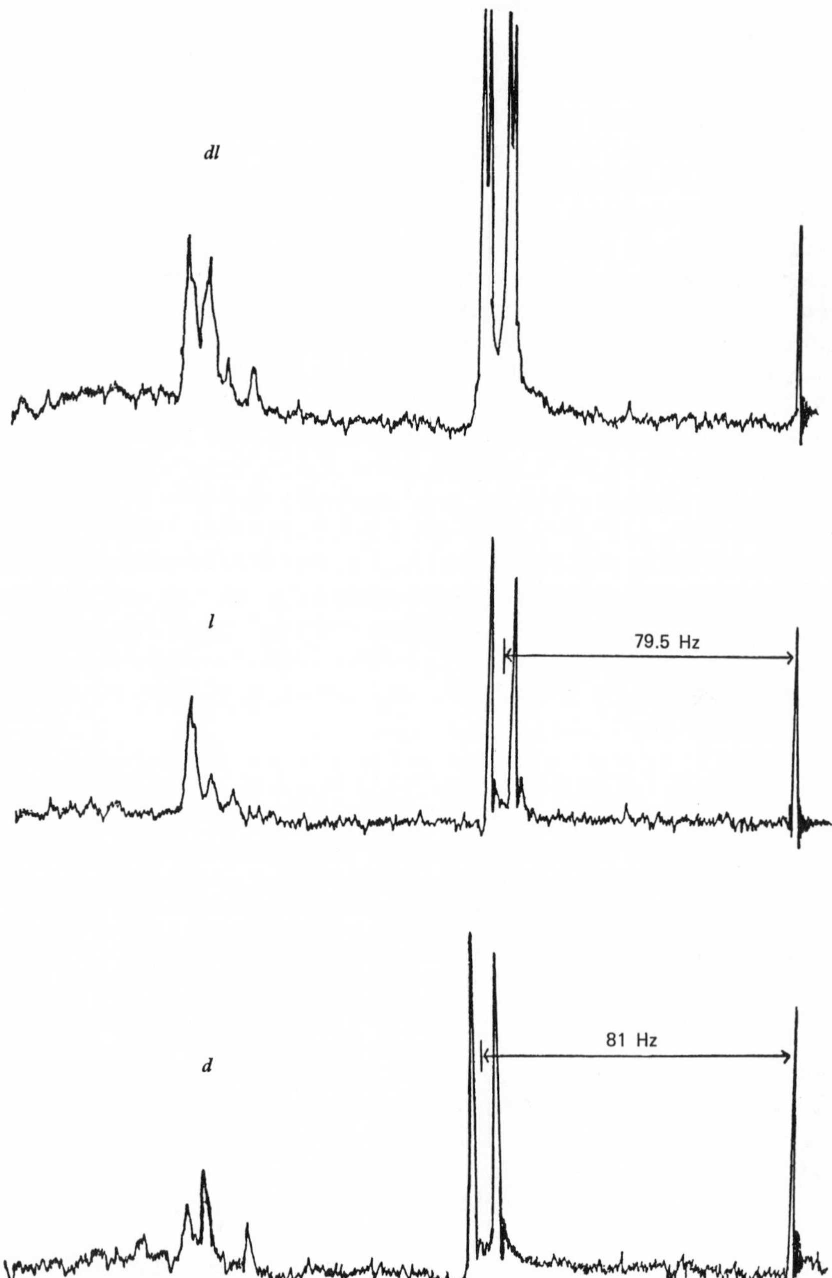

Fig. 6-25 Nmr spectra of racemic and optically active Na[Coox$_2$(*l*-pn)] in D_2O with NaTMS as internal reference [9].

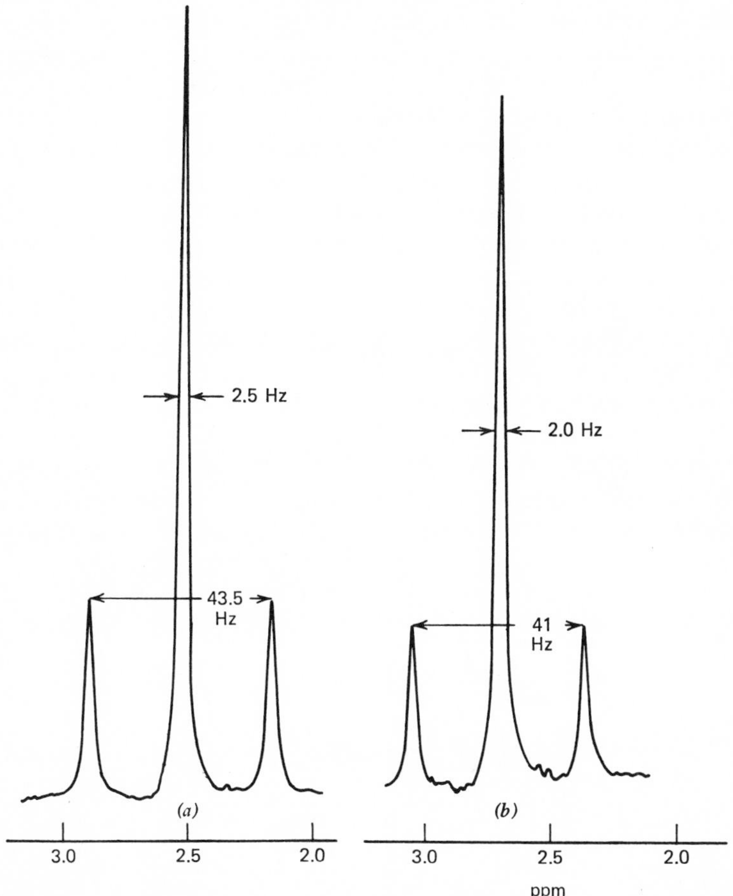

Fig. 6-26 Sixty-megahertz spectra of (*a*) $[Pt(ND_2CH_3)_4]Cl_2$, and (*b*) $[Pt(ND_2CH_2\text{-}CH_2ND_2)_2]Cl_2$ in D_2O [6].

the 60-MHz spectrum in 2 *M* perchloric acid shows two overlapping quintets and, on deuteration, two interacting triplets. Buckingham, Durham, and Sargeson [10], however, have reported a 100-MHz spectrum in which the signal appears as two overlapping seven-line multiplets (see Fig. 6-10). Sargeson [59] claims that this resembles an A_2B_2 system and that the same result is obtained when the NH_2 groups are deuterated in D_2O. He further suggests that the extensive fine structure for the D_2O complex derives from the retention of conformation, which is brought about by hydrogen bonding involving the D_2O and —NH_2 groups. Extensive fine structure has also been observed for *cis*-β-[Cotrien$(D_2O)_2]^{3+}$ [60]. As stated above, however, conformation retention is not necessary for the CH_2 protons to remain nonequivalent. The greater degree of fine structure for the D_2O complexes could perhaps be due to the chemical-shift difference between A,A′ and B,B′ being greater for D_2O than for other ligands.

The NH region usually shows at least three bands. Consider the complex in Fig. 6-27. In each ethylenediamine ring there are four differently oriented NH protons. H_A and H_B are attached to N trans to NH_2 and thus should be less shielded than H_C and H_D, which are attached to N trans to X. If the magnetic anisotropies of M—X and M—N ($\Delta\chi$) are positive, a hydrogen located directly above the M—X or M—N bond would be shielded relative to a hydrogen positioned along the M—X or M—N axis. Thus, H_A would be shielded by two M—X bonds and to a lesser extent by two M—N bonds; H_B would be primarily shielded by a M—X and a M—N bond and to a lesser extent by a second set of M—X and M—N bonds; H_C would also be shielded by a M—X and a M—N bond but, in contrast to H_B, would be influenced to a smaller degree by two other M—N bonds; and finally, H_D would be shielded first by two M—N bonds and second by a M—X and another M—N

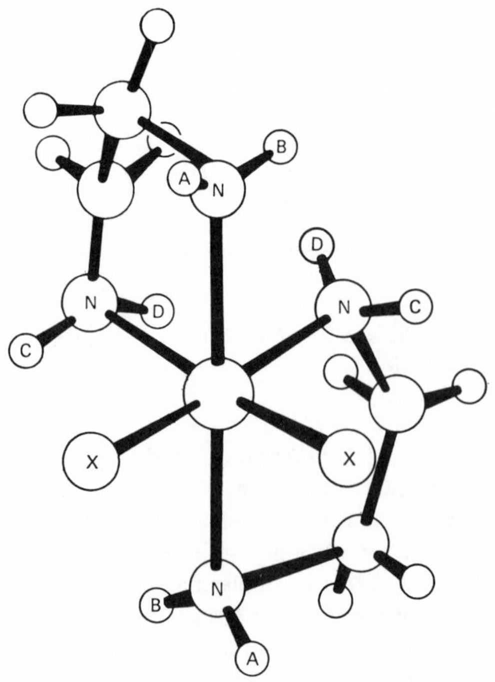

Fig. 6-27 Cis-$[Coen_2X_2]^{n+}$.

bond. In addition to these influences, H_D would be affected by the ring current in the other ethylenediamine chelate ring, because it lies directly above the ring. As a result of all these factors, it is not surprising to find that the N—H resonance pattern changes from compound to compound (see Fig. 6-10). The 100-MHz spectrum of *cis*-$[Coen_2Cl_2]^+$ shows NH absorption at 4.8, 5.1, and 6.3 ppm (2:2:4) with the low-field band showing some incipient splitting. For the corresponding diammine, three bands are again observed, but the four-proton band is at high field.

Sargeson [10,59] has suggested that the NH spectrum can be rationalized in terms of "three kinds of proton, (a) those that see essentially X groups (2 protons), (b) those that see only NH_2 groups (2 protons), and (c) those that see both X and NH_2 groups (4 protons)."

Lantzke and Watts [41] have analyzed the spectra in a different way. They have classified the NH protons according to whether the proton's time-averaged position is on the same side of a plane through the cobalt atom at right angles to the Co—X bond (proximal) or on the far side (distal). They claim that proximal and distal protons are deshielded and shielded by X, respectively. Further, because the environment of H_B is similar to that of the NH protons in the trans isomer, Lantzke and Watts have assigned H_B to the resonance that has the same chemical shift as the NH protons in the trans complex. Where the four NH resonances are observed, they have assigned the band at highest field to H_D because it is distal to both X ligands and is bonded to a nitrogen trans to X. The next resonance has been assigned to H_C, which is on the same nitrogen as H_D but is proximal to one X ligand. The remaining resonance is allocated to H_A.

The tris(ethylenediamine) and related complexes have also been studied extensively [6,27,56,63,67,69]. Woldbye [67] found that the nmr spectrum of tris(meso-2,3-diaminobutane)cobalt(III) in D_2O showed two doublets of equal intensity with δ values of 1.73 and 1.85 ppm. The equivalent *d*-2,3-diaminobutane [67] and *l*-propylenediamine [63] complexes show one doublet at approximately 1.85 ppm. The 1.85- and 1.73-ppm resonances were ascribed to the equatorial and axial methyl groups. These results have been taken to indicate that the conformational inversion in these complexes is slow, for if the inversion was rapid it was thought that the two methyl resonances in the meso-2,3-diaminobutane complexes would yield a time-averaged single doublet [59,67]. However, the assumption is in error; the two methyl groups are magnetically nonequivalent at all stages of the ring inversion; hence the observation of two methyl resonances does not allow a conclusion to be drawn concerning the rate of ring inversion. It is interesting to note that for $[Co(NH_3)_4(meso\text{-}bn)]^{3+}$, in which the two methyl groups are indistinguishable on ring inversion, the nmr spectrum observed in this laboratory and by Saito and his coworkers [68] shows only one sharp methyl doublet.

The NH and CH resonances for $[Coen_3]^{3+}$ are broad and poorly resolved at room temperature and at $-39°$ [63]. Warming the sample to $+70°$ tended to sharpen the resonances [69]. This supports the proposal that the conformational interchange is slow at room temperature. Deuteration of the NH_2 groups did not affect the CH band greatly: the half-width decreased from 21 to 18 Hz. If the conformational interchange is slow, the broad CH resonance must be an envelope of a complex system of multiplets. The CH resonances are much sharper for some other metal ions—rhodium, 3.5 Hz, platinum(IV) 1.3 Hz [6], for example—and if the inversion of the conformations is slow for these complexes, δ_{ae} (the chemical-shift difference between the axial and equatorial protons) must be very small.*

If the ring inversion is rapid on the nmr time scale, the observed nmr spectra are those of the equilibrium populations of conformers, and are determined by the *effective* chemical shift difference between non-equivalent protons together with the effective coupling constants between these protons. If each configuration were of equal energy, this effective chemical shift difference could result from the inherent magnetic non-equivalence of the geminal protons. Alternatively, if various configurations are not of equal energy, the effective chemical shift difference will depend on the populations of the various conformations. For example, if the D($\delta\delta\delta$) and D($\delta\delta\lambda$) forms are of equal energy and significantly more stable than the D($\delta\lambda\lambda$) and D($\lambda\lambda\lambda$) configurations, each proton will be in a δ conformation $\frac{5}{6}$ of the time and in a λ conformation, $\frac{1}{6}$ of the time, with the result that the effective chemical shift difference between the axial and equatorial protons will be $\frac{2}{3}$ of the *intrinsic* chemical shift difference for a fixed conformation. The relatively broad cobalt(III) methylene resonance could be partly due to a large intrinsic chemical shift difference. However, it could also result from the coupling of the protons with the cobalt nucleus.†

Sargeson [59,63] has suggested that the appearance of two NH resonances for $[Coen_3]^{3+}$ is due to the presence of molecules containing unfavorable conformations—D($\delta\delta\lambda$)-$[Coen_3]^{3+}$, for example. This conclusion followed the study of a series of ethylenediamine and propylenediamine complexes. D-[Coen$_2$(R-pn)]$^{3+}$, D-[Coen(R-pn)$_2$]$^{3+}$, and D-[Co(R-pn)$_3$]$^{3+}$ have in their

* Note added in proof: Elsbernd and Beattie have recently reported the nmr spectrum of nitrogen-deuterated $[Ruen_3]^{2+}$ which shows a very well resolved fine structure for the methylene proton resonances typical of an AA′BB′ multiplet [H. Elsbernd and J. K Beattie, *J. Am. Chem. Soc.*, **91**:4573 (1969)]. This structure persisted even at 100°. The spectrum has been analyzed to give the following coupling constants: $J_{H_A-H_B} = -12.5$, $J_{H_A-H_{A'}} = 9.4$, $J_{H_A-H_{B'}} = 3.8$, and $J_{H_B-H_{B'}} = 5.5$ Hz. [J. K. Beattie and H. Elsbernd, *J. Am. Chem. Soc.*, **92**:1946 (1970)]. These results are consistent with a rather strongly puckered conformation with ω about 63°.

† Note added in proof: Beattie has recently presented evidence to support this proposal (J. K. Beattie, Personal communication, 1970).

most stable configurations the conformations, $\delta\delta\lambda$, $\delta\lambda\lambda$, and $\lambda\lambda\lambda$, respectively, because R-propylenediamine prefers the λ conformation irrespectively of the overall chirality of the complex. These complexes all show two NH resonances in contrast to the L isomers, which, according to Sargeson, have the $\lambda\lambda\lambda$ configuration and which exhibit only one NH resonance. Fung [27] has criticized Sargeson's proposal on the basis that Yoneda and Morimoto [69] have shown that the two NH bands are, in fact, of the AB type, because the two bands have shoulders, and on increasing the field strength, the separation between the two main peaks increases while the separation between each main peak and its shoulder remains constant. From a study of the variation of the NH signals with temperature, Fung estimated that the ring-inversion activation energy is approximately $+10.5$ kcal mol^{-1}. However, it must be emphasized that, for these tris complexes, the protons in each NH_2 group do not become equivalent if the conformation interchange is fast. One of the protons is always above a chelate ring and feels the shielding effects associated with the ring, whereas the second proton is always directed away from the chelate ring.

From this survey of the five-membered diamine chelate-ring systems, it appears that it is possible to obtain stereochemical information from the NH resonances. Unfortunately, nuclear quadrupole relaxation effects due to ^{14}N, as well as NH—CH coupling, make the resonances broad. The bands could be sharpened by decoupling either the ^{14}N nucleus or the CH protons. Although, at present, only limited information has been gained from the CH absorptions, detailed analyses of the CH regions of high-resolution spectra of model complexes containing rigid chelate rings or of systems for which rapid conformation inversion does not lead to the protons becoming equivalent could yield useful data on the geometries of chelate rings.

Conformational information has been obtained by an analysis of the CH resonances for two five-membered diamine systems: bis(L-α,β-diaminopropionato)platinum(II) [22], and bis(R-propylenediamine)palladium(II) [68]. For both systems the ligands are expected to favor the conformation with the substituent equatorial. If this preference is substantial, the resultant H—H coupling scheme will reflect the geometry of the ring, whether or not the ring is inverting rapidly, because under no circumstances will the CH protons become equivalent and the time-averaged signal will closely approximate that of the preferred conformation.

The two possible chiralities for the L-α,β-diaminopropionate chelate ring are shown in Fig. 6-28. When the carboxylate group is equatorial, the Pt—N—C—H_X and Pt—N—C—H_A dihedral angles are both approximately 90°, whereas the Pt—N—C—H_B dihedral angle is close to 150°. If the Karplus relationship, Eq. (6-9), holds for this system, one would expect that J_{Pt-H_X} and J_{Pt-H_A} would be equal but much less than J_{Pt-H_B}. Further, when —COO^- is equatorial, H_X is approximately trans to H_A and gauche to H_B and therefore

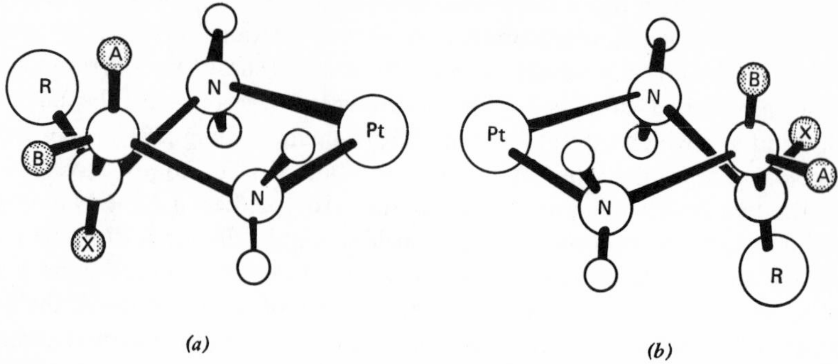

Fig. 6-28 Two chiralities for the (L-α, β-diaminopropionato)platinum(II) chelate ring where R is —COO$^-$.

$J_{H_A-H_X}$ would be much larger than $J_{H_B-H_X}$. On the other hand, if the axial conformation were the most preferred, the Pt—N—C—H_X and Pt—N—C—H_A dihedral angles would be about 150°, and the Pt—N—C—H_B angle about 90°, which would make J_{Pt-H_X} and J_{Pt-H_A} equal but greater than J_{Pt-H_B}. In addition, because H_X would be gauche to both H_A and H_B, $J_{H_A-H_X}$ and $J_{H_B-H_X}$ would be equal. The observed 60-MHz spectrum and Erickson's analysis are shown in Fig. 6-29. The coupling constants were estimated to be $J_{H_A-H_B} = -12.8$ Hz, $J_{H_A-H_X} = 9.8$ Hz, $J_{H_B-H_X} = 4.8$ Hz, $J_{Pt-H_X} = 26.4$ Hz, $J_{Pt-H_A} = 28.4$ Hz, $J_{Pt-H_B} = 56.4$. It is obvious from these results that —COOH is equatorial because H_A and H_B are not equally coupled with H_X and the ^{195}Pt—H_X and ^{195}Pt—H_A coupling is much weaker than that between Pt and H_B. Because the coupling constants were found to be independent of the state of ionization of the carboxylate group and also of the geometrical isomeric form of the bis complex, the geometry of the conformation must not be sig-

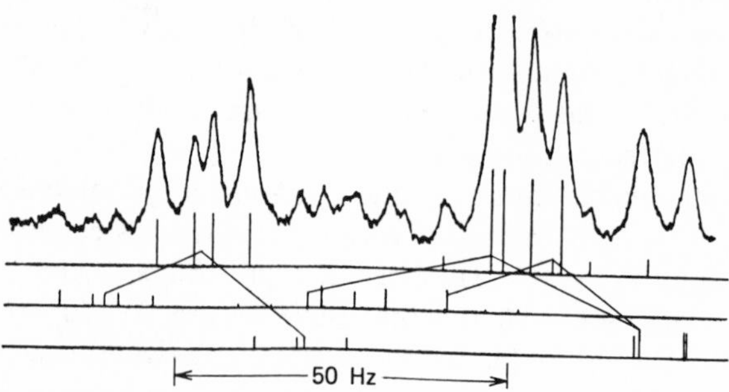

Fig. 6-29 Sixty-megahertz spectrum of $[Pt(L\text{-}DapH)_2]Cl_2$ in D_2O [22].

nificantly influenced by electrostatic interactions involving the negatively charged carboxylate group and by any interactions with the other chelate ring.

The nmr spectrum of $[Pd(R\text{-pn})_2]Cl_2$ in D_2O consists of a doublet at 1.22 ppm due to the methyl group, two quartets (2.53 {A} and 2.81 {B} ppm) from the two methylene protons, and a multiplet at 3.18 ppm due to the methine proton [68]. When the methyl resonance is irradiated, the multiplet collapses to a quartet {X}. If the chelate ring is λ, geminal, trans, and gauche coupling constants would be expected, but, if it is δ, only one geminal and two gauche couplings would be found. The spectrum has been analyzed to give $J_{H_A-H_X}$ = 9.9, $J_{H_B-H_X}$ = 4.2 and $J_{H_A-H_B}$ = −12.5 Hz. This is consistent with the λ conformation with A and B the geminal methylene protons (A, axial, and B, equatorial), and X the methine proton. As found for other systems, the axial proton resonated at higher field than the equatorial. The coupling constants obtained for this system are similar to those found for the L-α,β-diamino-propionate complex.

Erickson and his coworkers [22] have also studied a six-membered chelate ring system, bis(L-histidine)platinum(II). The two boat conformations shown in Fig. 6.30*a* and *b* were assumed to be the only two that had to be considered.

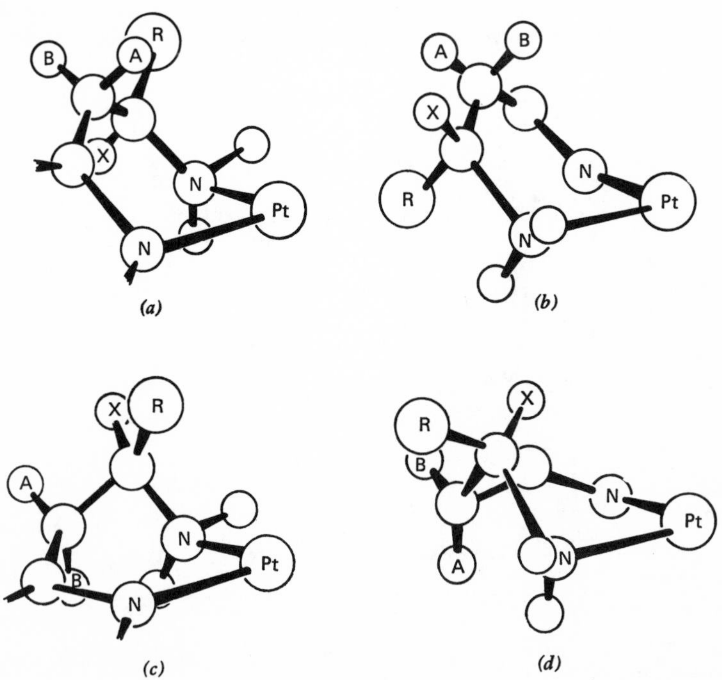

Fig. 6-30 (L-Histidine)platinum(II) chelate rings where R is the carboxylate group.

In these conformations, however, certain of the groups are in eclipsed orientations, and although no detailed conformational analysis has been carried out, it would appear that the alternative conformations, *c* and *d*, are more likely to have the lower energy. The complex has been studied at low pH, where the carboxylate group is protonated, and at about pH = 12, where the carboxylate and imidazole NH are ionized. An analysis of the nmr spectra, given in Fig. 6-31, shows that at low pH the couplings of H_A and H_B with H_X are equal and relatively weak (~4.6 Hz), whereas, at high pH, H_X is coupled much more strongly to one of the two methylene protons (~8 vs ~3 Hz). The low pH form has either structure *b* or *c* because both have H_X—C—C—H_A and H_X—C—C—H_B dihedral angles approximately equal and of the order

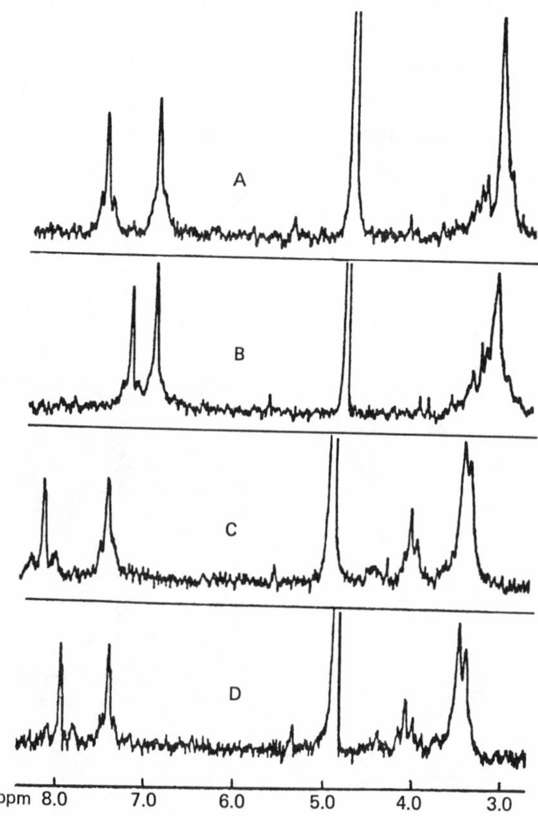

Fig. 6-31 Sixty-megahertz spectra of L-histidine complexes of platinum(II) in D_2O: (A) cis-[Pt(L-his)$_2$] at pH > 12; (B) *trans*-[Pt(L-his)$_2$] at pH > 12; (C) *cis*-[Pt(L-his)$_2$] at pH < 0; (D) *trans*-[Pt(L-his)$_2$] at pH < 0 [22].

of 60°. Both structures *a* and *d* could be used to explain the high pH data, because H_A is trans to H_X and H_B gauche to H_X in both conformations.

Other six-membered chelate ring systems have been studied [6], but in general no definitive evidence has been obtained regarding preferred conformations or, indeed, the rate of conformational interchange. Nevertheless, just as low temperature studies of cyclohexane and its derivatives have led to stereochemical information regarding that system, it is possible that similar studies with the complex six-membered ring systems might prove invaluable.

AMINO ACID COMPLEXES

Conformational analysis has predicted that the amino acid chelate-ring systems are, in general, less puckered than similar diamines, and the energy differences between the two chiralities for *C*- and *N*-substituted chelates are much smaller than for the corresponding diamines. The barrier to ring inversion is also likely to be much less than for the diamines, and, thus, the nmr spectra are much more likely to be time-averaged. Because the preference for one particular chirality is small, the resultant spectra do not reflect so closely the geometry of any one conformation. However, the analyses of the spectra are capable of indicating which chirality is preferred.

Erickson and his coworkers have interpreted the nmr of a number of platinum(II) complexes of amino acids, including sarcosine, aspartic and glutamic acids [22]. The spectra of the sarcosine complexes $[Pt(NH(CH_3)CH_2COO)Cl_2]^-$ and $[Pt(ND(CH_3)CH_2COO)Cl_2]^-$ are given in Fig. 6-32. The CH_2 signal of the former compound consists of eight lines typical of the ABX system, where X is the NH resonance at 5.97 ppm. Platinum (^{195}Pt) satellites are present but indistinct. The corresponding absorption in the second compound is basically an AB quartet with the ^{195}Pt side bands superimposed. The methyl proton signal, which appears as a doublet with its two doublet ^{195}Pt satellites, collapses to a singlet flanked by satellites in the spectrum of the deuterated complex. The spectra were analysed by Erickson and his coworkers [22] to give $\delta_{H_A} = 3.48$, $\delta_{H_B} = 3.98$, $\delta_{CH_3} = 2.71$, $\delta_{H_X} = 5.97$, $J_{H_AH_B} = 17.0$, $J_{H_AH_X} = 4.5$, $J_{H_BH_X} = 6.5$, $J_{Pt-H_A} = 38.5$, $J_{Pt-H_B} = 33.5$, $J_{Pt-CH_3} = 43.5$, and $J_{H_X-CH_3} = 6.0$, with the δ values in parts per million and the coupling constants in hertz. Although it is impossible to assign the CH resonances unequivocally to the individual methylene protons, molecular models show that H_A in Fig. 6-33 probably resonates at a higher field than H_B mainly because of the anisotropic shielding of the N—CH_3 bond. If this is correct, this proton is coupled more strongly with ^{195}Pt and more weakly with H_X than the other methylene proton. If a Karplus relationship is applicable to this system, these results support a conformation of the kind shown in Fig. 6-33, in which the Pt—N—C—H_A dihedral angle is closer to 180° than

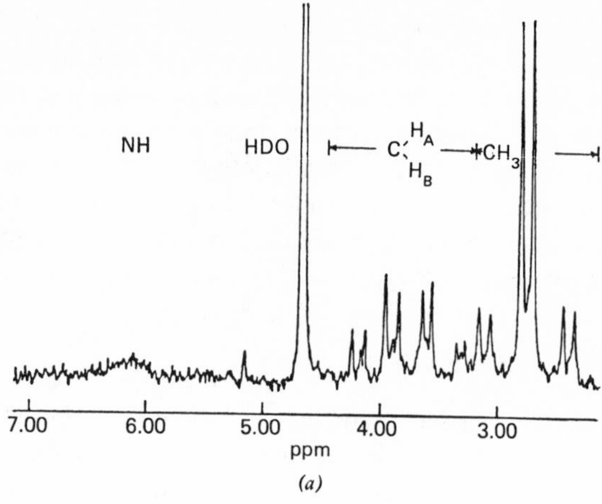

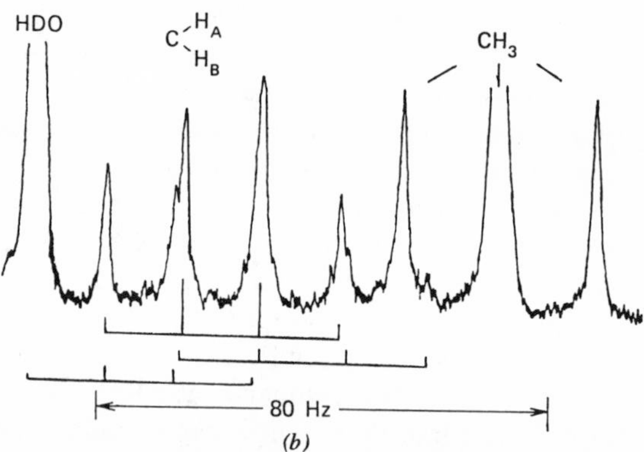

Fig. 6-32 Sixty-megahertz spectra of (*a*) $K[Pt(NH(CH_3)CH_2COO)Cl_2]$ in 0.01 *M* D_2SO_4 and (*b*) $K[Pt(ND(CH_3)CH_2COO)Cl_2]$ in D_2O [22].

Pt—N—C—H_B, and H_A—C—N—H_X is closer to 90° than H_B—C—N—H_X. It is to be noted that in this conformation the methyl group is more axial than equatorial. It must be remembered, however, that this conclusion depends on the assignment of H_A to the higher-field methylene resonance, and this assignment could be in error.

The spectra of 1:2 platinum(II) complexes of L-aspartate (NH_2—CH-(COO^-)—CH_2—COO^-) and L-glutamate (NH_2—CH(COO^-)—CH_2—CH_2—COO^-) in D_2O are given in Fig. 6-34. The α-proton signal for both complexes appears as a triplet flanked by the ^{195}Pt side bands. The ^{195}Pt—H_α coupling constants were found to be relatively small (20 Hz for aspartate and 26.7 Hz for glutamate) compared with values obtained for the Dap and

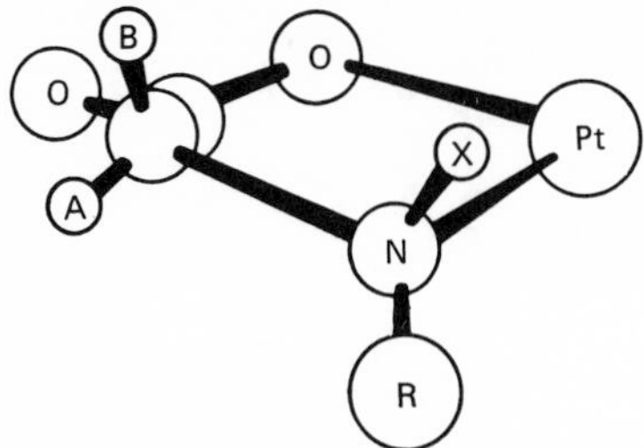

Fig. 6-33 Sarcosinatoplatinum(II) chelate ring with R = CH_3.

sarcosine complexes. This is consistent with the structure given in Fig. 6-35, which has the δ conformation with the substituent in an equatorial orientation.

Cobalt(III) complexes of amino acids have also been studied, but little conformational information has been obtained [10,11,12,20,32]. However, other important information has been forthcoming.

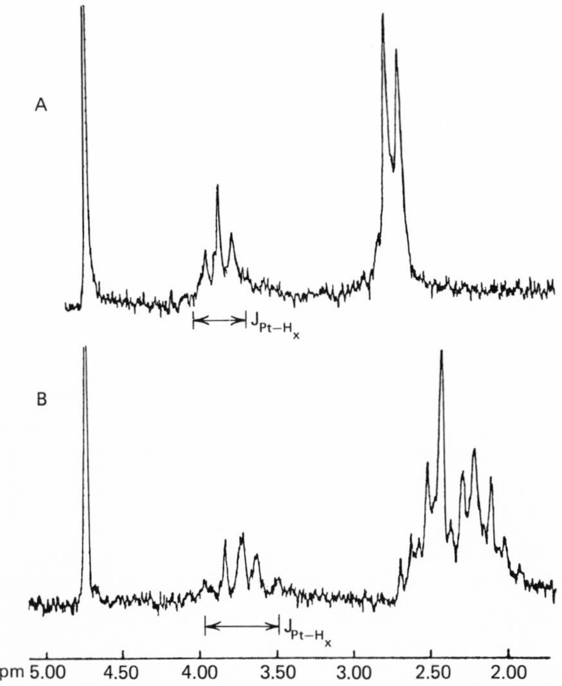

Fig. 6-34 Sixty-megahertz spectra of (A) $[Pt(L\text{-asp})_2]^{2-}$ and (B) $[Pt(L\text{-glu})_2]^{2-}$ in D_2O [22]

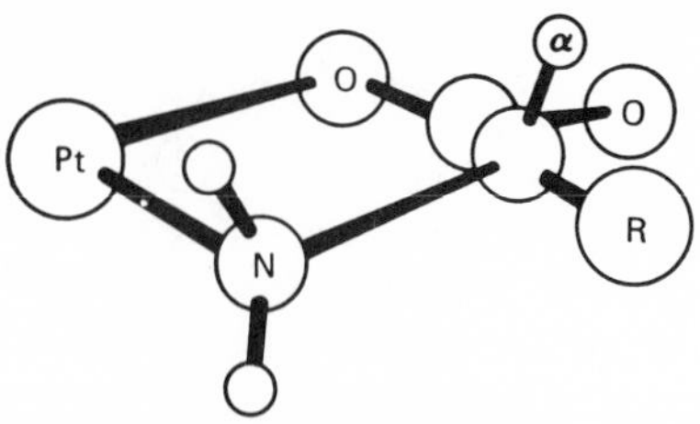

Fig. 6-35 (L-aspartato)- and (L-glutamato)platinum(II) chelate ring with R = $-CH_2$-COO^-, and $—CH_2CH_2COO^-$, respectively.

Denning and Piper [20] were able to assign absolute configurations to the two trans isomers of [Co(L-leu)$_3$] based on their nmr spectra. The spatial distributions of the $—CH_2—CH(CH_3)—CH_3$ groups in the two isomers are shown in Fig. 6-36. If the alkyl groups are in equatorial orientations, in the L isomer the C—C bond axes are oblique to the pseudo-C_3 axis, and the alkyl groups are, in the main, well away from other groups in the molecule. The two methyl groups of each ligand are not related by a C_2 axis, but because the group is no doubt freely rotating, they experience virtually the same average environment. Further, two of the ligands are similarly oriented about the pseudo-C_3 axis, and, therefore, although all the methyl protons are expected to have similar chemical shifts, it is possible that two sets of protons have δ values slightly different from the third. *l-trans*-[Co(L-leu)$_3$] shows this behavior

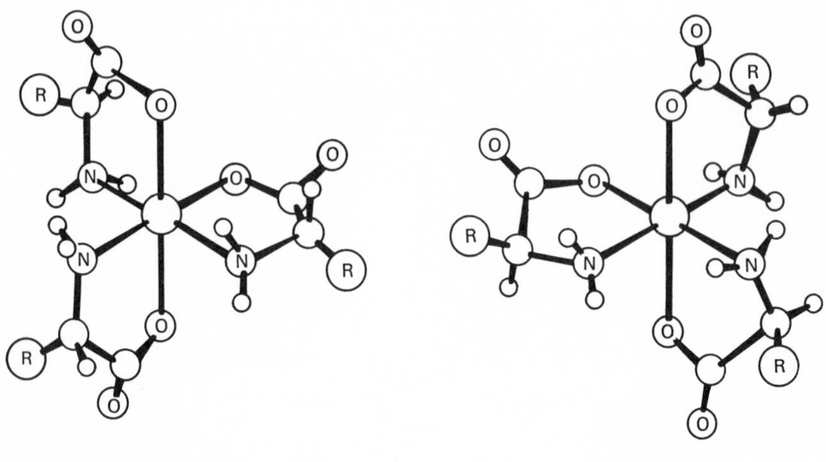

Fig. 6-36 D and L-trans-[Co(L-leu)$_3$] with R = $—CH_2CH(CH_3)_2$.

for its CH_3 resonances (see Fig. 6-37). Denning and Piper [20] analyzed the spectrum as arising from two doublets centered at 67 Hz downfield from TMS (60 MHz instrument) with almost identical chemical shifts, and one doublet centered at 63 Hz. The signals are broad, because the two methyl groups in each ligand are not completely equivalent. Denning and Piper [20] have suggested that in the D complex two of the alkyl groups interact, resulting in the loss of their freedom to rotate, whereas the third alkyl group, which is not involved in the "clash" and therefore is free to rotate, retains the essential degeneracy of its methyl resonances. This is consistent with the analysis of the CH_3 proton resonance spectrum shown in Fig. 6-37 for the dextro isomer. Two pairs of doublets with $J = 6.9$ Hz and $J = 5.3$ Hz are assignable to the four methyl groups of the interacting alkyl groups, and the two virtually degenerate doublets with $J = 5.7$ Hz are assignable to the two methyl groups of the other ligand.

Buckingham, Marzilli, and Sargeson [11] have used nmr to determine the conformational energy difference between D- and L-alanine and D- and L-valine in the complex D-$[Coen_2am]^{2+}$. The base-catalyzed exchange of H_α was used to equilibrate each system, and the isomeric ratio was determined for each system from the nmr of the equilibrium mixtures by determining the areas under the respective H_α signals. For valine with the nitrogen deuterated, the H_α signal appears as a doublet split by the proton on the tertiary carbon of the isopropyl group. The doublets for D- and L-valine in the D complex are centered at about 3.78 and 3.57 ppm, respectively, and, in the equilibrium

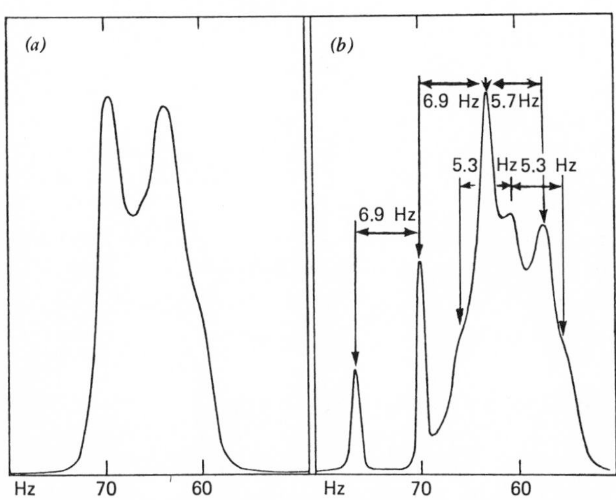

Fig. 6-37 Sixty-megahertz spectra of (*a*) *l-trans*-[Co(L-leu)$_3$] and (*b*) *d-trans*-[Co-(L-leu)$_3$] [20].

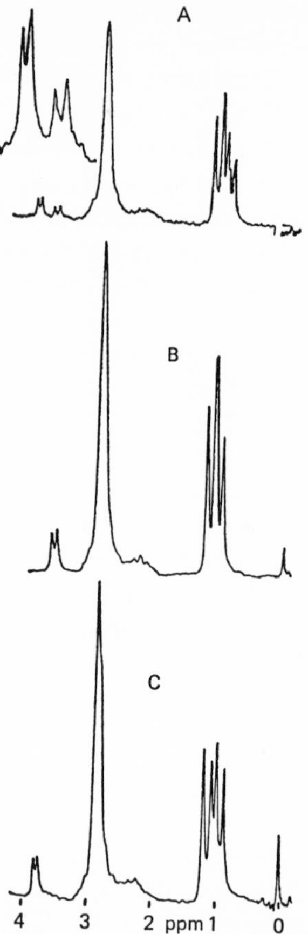

Fig. 6-38 Sixty-megahertz spectra of (A) the equilibrium mixture of D-$[Coen_2(L\text{-val})]^{2+}$ and D-$[Coen_2(D\text{-val})]^{2+}$; (B) D-$[Coen_2(L\text{-val})]^{2+}$ and (C) D-$[Coen_2(D\text{-val})]^{2+}$ [11].

mixture, their areas are in the ratio 3:2 (see Fig. 6-38). This amounts to a conformational energy difference of 0.30 kcal mol^{-1} in favor of D-valine. A similar study of the H_α quartets of the alanine complex showed that the conformational energy difference is negligibly small.

POLYAMINOCARBOXYLATE COMPLEXES

The nmr spectra of metal complexes of polyaminocarboxylates (e.g., EDTA) have been found to exhibit an unusual amount of fine structure, because some of the chelate rings are relatively rigid, and chemical-shift differences are more marked than for the simple systems we have already considered. This has resulted in perhaps the most successful application of nmr to the determination of the absolute configuration of metal complexes published so far, when Schoenberg, Cooke, and Liu [61] studied the isomers of (ethylenediamine-*N*,*N*′-di-L-α-propionato)ethylenediaminecobalt(III), [Co(L-EDDP)en]$^+$.

The structures of two trans and two cis isomers of Co(L-EDDP)en$^+$ are shown in Fig. 6-39. It should be noted that the L-cis isomer drawn has the RS configuration. Schoenberg and coworkers based their discussion on the SS configuration, but a preliminary conformational analysis suggests that the RS isomer, which has the meridional methyl group equatorial, is more stable than the SS, which has the methyl group axial. Nevertheless, the conclusions one can draw from the nmr are independent of which isomer is the more stable.

First, let us consider the trans complexes. The parent compound [Co(EDDA)en]$^+$, where EDDA is ethylenediamine-*N*,*N*′-diacetate, has two equivalent acetate rings related by a C_2 axis of symmetry. Each acetate group has two nonequivalent protons that couple to give an AB quartet (see Fig. 6-40) [43]. The high-field proton H_B is assigned to the proton that corresponds to R in the D-trans complex in Fig. 6-39, because it is more shielded by the

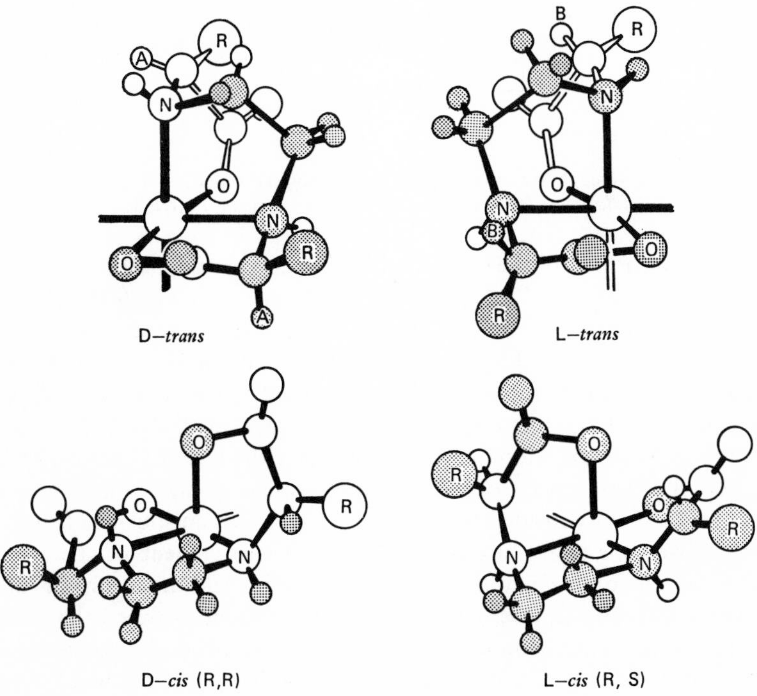

Fig. 6-39 Isomers of [Co(L-EDDP)en]$^+$.

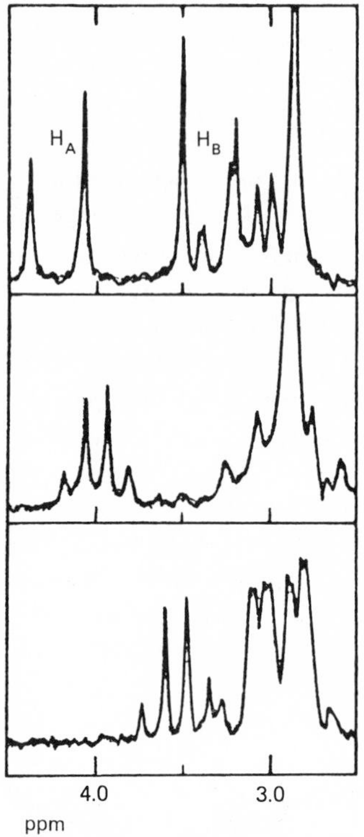

Fig. 6-40 Sixty-megahertz spectra of *trans*-[Co(EDDA)en]$^+$ (*top*), *d-trans*-[Co(L-EDDP)en]$^+$ (*middle*), and *l-trans*-[Co(L-EDDP)en]$^+$ (*bottom*) after deuteration [61].

magnetic anisotropy of the C—N bond of the ethylenediamine backbone than H_A. In the nmr of D- and L-*trans*-[Co(L-EDDP)en]$^+$ (see Fig. 6-40), H_α for the *d* isomer is found to resonate as a quartet at a field similar to that found for H_A and for the *l* isomer at approximately the field of the H_B proton. Assuming that the methyl substitution has not greatly affected the relative chemical shifts of H_A and H_B, this observation is consistent with the dextro isomer's having the D configuration and the levo isomer the L configuration. Further, the methyl group protons of the *d* isomer resonate at higher field than those of the *l*, and if one considers the effect of the magnetic anisotropy of the C—N bond above, this result strongly supports the assignment of the D and L configurations to the *d* and *l* isomers, respectively. Additional strong support for this conclusion can be derived from the coupling between the NH and H_α protons. In the D complex the H—N—C—H_α dihedral angle is approximately 10 to 20°, and significant coupling would be expected, whereas in the L complex, the angle is close to 90°, and little, if any, coupling would be observed. Because coupling is observed for the dextro isomer but not the levo, the *d* isomer must have the D configuration.

In the cis complexes the two propionate arms are not equivalent: one we classify as the apical arm, the other as the meridional arm. The two H_α resonances appear as overlapping quartets. The NH protons resonate at 6.9 and 6.0 ppm, with the NH associated with the apical ring at higher field because it is trans to $—COO^-$, in contrast to the other NH, which is trans to $—NH_2$. In the D isomer the H—N—C—H_α dihedral angle for the apical ring is approximately 90° and, therefore, $J_{NH—H_\alpha}$ would be very small. In the L isomer, however, this dihedral angle is in the vicinity of 10 to 20° and would lead to coupling between NH and H_α. Coupling between these protons is observed in the nmr spectrum of the levo isomer, which must therefore have the L configuration. Schoenberg and his coworkers suggest that the H—N—C—H_α dihedral angle for the meridional ring is about 55° in the L isomer, giving rise to a small value for the coupling constant, and, in the D isomer, has a value that would give rise to more significant coupling. However, the Dreiding molecular models of the cis complexes show that in the L the H—N—C—H_α angle is about 30° and, in the D, about 150°. If this approximates reality, coupling should be observable for both isomers. Unfortunately, the experimental data presented do not allow a conclusion to be drawn. Support for the assignment above can be obtained by considering the chemical shift of the apical H_α proton. The D-trans complex is converted to the L-cis by shifting the position of one of the apical rings into a meridional position, and thus the apical H_α of the L-cis isomer could be expected to resonate at a similar field to the apical H_α of the D-trans complex. The apical H_α signal appears at 3.53, 3.99, 3.95 for the L-trans, D-trans, and *l*-cis isomers, respectively, supporting the proposal that the *l*-cis compound has the L configuration.

Schoenberg, Cooke, and Liu [61] have also successfully studied the corresponding *l*-propylenediamine complexes, and the research group under Cooke has extracted considerable chemical-shift and coupling-constant data from the nmr spectra of related complexes without actually determining absolute configurations (see, for example, [43] and references to Cooke above). Complexes of ethylenediaminetetracetic acid and its substituted analogs have also been extensively studied, but unequivocal absolute configuration data have not been obtained yet (see, for example, [19,62]).

PSEUDOTETRAHEDRAL COMPLEXES

Very recently an important series of papers has been published by Holm and his research group on stereochemical aspects of some pseudotetrahedral complexes (see, for example, [54] and references therein). Of interest to this review

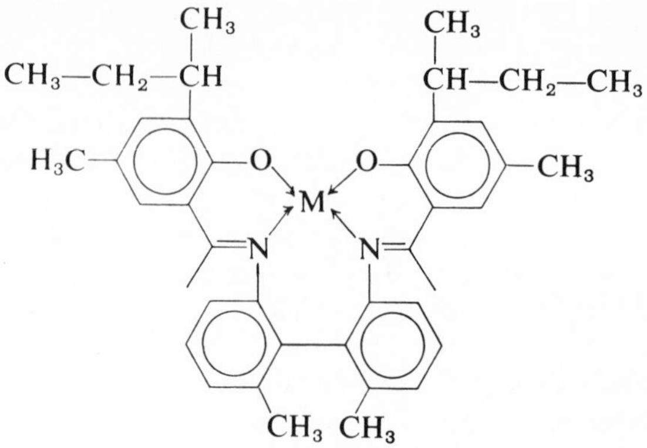

Fig. 6-41. 2,2′-Bis(3-*sec*-butyl-5-methylsalicylideneamino)-6,6′-dimethylbiphenyl.

is the nmr investigation of the diastereoisomers of the nickel(II) complexes of 2,2′-bis(salicylideneamino)-6,6′-dimethylbiphenyl {$(X\text{-sal})_2$bmp} (see Fig. 6-41) and related quadridentate Schiff bases [54]. These ligands can form dissymmetric tetrahedral complexes that are inert to racemization, although they are in rapid equilibrium with a diamagnetic square-planar configuration. This inertness arises, because the ligand, which can itself be resolved, is inert to inversion because of the restricted rotation of the substituted biphenyl. The two enantiomeric configurations are shown in Fig. 6-42 and are labeled S and R (Δ and Λ [54]) according to the rules given in Chap. 2. Because the configurations of *d*- and *l*-2,2′-diamino-6,6′-dimethylbiphenyl are known to be R and S, respectively [50], the absolute configurations of the complexes formed by the Schiff bases of the diamine are uniquely defined.

The nickel(II) complex of racemic 2,2′-bis(3-*dl*-*sec*-butyl-5-methylsalicylideneamino)-6,6′-dimethylbiphenyl has six isomers:

$$\mathrm{R}(dd) = \mathrm{S}(ll)$$
$$\mathrm{S}(dd) = \mathrm{R}(ll)$$
$$\mathrm{R}(dl) = \mathrm{S}(dl) \qquad \text{meso}$$

Fig. 6-42 Absolute configurations.

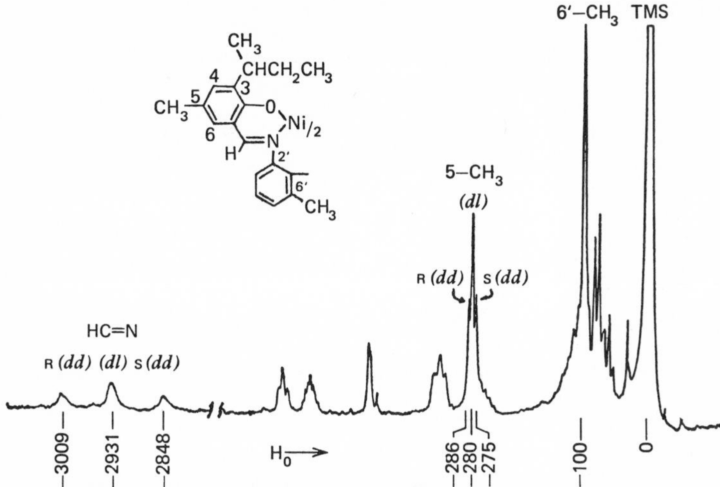

Fig. 6-43 Sixty-megahertz nmr spectrum of the "active" and meso diastereoisomers of [Ni(3-s-bu-5-Me-sal)$_2$bmp] in $CDCl_3$ solution at 27°. Frequencies (hertz) are the chemical shifts [54].

where *d* and *l* refer to the $[\alpha]_D$ of the 3-*sec*-butyl-5-methyl-salicylaldehyde. Because the absolute configuration of the complex is inert to racemization, it is possible, in principle, to observe separate signals for the three pairs of enantiomers. The nmr spectrum of this complex is shown in Fig. 6-43. The 5-methyl and azomethine signals that occur in the diamagnetic free ligand at 132 and 497 Hz respectively, downfield from TMS, are displaced to lower field because of the contact shift with the paramagnetic nickel(II). These two signals are each split into three components with an approximate 1:2:1 intensity ratio. The three components were identified as arising from R(*dd*)S(*ll*), R(*dl*)-S(*dl*), and S(*dd*)R(*ll*) in this order with the last at highest field. From this and a previous study [23], Holm and his coworkers concluded that the chemical-shift differences between diastereoisomeric nickel(II) complexes are "essentially contact shift differences which arise from observable inequalities in the free energy changes for the planar-tetrahedral interconversion of the active and meso forms."

GENERAL READING

1. Jackman, L. M., "Applications of Nuclear Magnetic Resonance Spectroscopy in Organic Chemistry," Pergamon, London, 1959.
2. Pople, J. A., W. G. Schneider, and H. J. Bernstein, "High Resolution Nuclear Magnetic Resonance," McGraw-Hill, New York, 1959.
3. Emsley, J. W., J. Feeney, and J. H. Sutcliffe, "High Resolution Nuclear Magnetic Resonance Spectroscopy," vols. 1 and 2, Pergamon, London, 1965.

REFERENCES

4. Abraham, R. J., and K. A. McLauchlan, *Mol. Phys.*, **5**:513 (1962).
5. Anet, F. A. L., *Canad. J. Chem.*, **39**:789 (1961).
6. Appleton, T., and J. R. Hall, personal communication, 1969.
7. Booth, H., and G. C. Gidley, *Tetrahedron Letters*, **1964**:1449.
8. Bothner-by, A. A., and C. Naar-Colin, *J. Am. Chem. Soc.*, **84**:743 (1962).
9. Brushmiller, J. G., and L. G. Stadtherr, *Inorg. Nucl. Chem. Letters*, **3**:525 (1967).
10. Buckingham, D. A., L. Durham, and A. M. Sargeson, *Aust. J. Chem.*, **20**:257 (1967).
11. Buckingham, D. A., L. G. Marzilli, and A. M. Sargeson, *J. Am. Chem. Soc.*, **89**:5133 (1967).
12. Buckingham, D. A., S. F. Mason, A. M. Sargeson, and K. R. Turnbull, *Inorg. Chem.*, **5**:1649 (1966).
13. Buckingham, D. A., I. I. Olsen, and A. M. Sargeson, *Aust. J. Chem.*, **20**:597 (1967).
14. Chakravorty, A., and R. H. Holm., *Inorg. Chem.*, **3**:1521 (1964).
15. Chakravorty, A., and K. C. Kalia, *Inorg. Chem.*, **6**:690 (1967).
16. Chu, M. W. S., D. W. Cooke, and C. F. Liu, *Inorg. Chem.*, **7**:2543 (1968).
17. Clifton, P., and L. Pratt, *Proc. Chem. Soc.*, **1963**:339.
18. Cooke, D. W., *Inorg. Chem.*, **5**:1141 (1966).
19. Day, R. J., and C. N. Reilley, *Anal. Chem.*, **37**:1326 (1965).
20. Denning, R. G., and T. S. Piper, *Inorg. Chem.*, **5**:1056 (1966).
21. Eidson, A. F., and C. F. Liu, Paper presented to the 156th ACS Meeting, Atlantic City, 1968.
22. Erickson, L. E., J. W. McDonald, J. K. Howie, and R. P. Clow, *J. Am. Chem. Soc.*, **90**:6371 (1968).
23. Ernst, R. E., M. J. O'Connor, and R. H. Holm, *J. Am. Chem. Soc.*, **89**:6104 (1967).
24. Fay, R. C., and T. S. Piper, *J. Am. Chem. Soc.*, **84**:2303 (1962).
25. Fay, R. C., and T. S. Piper, *J. Am. Chem. Soc.*, **85**:500 (1963).
26. Froebe, L. R., M.S. dissertation, Univ. of North Dakota, 1967.
27. Fung, B. M., *J. Am. Chem. Soc.*, **89**:5788 (1967).
28. Gutowsky, H. S., G. G. Belford, and P. E. McMahon, *J. Chem. Phys.*, **36**:3353 (1962).
29. Gutowsky, H. S., and C. Juan, *Disc. Faraday Soc.*, **34**:115 (1962).
30. Gutowsky, H. S., M. Karplus, and D. M. Grant, *J. Am. Chem. Soc.*, **83**:4726 (1961).
31. Gutowsky, H. S., D. W. McCall, and C. P. Schichter, *J. Chem. Phys.*, **21**:279 (1953).
32. Halpern, B., A. M. Sargeson, and K. R. Turnbull, *J. Am. Chem. Soc.*, **88**:4630 (1966).
33. Henney, R. C., H. F. Holtzclaw, and R. C. Larson, *Inorg. Chem.*, **5**:940 (1966).
34. Jensen, F. R., D. S. Noyce, C. H. Sederholm, and A. J. Berlin, *J. Am. Chem. Soc.*, **82**:1256 (1960).
35. Jolly, W. L., A. D. Harris, and T. S. Briggs, *Inorg. Chem.*, **4**:1064 (1965).
36. Kalia, K. C., and A. Chakravorty, *Inorg. Chem.*, **7**:2016 (1968).
37. Karplus, M., *J. Chem. Phys.*, **30**:11 (1959).
38. Karplus, M., *J. Chem. Phys.*, **33**:1842 (1960).
39. Karplus, M., *J. Am. Chem. Soc.*, **84**:2458 (1962).
40. Karplus, M., *J. Am. Chem. Soc.*, **85**:2870 (1963).

41. Lantzke, I. R., and D. W. Watts, *Aust. J. Chem.*, **20**:35 (1967).
42. Lauterbur, P. C., and R. J. Kurland, *J. Am. Chem. Soc.*, **84**:3405 (1962).
43. Legg, J. I., and D. W. Cooke, *Inorg. Chem.*, **4**:1576 (1965).
44. Legg, J. I., and D. W. Cooke, *Inorg. Chem.*, **5**:594 (1966).
45. Lemieux, R. U., and J. W. Lown, *Canad. J. Chem.*, **42**:893 (1964).
46. McConnell, H. M., *J. Chem. Phys.*, **27**:226 (1957).
47. McConnell, H. M., and C. H. Holm, *J. Chem. Phys.*, **27**:314 (1957).
48. McConnell, H. M., and R. E. Robertson, *J. Chem. Phys.*, **29**:1361 (1958).
49. McDonald, C. C., and W. D. Phillips, *J. Am. Chem. Soc.*, **85**:3736 (1963).
50. McGinn, F. A., A. K. Lazarus, M. Siegel, J. E. Ricci, and K. Mislow, *J. Am. Chem. Soc.*, **80**:476 (1958).
51. Marzilli, L. G., and D. A. Buckingham, *Inorg. Chem.*, **6**:1042 (1967).
52. Meinwald, J., and A. Lewis, *J. Am. Chem. Soc.*, **83**:2769 (1961).
53. Moritz, A. G., and N. Sheppard, *Mol. Phys.*, **5**:361 (1962).
54. O'Connor, M. J., R. E. Ernst, and R. H. Holm, *J. Am. Chem. Soc.*, **90**:4561 (1968).
55. Petrakis, L., and C. H. Sederholm, *J. Chem. Phys.*, **35**:1243 (1961).
56. Powell, D. B., and N. Sheppard, *J. Chem. Soc.*, **1959**:791.
57. Ramsay, N. F., *Phys. Rev.*, **91**:303 (1953).
58. Ramsay, N. F., and E. M. Purcell, *Phys. Rev.*, **85**:143 (1952).
59. Sargeson, A. M., Conformations of Coordinated Chelates, in R. L. Carlin (ed.), "Transition Metal Chemistry," vol. 3, pp. 303–343, Marcel Dekker, New York, 1966.
60. Sargeson, A. M., and G. H. Searle, *Inorg. Chem.*, **6**:787 (1967).
61. Schoenberg, L. N., D. W. Cooke, and C. F. Liu., *Inorg. Chem.*, **7**:2386 (1968).
62. Smith, B. B., and D. T. Sawyer, *Inorg. Chem.*, **7**:2020 (1968).
63. Spees, S. T., L. J. Durham, and A. M. Sargeson, *Inorg. Chem.*, **5**:2103 (1966).
64. Spiesecke, H., and W. G. Schneider, *J. Chem. Phys.*, **35**:722 (1961).
65. Williamson, K. L., and W. S. Johnson, *J. Am. Chem. Soc.*, **83**:4623 (1961).
66. Williamson, K. L., Y. F. Li, F. H. Hall, and S. Swager, *J. Am. Chem. Soc.*, **88**:5678 (1966).
67. Woldbye, F., "Studier over Optisk Aktivitet," Polyteknisk Forlag, Copenhagen, 1969.
68. Yano, S., H. Ito, Y. Koike, J. Fujita, and K. Saito, *Chem. Commun.*, **1969**:460.
69. Yoneda, H., and Y. Morimoto, *Bull. Chem. Soc. Japan*, **39**:2180 (1966).

7

Miscellaneous Techniques

INFRARED SPECTROSCOPY

Although infrared (IR) spectroscopy provided the first experimental evidence to support Theilacker's original proposal that ethylenediamine chelate rings are not puckered, the technique's contribution to absolute-configuration determination has not progressed much further. It has been successfully used, however, to distinguish geometrical isomers, and this aspect of the subject is covered in Chap. 8.

Because of solubility problems, most of the complexes of interest have been studied in the solid, polycrystalline phase either in the form of a mull or disk (e.g., polythene, KBr). Such studies are thwart with danger:

1. The intramolecular interactions that might allow isomers to be distinguished are often of the same order of magnitude as intermolecular crystal effects.
2. The expected molecular-symmetry selection rules, which are usually the basis for differentiating isomers, do not necessarily apply because the crystal-site symmetry could split otherwise degenerate modes and make otherwise IR inactive vibrations allowed.
3. Interactions, such as hydrogen bonding, can shift the frequencies of some modes, giving rise to difficulties in the assignment of the peaks.

The various rotamers of a 1,2-disubstituted ethane ligand, *trans*(C_{2h}), *cis*(C_{2v}), and *gauche*(C_2), have different symmetries and would therefore be expected to have different numbers of IR-active fundamentals (see Fig. 7-1). The IR study of such rotamers has been reviewed by Mizushima [26] and Sheppard [44]. Let us consider 1,2-dithiocyanatoethane both as a free ligand

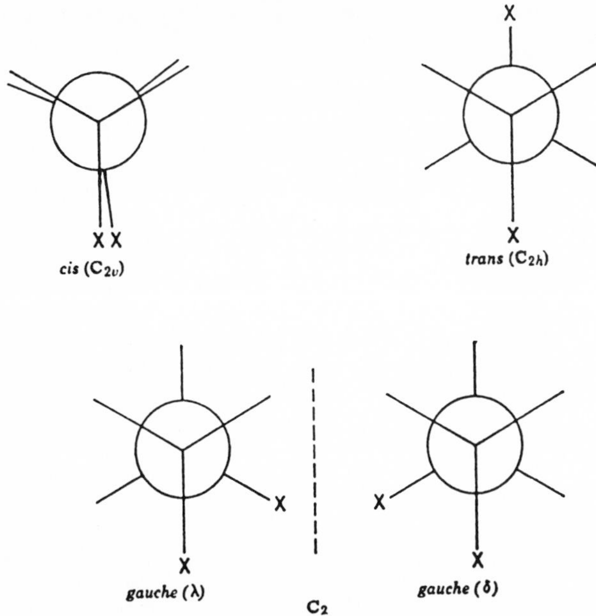

Fig. 7-1 Rotamers of 1,2-disubstituted ethanes.

and as a chelate in a complex of the type $PtL(Cl)_2$. The Raman and IR frequencies of the solid ligand do not correspond except for an accidental degeneracy of the symmetric and antisymmetric —C≡N frequencies (see Table 7-1). Therefore, because the trans rotamer is the only isomer with a center of symmetry, the ligand must have the trans conformation in the solid state [27,41]. This conclusion was confirmed by an X-ray analysis [12]. The trans modes appear at almost the same frequencies in a chloroform solution of the ligand together with a large number of other bands. These latter peaks also appear in the solid spectrum of $PtL(Cl)_2$, but the trans modes are missing.

The normal frequencies of 1,2-dichloroethane have been calculated as an eight-body problem, and all the observed frequencies were assignable to normal vibrations of trans and gauche molecules. It was found that, for the chloro compound, only the trans occurred in the solid state, but both trans and gauche conformations were present in the liquid and gaseous phases. Further, CH_2 rocking modes were found at 768 cm^{-1} for the trans and at 880 and 941 cm^{-1} for the gauche. The cis was expected to have an IR-active CH_2 rocking vibration at about 740 cm^{-1} (B_2,C_{2v}) and an inactive mode at about

Table 7-1 Infrared and Raman frequencies assigned to the fundamental vibrations of 1,2-dithiocyanatoethane and 1,2-dithiocyanatoethanedichloroplatinum(II) [27]

Raman (solid)	Infrared: Solid	Infrared: Coordination complex	Infrared: $CHCl_3$ soln.	Assignment
292(4)				*trans*-δ(C—C—S) A_g
	660(*m*)		660(*m*)	*trans*-ν(C—S) B_u
673(2*b*)				*trans*-ν(C—S) A_g
	680(*m*)		677(*m*)	*trans*-ν(C—S) B_u
722(6)				*trans*-ν(C—S) A_g
	749(*s*)		[a]	*trans*-ρr(CH_2) A_g
		847(*s*)	845(*m*)	*gauche*-ρr(CH_2) B
		929(*m*)	918(*m*)	*gauche*-ρr(CH_2) A
1037(2)				*trans*-ν(C—C) A_g
		1052(*m*)	[a]	*gauche*-ν(C—C) A
		1110(*m*)	1100(*w*)	*gauche*-ρt(CH_2) A or B
	1145(*m*)		1140(*m*)	*trans*-ρt(CH_2) A_u
	1220(*s*)		1215(*s*)	*trans*-ρw(CH_2) B_u
		1280(*m*)	1285(*s*)	*gauche*-ρw(CH_2) A or B
1291(3)				*trans*-ρw(CH_2) A_g
		1410(*s*)	1419(*s*)	*gauche*-δ(CH_2) A or B
	1423(*s*)			*trans*-δ(CH_2) B_u
1422(1)			1423(*s*)	*trans*-δ(CH_2) A_g
	2155(*s*)		2170(*s*)	*trans*-ν(C≡N) B_u
2160(7)				*trans*-ν(C≡N) A_g
		2165(*s*)	2170(*s*)	*gauche*-ν(C≡N) A and B
2947(2)			[a]	ν(C—H)
	2960(*w*)		[a]	ν(C—H)
2986(2*b*)			[a]	ν(C—H)
3010(1*b*)		2993(*sh*)	[a]	ν(C—H)
		3140(*s*)	[a]	ν(C—H)

[a] The frequency region where the absorption peak cannot be measured accurately because of the absorption of chloroform.

937 cm^{-1} (A_2,C_{2v}). Because only CH_2 rocking vibrations are found in the 740 to 1,000 cm^{-1} region, the 749 cm^{-1} vibration in solid 1,2-dithiocyanatoethane has been assigned to the trans CH_2 rocking mode, and the 845 and 918 cm^{-1} frequencies in the solution and the 847 and 929 cm^{-1} modes in the complex have been assigned to CH_2 rocking vibrations of the gauche conformation. This assignment was confirmed by a normal coordinate analysis.

Similarly, 1,2-dimethylthioethane has been shown to exist as the trans conformer in the solid but as both gauche and trans conformers in solution and as gauche in complexes of the type ML(Cl)$_2$, where M is Pt^{2+}, Pd^{2+}, Hg^{2+}, Cd^{2+}, and Ni^{2+} [45,47]. An X-ray analysis of $[Cu(NC—CH_2—CH_2—CN)_2]NO_3$ has shown that 1,2-dicyanoethane also has a gauche conformation, but in this complex the ligand is found to coordinate to two different copper atoms by way of the two —CN groups (see Fig. 7-2) [19]. The IR spectrum is consistent with the gauche conformation [25].

The ligand of greatest interest is, of course, ethylenediamine. The simplicity of the IR of the salt, $en(HCl)_2$, suggests the trans conformation for the compound [37]. When ethylenediamine is chelated, it is sterically impossible for it to have the trans conformation. From a survey of the IR of a number of ethylenediamine complexes of different metal ions, Powell and Sheppard [38] proposed that there were two kinds of chelate conformations, classified as A and B, one of which was gauche and the other cis. This was later retracted when X-ray data showed that both classes had the gauche conformation [40].

Brodersen [2] has analyzed the IR spectrum of solid $HgenCl_2$ and concluded from the simplicity of the spectrum that the ligand has the trans conformation. Because this conformation cannot exist if the ligand is chelated, the result implied that the complex contained a bridging ethylenediamine. This conclusion was shown to be correct by an X-ray analysis [2]. Powell and Sheppard [39] have studied the IR of a bridging ethylenediamine in the complex $(C_2H_4)Cl_2Pt—NH_2CH_2CH_2NH_2—Pt(Cl)_2(C_2H_4)$. After the ethylene modes

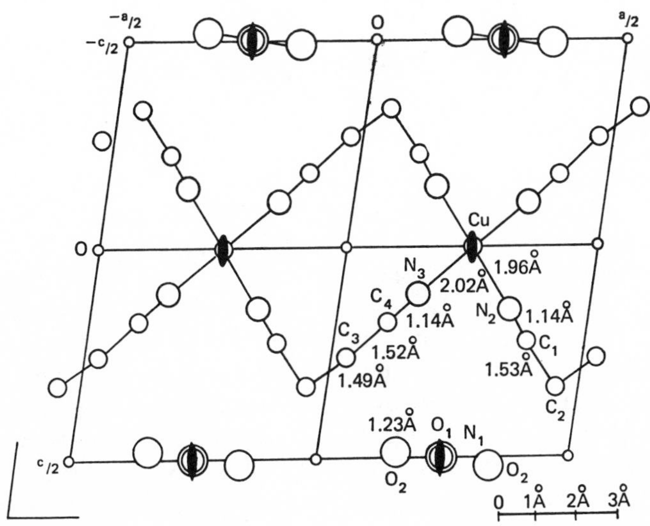

Fig. 7-2 Structure of succinonitrilecopper(I) determined by X-ray diffraction: projection along (010) [19].

were subtracted, the spectrum was found to be much simpler than that of $PtenCl_2$, which contains chelated ethylenediamine with the gauche conformation (see Fig. 7-3). They concluded that the bridging ethylenediamine had the trans configuration. Newman and Powell [34] compared the IR of zinc, cadmium, and mercury complexes of the type $MenCl_2$ to that of (C_2H_4)-$Cl_2Pt—NH_2CH_2CH_2NH_2—Pt(Cl)_2(C_2H_4)$ and concluded, in agreement with Brodersen, that the bridging ethylenediamine has the trans configuration.

In Fig. 7-4, the IR of three silver(I)-ethylenediamine compounds are compared with the spectra of $HgenCl_2$, which is known to have a trans bridging group, and $[Znen_2]Cl_2$, which has gauche chelated ethylenediamine [35]. $(AgCl)_2en$ and $[Agen]_2[PtCl_4]$ are found to have spectra similar to $HgenCl_2$ with two strong bands at about 1,010 cm^{-1} [ν_{asym}(C—N)] and 1,080 cm^{-1} [$\delta(NH_2)$]. Newman and Powell [35] concluded that both have bridging ethylenediamines and suggested that $(AgCl)_2en$ has the structure, $Cl—Ag—NH_2$-$CH_2CH_2NH_2—Ag—Cl$ and that $[Agen]_2[PtCl_4]$ most probably contains Ag—en—Ag—en— chains. The spectrum of AgenCl was more complex, resembling that of $[Znen_2]Cl_2$ with three strong bands near 1,000 cm^{-1} and a doublet near 840 cm^{-1}, typical of a gauche ethylenediamine. Two structures were postulated:

$CH_2—CH_2$ (2+)
H_2N NH_2
↓ ↓
Ag Ag
↑ ↑
H_2N NH_2
$CH_2—CH_2$

and

NH_2 Cl NH_2
H_2C CH_2
Ag Ag
H_2C CH_2
NH_2 Cl NH_2

Newman and Powell [35] favored the first of these structures, which had previously been proposed by Schwarzenbach [43] for $[Ag_2en_2]SO_4$.

Trans-bis(ethylenediamine) complexes have two possible configurations, meso ($\delta\lambda$) or chiral ($\delta\delta,\lambda\lambda$). IR has been used to differentiate between them. The meso form has a center of symmetry and should give rise to fewer IR-active fundamentals than the chiral, which does not have a center of symmetry. Further, the IR frequencies of the meso should not coincide with its Raman frequencies. Both IR and Raman spectra of $[Pten_2]^{2+}$ were interpreted by Mathieu to indicate that the compound was in the meso form [24]. Quagliano and Mizushima [41] found that the IR spectrum of *trans*-$[Coen_2Cl_2]Cl$ was also consistent with the meso configuration in agreement with an X-ray analysis [33].

It has been found that an equatorial group X attached directly to a cyclohexane ring shows a C—X stretching vibration at a higher frequency than the

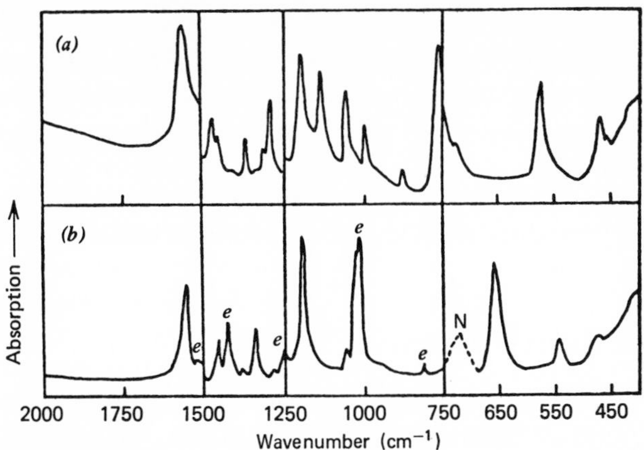

Fig. 7-3 IR spectra of (*a*) $PtenCl_2$ and (*b*) $Pt_2Cl_4(C_2H_4)_2en$; *e* = ethylene band; *N* = nujol band [39].

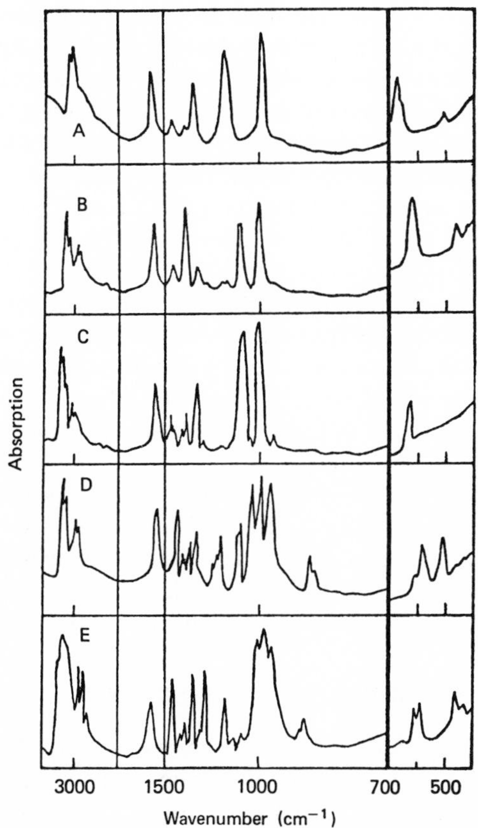

Fig. 7-4 IR spectra of (A) $HgenCl_2$, (B) $(AgCl)_2en$, (C) $[Agen]_2[PtCl_4]$, (D) AgenCl, and (E) $[Znen_2]Cl_2$ [35].

corresponding axial group [10]. In theory it should be possible to determine the axial-equatorial equilibrium constant for a labile system by measuring the relative intensities of the C—X stretching bonds. In practice, however, it is necessary to know the molecular-extinction coefficients for both isomers, and these have been determined only for a limited number of organic systems. Nevertheless, the technique could possibly be advantageously applied to some chelate-ring systems.

DELÈPINE'S ACTIVE RACEMATE METHOD

When a chiral compound D-A′ is mixed with a closely related racemate DL-A in solution, a *mixed* racemate D-A′ L-A sometimes precipitates as the least soluble material. This is called an "active racemate" [5]. The term "active" refers to the fact that the compound most probably shows optical activity, because the rotations of D-A′ and L-A are unlikely to be canceled completely. The compound that enters the racemate and the isomer that it replaces have the same configuration. Thus, this method enables configurations to be related. It is applicable only to systems in which the least soluble phase is a racemic compound, in contrast to a racemic mixture, in which single crystals contain either all *d* or all *l* isomers, or a racemic solid solution, in which the crystals contain variable amounts of the *d* and *l* forms.

This method has been used by Delèpine for a number of systems [6] and recently by Elsbernd and Beattie for isomers of $[Ruen_3]^{3+}$ [11]. In the latter study, when d_{350}-$[Ruen_3]I_3$ and *dl*-$[Coen_3]I_3 \cdot H_2O$ were dissolved in trifluoroacetic acid (under a nitrogen atmosphere) and the volume reduced under vacuum to about one-third of the original volume, dark purple, needle-shaped crystals containing d_{350}-$[Ruen_3]^{3+}$ and *d*-$[Coen_3]^{3+}$ were formed. The filtrate showed the presence of *l*-$[Coen_3]^{3+}$. Similarly, when *l*-$[Coen_3]I_3 \cdot H_2O$ and *dl*-$[Ruen_3]I_3$ were dissolved in trifluoroacetic acid and the procedure above repeated, the least soluble product contained the levo isomers of the two compounds. Further, when a saturated solution of *l*-$[Coen_3]I_3$ was added to solutions of l_{350}-$[Ruen_3]^{3+}$ and d_{350}-$[Ruen_3]^{3+}$ separately, a precipitate formed in the solution with the two levo isomers, but no precipitate appeared in the other solution. From this it can be concluded that, because *l*-$[Coen_3]^{3+}$ has the L configuration, l_{350}-$[Ruen_3]^{3+}$ probably has the D configuration.

DIASTEREOISOMER SOLUBILITY CRITERION

When one optical isomer of a chiral ionic compound, say, *d*-A^+X^-, is added to a solution containing a racemate of an oppositely charged species *dl*-B^-, the two possible diastereomeric salts *d*-A^+*d*-B^- and *d*-A^+*l*-B^- are capable of showing different solubilities, and it is possible, if the right conditions are chosen, that one form will be preferentially precipitated. This is the principle of Pasteur's *second method* of resolution. Werner proposed that, for a series of related racemic compounds, the isomers that were preferentially precipitated by the same resolving agent would have related absolute configurations [49]. Jaeger, however, has pointed out that this rule would hold only if the least soluble diastereoisomers were isomorphous [17].

The relative solubilities of the diastereoisomers are governed by at least the following three factors:

1. In the presence of the resolving agent, the activities of the two isomers are not equilvalent [20], and this could give rise to a difference in solubilities [3,4].
2. Specific interactions in the solid state of the kind that caused the difference in activities in solution.
3. Differences in energy arising from the close packing of the ions in the crystalline state. The interactions of the first two kinds are not fully understood. Intuitively, however, one would predict that, for two closely related complexes, the interactions in solution would lead to the same configurations being preferred. In the solid state, such a conclusion could be drawn only if the crystal structures were similar. The third factor would require that the structures of the least soluble diastereoisomers are isomorphous.

An added complication arises if the diastereoisomer that precipitates is not decided by solubility differences but by the rate of nucleation. This can be checked by studying whether timing is important for the resolution.

The absolute configurations of the complexes $[Men_3]^{3+}$, where M is cobalt, rhodium, iridium, and chromium, have been related by this method [22]. With *d*-tartrate as the resolving agent, the diastereoisomers *d*-$[Coen_3]Cl \cdot$ *d*-tart$\cdot 5H_2O$ [50] and *l*-$[Rhen_3]Cl \cdot$*d*-tart$\cdot 4H_2O$ [17] are less soluble.* With *d*-nitro-camphor, *d*-$[Cren_3]^{3+}$ [52], *l*-$[Rhen_3]^{3+}$ [51], and *l*-$[Iren_3]^{3+}$ [53] were found to form the less soluble diastereoisomers. Thus, if this rule holds, *d*-$[Coen_3]^{3+}$, *d*-$[Cren_3]^{3+}$, *l*-$[Rhen_3]^{3+}$, and *l*-$[Iren_3]^{3+}$ have related configurations.

Garbett and Gillard have observed that with complexes of the type *cis*-$[Men_2Cl_2]^+$ the less soluble diastereoisomers with *d*-bromocamphorsulfonate are *d*-$[Coen_2Cl_2]^+$, l_{535}-$[Rhen_2Cl_2]^+$, and *d*-$[Cren_2Cl_2]^+$ and, with antimony *d*-tartrate, *d*-$[Coen_2Cl_2]^+$ and *d*-$[Cren_2Cl_2]^+$. They have concluded

* In some cases differences in the number of waters of crystallization need not invalidate this rule, because the additional water molecules could be occupying holes in the crystal lattice and need not be significantly affecting the energetics of crystallization. In this case, however, the former complex has been shown to be triclinic and the latter monoclinic [17].

that these isomers have the same configuration [14]. Similarly, with *d*-camphorsulfonate, *l*-$[Coen_2(NO_2)_2]^+$ and *d*-$[Iren_2(NO_2)_2]^+$ and, with dibenzoyl *d*-tartrate, *d*-$[Coen_2F_2]^+$ and *d*-$[Cren_2F_2]^+$ were found to form the less soluble diastereoisomers and separately were thought, therefore, to have the same configuration. Although Garbett and Gillard pointed out the necessity for the crystals to be isomorphous before a comparison can be objectively made, no evidence was given for isomorphism.

Garbett and Gillard have criticized prior comparisons of configuration in the literature for the series of complexes *cis*-$[Coen_2XY]^{n+}$ [14]. They have pointed out that complexes with $n = 1$, 2, and 3 are unlikely to form isomorphous salts with a resolving agent. They observed that for the closely related complexes $[Coen_2Cl_2]^+$, $[Coen_2BrCl]^+$, and $[Coen_2Br_2]^+$ the dextro isomers had similar solubilities with three distinct resolving agents. Even such a correlation was not without risk, however, because, for the complexes $[Coen_2(OH_2)X]^{2+}$, the less soluble diastereoisomers with *d*-camphorsulfonate and with antimony *d*-tartrate contain the *d* forms of both bromo- and chloro-cations, but, with *d*-bromocamphorsulfonate, the less soluble diastereoisomers contain the *d*-chloro and *l*-bromo complexes.

Tris(1,10-phenanthroline) and tris(2,2′-bipyridine) complexes have been resolved for a series of divalent metal ions (see Ref. 42). With antimony *d*-tartrate as the resolving agent, *d*-$[Niphen_3]^{2+}$, *l*-$[Fephen_3]^{2+}$, *d*-$[Ruphen_3]^{2+}$, and d_{546}-$[Osphen_3]^{2+}$ form the least soluble diastereoisomers. With the same resolving agent, *l*-$[Nibipy_3]^{2+}$, *d*-$[Febipy_3]^{2+}$, *l*-$[Rubipy_3]^{2+}$, and l_{546}-$[Osbipy_3]^{2+}$ are precipitated as their iodide antimony *d*-tartrate salts. The enantiomers are precipitated with *d*-tartrate. Sargeson, however, has pointed out that the validity of any assignment of configuration based on this data is questionable because the relative solubilities of the diastereoisomers depend, in some instances, on the concentration of the resolving agent that has been used [42]. Kennard and Seccombe have studied the unit cell found for *l*-$[Fephen_3]$ $(d\text{-Sbtart})_2 \cdot 8H_2O$ [46] to see if the geometry is such that the L configuration could be as comfortably accommodated as the D [18]. Their calculations showed that, for the L, certain nonbonded interaction distances were such as to preclude this possibility.

Some complexes formed with chiral ligands are inherently diastereomeric, and, under the right conditions, the isomers should be separable without the aid of a resolving agent. If the ligand is relatively stereospecific, the isomer that precipitates first need not be the least soluble, because the isomers are likely to be present in different concentrations. Conversely, the isomer that precipitates first need not be the most preferred isomer.

Liu and Douglas have studied the diastereoisomers formed by complexes of the type $[Coen_2am]^{2+}$ with a number of resolving agents [21]. With either *d*-bromocamphorsulfonate or *d*-tartrate, complexes with achiral or D-amino acids were precipitated as their l_{546} isomers, and the complexes with L-amino acids were precipitated as their d_{546} isomers. Surprisingly, with *l*-tartrate, the same isomers were precipitated as for *d*-tartrate. This is possible because the diastereoisomers for the two resolving agents are not enantiomeric. As expected, however, when *d*-tartrate was reacted with *dl*-$[Coen_2\text{L-leu}]^{2+}$ and *l*-tartrate with *dl*-$[Coen_2\text{D-leu}]^{2+}$, the two least soluble diastereoisomers were found to be enantiomeric. With antimony *d*-tartrate, the dextro isomers of the complexes with D- and L-amino acids were precipitated in contrast to the results for the other resolving agents. This emphasizes the difficulties associated with this method of assigning configurations.

In conclusion it is apparent that not much weight can be given to the assignment of configuration by this method. If applied with caution, however, it could be used to provide corroborative evidence for other methods of configuration assignment. If possible, a number of resolving agents should be used to test for consistency, and the crystals formed for the series of compounds should be studied for isomorphism.

CHEMICAL CORRELATIONS

Relatively few unambiguous assignments of relative configuration have been made for octahedral metal complexes by means of chemical reactions. For such chemical correlations to be rigorous, there should be no fission of metal-ligand bonds during the relevant reaction. It is usually assumed, however, when metal-ligand bonds are broken that, in the absence of evidence to the contrary, the reactions proceed with retention of configuration.

A number of reactions of *cis*-bis(ethylenediamine)cobalt(III) complexes will be considered here.

(i) $$[L_5Co{-}NCS]^{n+} \xrightarrow{[O]} [L_5Co{-}NH_3]^{(n+1)+}$$

Werner carried out this reaction originally with chlorine or hydrogen peroxide as the oxidizing agent [48]. Garbett and Gillard, however, have found that a better yield is obtained with iodate [13]. The reaction proceeds without the rupture of the Co—N bond, and, therefore, the product has the same configuration as the starting material. The following complexes have been related in this way [13]:

$$d_{546}\text{-}[Coen_2Cl(NCS)]^+ \longrightarrow d\text{-}[Coen_2Cl(NH_3)]^{2+}$$
$$d\text{-}[Coen_2OH_2(NCS)]^{2+} \longrightarrow d\text{-}[Coen_2OH_2(NH_3)]^{3+}$$
$$d\text{-}[Coen_2NO_2(NCS)]^+ \longrightarrow d\text{-}[Coen_2NO_2(NH_3)]^{2+}$$
$$d\text{-}[Coen_2NH_3(NCS)]^{2+} \longrightarrow d\text{-}[Coen_2(NH_3)_2]^{3+}$$
$$d\text{-}[Coen_2(NCS)_2]^+ \longrightarrow d\text{-}[Coen_2(NH_3)_2]^{3+}$$

(ii) $$[L_5Co{-}\overset{*}{O}H_2]^{n+} \xrightarrow{NO_2^-} [L_5Co{-}\overset{*}{O}NO]^{(n-1)+} \longrightarrow [L_5Co{-}NO_2]^{(n-1)+}$$

When nitrite is reacted at pH 4 in the cold with an aqua complex of cobalt(III), the nitrito complex is formed without the breakage of the Co—O bond [29]. The nitrito rearranges intramolecularly to the nitro form without change in the configuration [28]. This substitution and subsequent rearrangement have been used for the system

$$d\text{-}[\mathrm{Coen_2X(OH_2)}]^{n+} \xrightarrow{\mathrm{NO_2^-}} d\text{-}[\mathrm{Coen_2X(ONO)}]^{(n-1)+} \longrightarrow d\text{-}[\mathrm{Coen_2X(NO_2)}]^{(n-1)+}$$

with X = OH_2, ONO^-, NO_2^-, NCS^-, Cl^-, and NH_3 to relate the configurations of the aqua, nitrito, and nitro complexes [13].

(iii) $$[\mathrm{L_4XCo{-}\overset{*}{O}H_2}]^{n+} \xrightarrow{\mathrm{HCO_3^-}} [\mathrm{L_4XCo{-}\overset{*}{O}{-}CO{-}OH}]^{(n-1)+} + [\mathrm{L_4XCo{-}\overset{*}{O}{-}CO{-}O}]^{(n-2)+} \downarrow [\mathrm{L_4Co}(\mathrm{O})(\overset{*}{\mathrm{O}})\mathrm{C=O}]^{(n-1)+} + \mathrm{X^-}$$

When an aqua complex of cobalt(III) is reacted with bicarbonate, it has been shown that the bicarbonate attacks the coordinated water, losing one of its original oxygen atoms in the process [16,36]. If X is a halide and L an amine, the carbonate chelates by the removal of X by way of an intramolecular mechanism. Optically pure d-$[\mathrm{Coen_2Cl(OH_2)}]^{2+}$ has been converted by this method with full retention of configuration to optically pure d-$[\mathrm{Coen_2CO_3}]^+$ [42].

(iv) $$[\mathrm{L_5Co{-}X}]^{n+} \xrightarrow{\mathrm{H_2O}} [\mathrm{L_5Co{-}OH_2}]^{(n+1)+} + \mathrm{X^-}$$

Acid hydrolysis of some complexes of the type $[\mathrm{Coen_2X_2}]^{n+}$, X = Cl^-, for example, has been shown to take place with retention of configuration [23].

The reactions above have enabled the configurations of an extensive series of *cis*-bis(ethylenediamine)cobalt(III) complexes to be related.

A small number of relatively unique correlations have been made for tris-(bidentate) complexes of cobalt(III); for example, on oxidation with cold neutral potassium permanganate, *l*-tetrakis(ethylenediamine)-μ-tartrato-dicobalt(III) is converted to *l*-oxalatobis(ethylenediamine)cobalt(III) and, according to Gillard and Price [15], therefore has the same configuration as the latter complex:*

* *l*-$[\mathrm{Coen_2ox}]^+$ has been assigned the L configuration from CD studies.

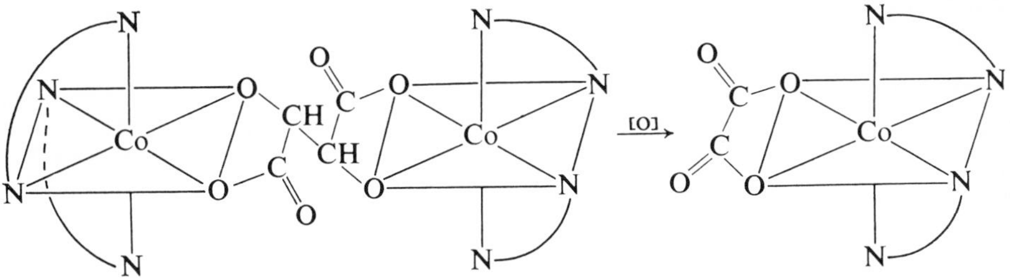

The levo isomers of salicylatobis(ethylenediamine)cobalt(III) and oxalatobis-(ethylenediamine)cobalt(III) have been related by a similar reaction [1].

It is hoped that, with the present emphasis in the literature on mechanistic aspects of inorganic reactions, chemical correlations of configuration will become more common and more soundly based.

OTHER METHODS

Dwyer and his coworkers have found that the dextro isomers of the tris(1,10-phenanthroline) complexes of nickel(II), ruthenium(II), and osmium(II) are more potent than their enantiomers when injected intraperitoneally into mice [7] or when reacted with toad rectus abdominis and innervated rat diaphragm muscle preparations [9], suggesting that they have the same configuration. Similar investigations with isolated biological preparations could provide a powerful method of configuration assignment because stereospecific effects are well known for a variety of systems.

Although configurational activity [8] has not been used directly to assign configuration, it shows promise of providing a useful procedure. If a dissymmetric compound A is labile to racemization, the equilibrium between the *d* and *l* isomers for the racemate could be shifted to favor one of the enantiomers, say, *d* when placed in a solution containing one isomer of a chiral molecule B. If the study is repeated for a very closely related series of compounds, B, B′, B″, and so on—$[ML_3]^{n+}$ for a series of M^{n+}, for example—the configurations of these complexes could be related, because it is expected that the complexes with the same configuration each shift the equilibrium in the same direction. At present this should be treated as an unsupported postulate, but, because a considerable amount of research is currently being undertaken in this field, its generality should shortly be extensively tested.

REFERENCES

1. Beaumont, A. G., and R. D. Gillard, *Chem. Commun.*, **1969**:438.
2. Brodersen, K., *Z. Anorg. Allgem. Chem.*, **298**:142 (1959).
3. Broomhead, J. A., Ph.D. thesis, Australian National University, Canberra, 1960.

4. Broomhead, J. A., *Nature*, **211**:741 (1966).
5. Delèpine, M., *Bull. Soc. chim. France*, (4) **29**:656 (1921).
6. Delèpine, M., *Bull. Soc. chim. France*, (5) **1**:1256 (1934).
7. Dwyer, F. P., E. C. Gyarfas, and J. H. Koch, *Aust. J. Biol. Sci.*, **9**:371 (1956).
8. Dwyer, F. P., E. C. Gyarfas, and M. F. O'Dwyer, *Nature*, **167**:1036 (1951).
9. Dwyer, F. P., E. C. Gyarfas, R. D. Wright, and A. Shulman, *Nature*, **179**:425 (1957).
10. Eliel, E. L., N. L. Allinger, S. J. Angyal, and G. A. Morrison, "Conformational Analysis," pp. 143, 144, Interscience, New York, 1965.
11. Elsbernd, H., and J. K. Beattie, *Inorg. Chem.*, **8**:893 (1969).
12. Foss, O., *Acta Chem. Scand.*, **10**:136 (1956).
13. Garbett, K., and R. D. Gillard, *Coord. Chem. Revs.*, **1**:179 (1966).
14. Garbett, K., and R. D. Gillard, *J. Chem. Soc., A.*, **1966**:802.
15. Gillard, R. D., and M. G. Price, *Chem. Commun.*, **1969**:67.
16. Hunt, J. P., A. C. Rutenberg, and H. Taube, *J. Am. Chem. Soc.*, **74**:268 (1952).
17. Jaeger, F. M., "Optical Activity and High Temperature Measurement," pp. 92, 93, McGraw-Hill, New York, 1930.
18. Kennard, C. H. L., and R. Seccombe, personal communication, 1969.
19. Kinoshita, Y., I. Matsubara, and Y. Saito, *Bull. Chem. Soc. Japan*, **32**:745 (1959).
20. Kirschner, S., N. Ahmad, and K. Magnell, *Coord. Chem. Revs.*, **3**:201 (1968), and references therein.
21. Liu, C. T., and B. E. Douglas, *Inorg. Chem.*, **3**:1796 (1964).
22. MacDermott, T. E., and A. M. Sargeson, *Aust. J. Chem.*, **16**:334 (1963).
23. Mathieu, J.-P., *Bull. Soc. chim. France*, (5) **4**:687 (1937).
24. Mathieu, J.-P., *J. chim. phys.*, **36**:308 (1939).
25. Matsubara, I., *Bull. Chem. Soc. Japan*, **34**:1710 (1961).
26. Mizushima, S., "Structure of Molecules and Internal Rotation," Academic, New York, 1954.
27. Mizushima, S., I. Ichishima, I. Nakagawa, and J. V. Quagliano, *J. Phys. Chem.*, **59**:293 (1955).
28. Murmann, R. K., *J. Am. Chem. Soc.*, **77**:5190 (1955).
29. Murmann, R. K., and H. Taube, *J. Am. Chem. Soc.*, **78**:4886 (1956).
30. Nakagawa, I., *J. Chem. Soc. Japan*, **74**:848 (1953).
31. Nakagawa, I., *J. Chem. Soc. Japan*, **75**:178 (1954).
32. Nakagawa, I., and S. Mizushima, *J. Chem. Phys.*, **21**:2195 (1953).
33. Nakahara, A., Y. Saito, and H. Kuroya, *Bull. Chem. Soc. Japan*, **25**:331 (1952).
34. Newman, G., and D. B. Powell, *J. Chem. Soc.*, **1961**:477.
35. Newman, G., and D. B. Powell, *J. Chem. Soc.*, **1962**:3447.
36. Posey, F. A., and H. Taube, *J. Am. Chem. Soc.*, **75**:4099 (1953).
37. Powell, D. B., *Spectrochim. Acta*, **16**:241 (1960).
38. Powell, D. B., and N. Sheppard, *J. Chem. Soc.*, **1959**:791.
39. Powell, D. B., and N. Sheppard, *J. Chem. Soc.*, **1959**:3089.
40. Powell, D. B., and N. Sheppard, *J. Chem. Soc.*, **1961**:1112.
41. Quagliano, J. V., and S. Mizushima, *J. Am. Chem. Soc.*, **75**:6048 (1953).
42. Sargeson, A. M., Optical Phenomena in Metal Chelates, in F. P. Dwyer and D. P. Mellor (eds.), "Chelating Agents and Metal Chelates," pp. 183–235, Academic, New York, 1964.
43. Schwarzenbach, G., *Helv. Chim. Acta*, **36**:23 (1953).
44. Sheppard, N., in H. W. Thompson (ed.), "Advances in Spectroscopy," vol. I, p. 288, Interscience, New York, 1959.

45. Sweeney, D. M., S. Mizushima, and J. V. Quagliano, *J. Am. Chem. Soc.*, **77**:6521 (1955).
46. Templeton, D. H., A. Zalkin, and T. Ueki, *Acta Cryst.*, **21**:A154 (supplement) (1966).
47. Welti, D., and D. Whittaker, *J. Chem. Soc.*, **1962**:4372.
48. Werner, A., *Z. Anorg. Allgem. Chem.*, **22**:91 (1900).
49. Werner, A., *Bull. Soc. chim. France*, **11**:1 (1912).
50. Werner, A., *Ber.*, **45**:121 (1912).
51. Werner, A., *Ber.*, **45**:1229 (1912).
52. Werner, A., *Ber.*, **45**:3061 (1912).
53. Werner, A., and A. Smirnoff, *Helv. Chim. Acta*, **3**:472 (1920).

8

Geometrical Isomerism

Various physical techniques have been applied to the study of geometrical isomers. The application of nmr to this problem was discussed at some length in Chap. 6 and will not be discussed here further, except to reemphasize that it is capable of yielding unambiguous assignments of configuration for a large number of the systems of interest. Normal X-ray structural analysis can also be applied to the solution of this problem. Its application will not be discussed here because no special technique or theory is required. However, the reader is referred to the general texts on X-ray analysis given in Chap. 4 and especially to an article by Saito [42].

INFRARED SPECTROSCOPY

The differentiation of geometrical isomers by IR spectroscopy usually depends on the premise that the higher symmetry of one isomer leads to a simpler spectrum for this compound compared with that of the other. This is true only if the crystal-site symmetries and other interactions in the solid state, such as hydrogen bonding, do not interfere. If the investigation is to be carried out in the solid phase, it is advisable to study the complexes with a range of counterions, including large ions, such as $[PtCl_4]^{2-}$, which are known to lead to a breakdown of hydrogen bonding.

The alternative procedure is purely empirical. The IR of a large series of closely related complexes with known structures are measured, and regions of the spectra are selected where the geometrical isomers are consistently found to show different peak patterns. These distinctive features of the cis and trans isomers are then sought in the IR spectrum of the compound under

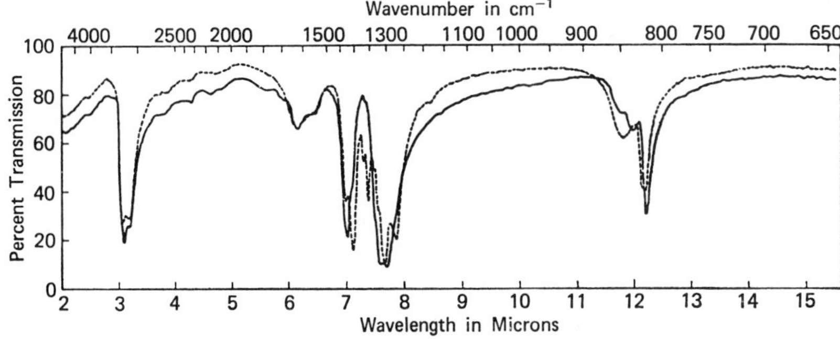

Fig. 8-1 IR spectra of the isomers of $[Co(NH_3)_4(NO_2)_2]$ Cl in KBr disks: *trans* (——); *cis* (– – –) [15].

investigation. Some examples of systems studied by these methods are discussed below.

$[Co(NH_3)_4X_2]$. Faust and Quagliano [15] observed that *cis-* and *trans-* $[Co(NH_3)_4(NO_2)_2]^+$ have distinctly different IR spectra in the 1200 to 1500 cm^{-1} region where the symmetric NH_3 deformation and the NO stretching modes absorb, with the trans having the simpler spectrum (see Fig. 8-1). Merritt and Wiberly [33] studied $[Co(NH_3)_4Cl_2]Cl$, but although the positions of the peaks varied slightly from cis to trans, the differences reported were too small to provide a sound basis for differentiating between them. Rigg and Sherwin [40] have claimed that the cis compound studied was in fact $[Co(NH_3)_4Cl(OH_2)]^{2+}$ and have reported data for the *cis*-dichloro compound but failed to report any definitive differences between the two isomers.

$[Coen(NH_3)_2X_2]$. The analysis of IR spectra of ethylenediamine complexes is complicated by a strong vibrational coupling in the chelate ring. A reliable assignment of the normal modes of a cobalt(III)-ethylenediamine complex based on a normal coordinate analysis has been achieved only for *trans*-$[Coen_2Cl_2]^+$ [35,36] (see Fig. 8-2). These assignments have been used in the subsequent discussion.

The three geometrical isomers of $[Coen(NH_3)_2Cl_2]^+$ (**X, XI, XII**) have been studied by Rigg and Sherwin above 650 cm^{-1} [40]. Distinct differences were reported in the CH_2 rocking region: the *trans*-dichloro compound had a single weak band at 876 cm^{-1}, whereas the two *cis*-dichloro compounds each showed two bands (*cis*-dichloro-*trans*-diammine: 892 and 878 cm^{-1}; *cis*-dichloro-*cis*-diammine: 896 and 880 cm^{-1}). Rigg and Sherwin pointed out that this observation is difficult to rationalize in terms of symmetry selection rules because both *trans*-dichloro and *trans*-diammine have the same

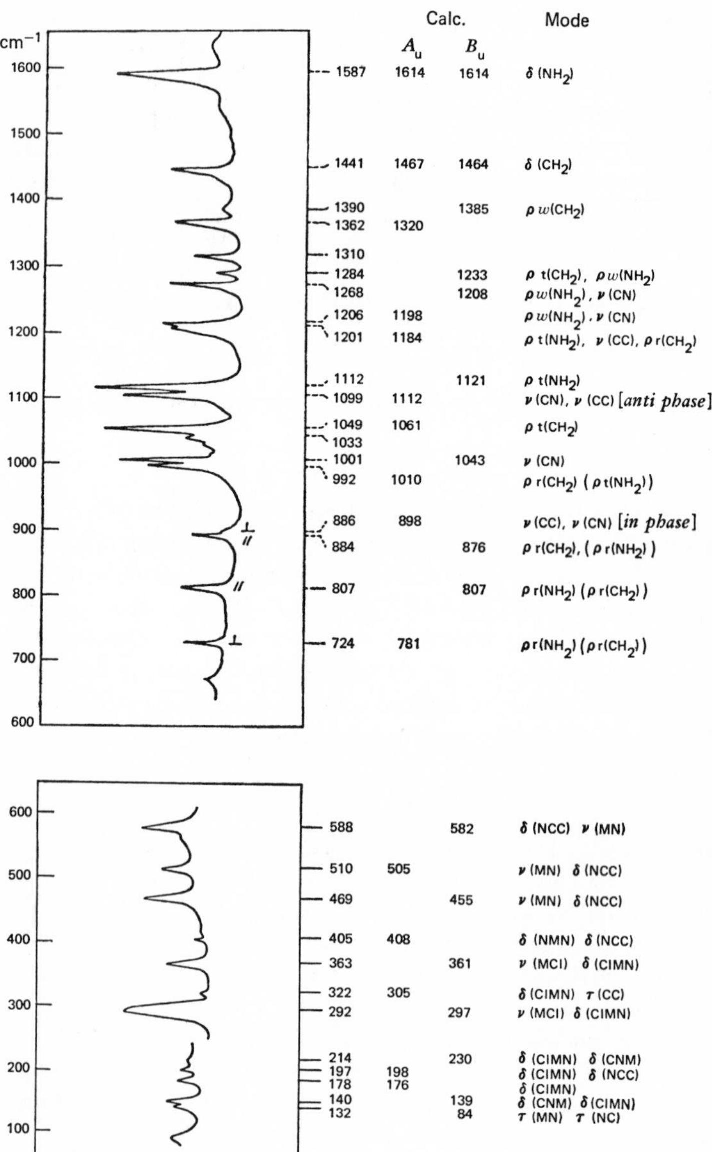

Fig. 8-2 IR spectrum of *trans*-$[Coen_2Cl_2]$ Br and the assignment of transitions [35, 36].

molecular symmetry. It is possible that one of the two vibrations observed for the *cis*-dichloro compounds is not a CH_2 rocking mode. Nakagawa [35] has calculated that, for *trans*-$[Coen_2Cl_2]^+$, an A_u symmetry band would appear at 898 cm^{-1} because of C—C and C—N in phase stretches, in addition to the 876 cm^{-1} CH_2 rocking mode B_u which is coupled with a NH_2 rocking mode.

Rigg and Sherwin [40] also observed differences in the 1300 cm^{-1} NH_3 deformation region. The two isomers with the NH_3 groups cis showed four bands, whereas the *trans*-diammine complex showed only two (see Fig. 8-3). It could well be informative to study these three isomers in the Co—Cl stretch region because it is known that the M—X stretching modes are structure-sensitive (e.g., see [12,20]).

[Coen₂X₂]. The trans isomer has been found to have the $\delta\lambda$ configuration for a number of complexes (e.g., see Chapter 4). In this form it has C_{2h} symmetry, in which only the *u* vibrations are IR-active. The cis complexes can have a symmetry no higher than C_2, in which all the modes are IR-active. This should allow a clear distinction to be drawn between the IR of the two isomers.

Merritt and Wiberly [33] studied a small number of cis and trans isomers and observed a shift of the band in the 1500 to 1600 cm^{-1} region to higher

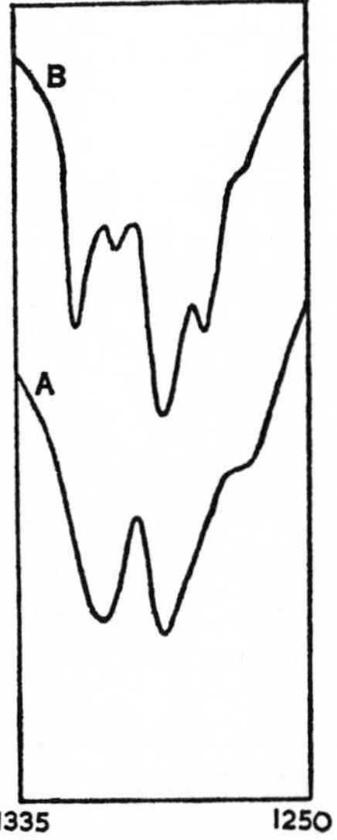

Fig. 8-3 IR spectra of (A) *cis*-dichloro-*trans*-diammine(ethylenediamine)cobalt(III) chloride, and (B) *trans*-dichloro-*cis*-diammine(ethylenediamine)cobalt(III) chloride [40].

energy on the change from cis to trans. These authors observed differences in the NH_2 rocking frequencies (820 to 780 cm^{-1}). Baldwin [5], however, has shown that, in a series of compounds of the type $[Coen_2AB]X$, the position and shape of the bands in this region depend on A, B, and X as well as on the geometrical configuration. Chamberlain and Bailar [11] reported differences in the 1120 to 1150 cm^{-1} region and proposed that, for the trans, there was a single band, which was split for the cis. Baldwin [5] has also criticized this method of differentiation because this region cannot be used for oxyanions and also the band structures were found to depend on the anion.

The CH_2 rocking region (850 to 900 cm^{-1}) was found to be one of the best regions for distinguishing the two isomers [5]. The cis isomers showed two bands, in contrast to the trans's one, and this result was independent of the unidentate ligands and the anion. The NH_2 asymmetric deformation modes (~ 1600 cm^{-1}) were also observed to be sensitive to the geometrical isomeric form [5,34]. Morris and Busch [34] found for the dinitro complex:

trans	1610 cm^{-1}
cis	1575 and 1617 cm^{-1}

and for the dichloro complex:

trans	1596 cm^{-1}
cis	1561 and 1630 cm^{-1}

Subsequently, Hughes and McWhinnie [20] applied each of these methods to a series of 13 cis and 8 trans complexes and found that the CH_2 rocking vibrations gave the most consistent results, but the method is handicapped by the fact that the transitions are weak. They observed that, in the 620 to 455 cm^{-1} region, no trans isomer had more than three strong bands, in contrast to the cis, which had at least four strong bands. Further, for the halogen complexes, the M—X stretching modes were at higher frequency for the trans than for the cis.

$[Rhen_2X_2]$ and $[Iren_2X_2]$. The cis and trans isomers of $[Rhen_2Cl_2]^+$ and $[Iren_2Cl_2]^+$ can clearly be differentiated through their IR spectra [23] (Fig. 8-4). In both the NH_2 asymmetric deformation (~1600 cm^{-1}) and the CH_2 rocking (850 to 900 cm^{-1}) regions, the cis isomers show two peaks, in contrast to the single sharp peak found for the trans.

$[Cren_2X_2]$. The observation that the asymmetric NH_2 deformation modes of cis complexes occur at lower frequency than the trans has been used to distinguish the isomers of bis(ethylenediamine)chromium(III) [30]. It was pointed out, however, that if the anion was polyatomic, the differentiation was very difficult. Hughes and McWhinnie [21] studied the low-frequency IR

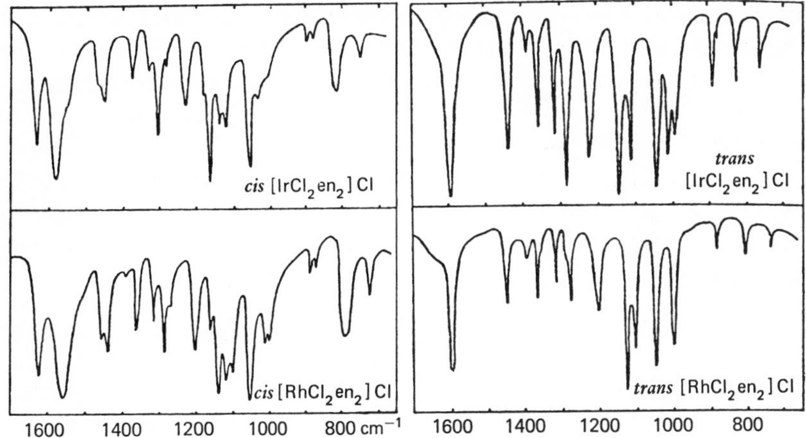

Fig. 8-4 IR spectra of *cis*- and *trans*-$[Rhen_2Cl_2]^+$ and $[Iren_2Cl_2]^+$ [23].

spectra (667 to 222 cm^{-1}) of a series of complexes and found marked differences among the isomers. Four strong bands were found for the cis: one at 545 to 535 cm^{-1}, which was strongly split; two other bands at ~480 and 430 cm^{-1}; and a much weaker band at 405 cm^{-1}. In contrast, the trans isomers showed only three bands: one at about 540 cm^{-1}, which was sometimes split; and the other two at about 490 and 440 cm^{-1} (see Fig. 8-5). The method was complicated by the SCN bending modes at about 475 cm^{-1} and the Cr—O stretching vibrations at about 470 cm^{-1}, but because they appear

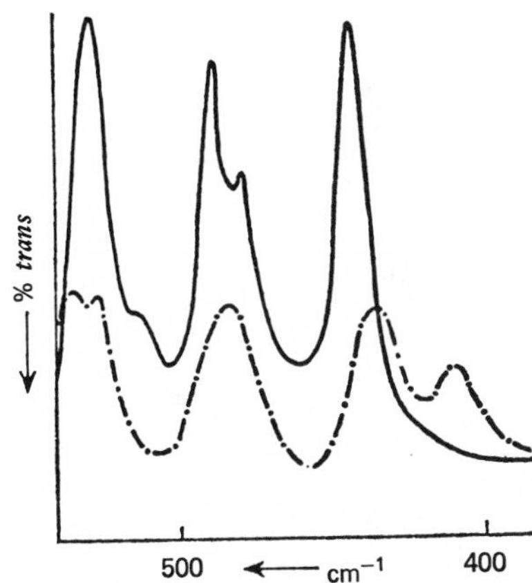

Fig. 8-5 IR spectra of *trans*-$[Cren_2Br_2]HgBr_3$ (——) and *cis*-$[Cren_2Br_2]Br$ (–·–·) [21].

as only shoulders, they do not affect the basic pattern. On deuteration, all four cis bands move to lower energy, whereas only the 545 to 535 cm^{-1} band moves for the trans isomer. The differences in the M—X stretching frequencies for the halogen complexes were not so marked for the chromium as they were for the cobalt complexes.

Multidentate amine complexes. Buckingham and Jones [9] have studied an extensive range of *cis*-α and *cis*-β triethylenetetramine complexes of cobalt(III) and have compared their IR with those of *cis*-α-$[RhtrienCl_2]^+$, *cis*-α-$[CrtrienCl_2]^+$, and *trans*-$[CotrienCl_2]^+$. They found the following:

1. The IR of *cis*-α-$[CotrienCl_2]^+$ resembles that of *cis*-$[Coen_2Cl_2]^+$ in many ways.

2. In the N—H stretching region (3000 to 3300 cm^{-1}) the *cis*-α complexes show three strong bands that are shifted to lower energy as a group on deuteration; the *cis*-β show a more complicated spectrum, with usually four, but

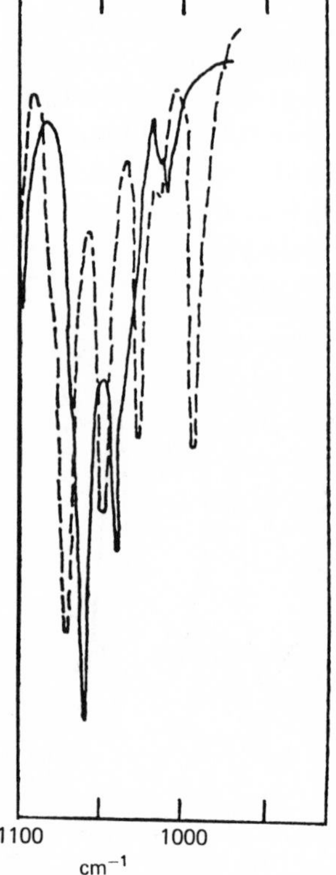

Fig. 8-6 IR spectra of *cis*-α (——) and *cis*-β (– – –) $[CotrienCl_2]\,Cl\cdot H_2O$ [9].

sometimes five, strong absorptions that are also moved to lower energy on deuteration; *trans*-[CotrienCl$_2$]$^+$ gives four medium to strong absorptions.

3. In the 990 to 1090 cm^{-1} region *cis*-α compounds show two strong absorptions, whereas the *cis*-β have four (see Fig. 8-6).

4. The asymmetric NH$_2$ deformation modes at about 1,600 cm^{-1} give rise to one main absorption for the *cis*-α (composite) and *trans*, whereas *cis*-β complexes give, in general, two, with the lower-energy absorption composite.

5. The CH$_2$ bending region of 1430 to 1490 cm^{-1} becomes less complicated and more intense in the order *cis*-α, *cis*-β, *trans*.

6. In the CH$_2$ rocking region the *cis*-β show four absorptions, whereas the *cis*-α series show only two.

From this survey they concluded that the isomer of [RhtrienCl$_2$]$^+$ prepared by the method of Johnson and Basolo [22] and the isomer of [CrtrienCl$_2$]$^+$ reported by Kling and Schlaefer [26] have the *cis*-α configuration. Buckingham and Jones [9] also applied their method of differentiation to the green "trans" and violet "cis" [CotrienBr$_2$]Br isomers prepared by Selbin and Bailar [44] and concluded that the "cis" form has the *cis*-β configuration, whereas the "trans" product is mainly *cis*-α contaminated with the *cis*-β form.

House and Garner applied the observations above to the differentiation of the isomers of triethylenetetraminechromium(III) [18] and to the isomers of chlorotetraethylenepentaminecobalt(III) and -chromium(III) [19]. Collman and Schneider have also used IR to relate the configurations of the cobalt(III) and rhodium(III) complexes of 1,4,7,10-tetraazacyclododecane (cyclen) [13].

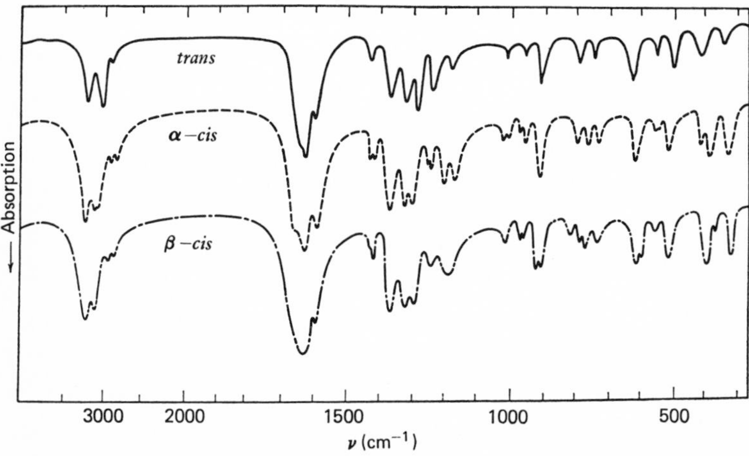

Fig. 8-7 IR spectra of *cis*- (two crystalline forms) and *trans*-[Pt(gly)$_2$] [37].

Amino acid complexes. Condrate and Nakamoto have carried out a normal coordinate analysis of a divalent metal–α-aminocarboxylate chelate ring [14]. The assignment of bands for the IR spectra of the geometrical isomers of $[Pt(gly)_2]$ (see Fig. 8-7) based on this analysis is given in Table 8-1. [24]. As might be expected from symmetry considerations, the cis isomer exhibits more bands than the trans for the platinum(II) [14,24,43] and copper(II) complexes [14,28,41,46]. The two geometrical isomers of $[Co(gly)_3]$ have also been distinguished by IR spectroscopy on the basis that the purple dihydrate has a more complicated spectrum than the red monohydrate and therefore is the lower-symmetry trans isomer [37,43] (see Fig. 8-8).

Table 8-1 Observed frequencies of several bidentate Pt(II) glycinato complexes, cm^{-1} [24]

	$Pt(gly)_2$			
K $PtCl_2(gly)$	β-cis	α-cis	trans	Band assignment
1645	1640	1639	1643	$\nu(C{=}O) + \nu(C{-}O)$
1575	1585[a]	1605	1610	$\delta(NH_2)$
		1425		
1412	1435	1419	1441	$\delta(CH_2)$
1355	1372	1370	1374	$\nu(C{-}O) + \nu(C{-}C) + \nu(C{=}O)$
1310	1317	1325	1332	$\nu(C{-}N) + \rho w(CH_2)$
1303[a]	1300	1300	1292[a]	
	1210	1255	1245	$\rho w(NH_2) + \rho t(NH_2) + \rho t(CH_2)$
1160	1188	1217		
		1032		
1025	1025	1023	1023	$\rho w(NH_2) + \nu(C{-}N)$
950	970	973		
	960	960	962	$\nu(C{-}C)$
	925	916		
918	910	911	917	$\rho r(CH_2)$
	834	795		
755	772	768	792	$\rho r(NH_2)$
715	735	738	745	$\delta(C{=}O) + \nu(C{-}C)$
	617	616	614	
585	595[a]	595[a]		$\pi(C{=}O)$
550	553	554	549	$\nu(Pt{-}N)$
510	510	510	495	Ring def + $\rho r(CO_2)$
	498[a]			
388	398	405[a]	411	$\nu(Pt{-}O) + \nu(Pt{-}N)$
		400		
	337	337	335	$\delta(C{-}C{-}N)$
320				$\nu(Pt{-}Cl)$

[a] Shoulder.

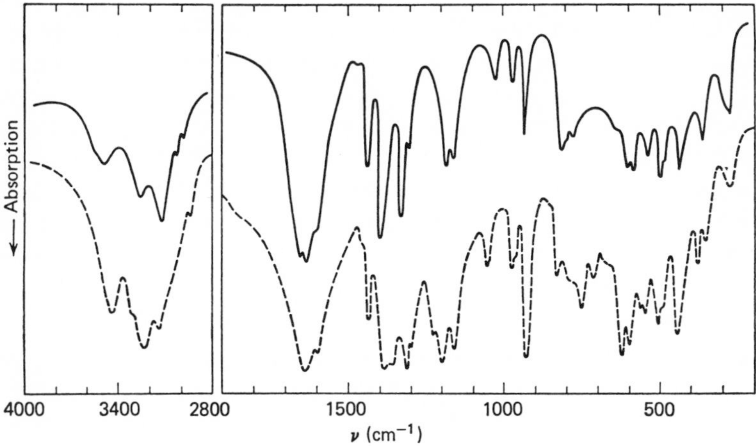

Fig. 8-8 IR spectra of *trans*- (- - -), and *cis*- (——) $[Co(gly)_3]$ [37].

Considerable care is needed in applying this method to the amino acid complexes, because the molecular symmetry for both isomers is usually very low, the hydrogen-bonding effects for the two isomers are often different, and sometimes in the solid state the various chelate rings in the one complex have different geometries. Further, the complexity of the IR spectra for amino acids other than glycine is usually so marked that it would be difficult to use this method objectively to distinguish geometrical isomers.

ELECTRONIC ABSORPTION SPECTROSCOPY

Perhaps the most commonly applied method for distinguishing geometrical isomers of metal complexes is based on the *d–d* transitions. As stated in Chap. 5, the gross features of these absorption bands can be interpreted in terms of three general chromophores: cubic (field: $x = y = z$), tetragonal (field: $x = y \neq z$), and rhombic (field: $x \neq y \neq z$). The low-field splitting of the triply degenerate excited states of metal ions, such as cobalt(III), chromium(III), and nickel(II), is often sufficiently different for the various geometrical isomers to allow an immediate assignment from a study of the spectra of the isomers. For other systems a more detailed analysis is necessary. The crucial factor governing the splitting is the difference in the ligand fields in the xy plane and along the z axis for the tetragonal field and along the three axes passing through the donor atoms for the rhombic—Eq. (5-34).

Basolo, Ballhausen, and Bjerrum first applied the method to some cobalt-(III) complexes, including the two geometrical isomers of tris(glycinato)-cobalt(III) [7]. *cis*-[Co(gly)$_3$] has a cubic chromophore, and thus the $^1A_{1g} \rightarrow {}^1T_{1g}$ and $^1A_{1g} \rightarrow {}^1T_{2g}$ absorption bands are expected to be Gaussian in shape, showing no signs of a low-field splitting. In contrast, *trans*-[Co(gly)$_3$] has a rhombic chromophore, and the absorption band, if not distinctly split, should be markedly broadened. The two isomers indeed show these features (see Fig. 8-9), and on this basis the red form has been assigned the cis structure and the purple the trans.

The cis and trans isomers of $[Co(NH_3)_4X_2]^{n+}$ and $[Coen_2X_2]^{n+}$ are usually distinguishable in this way. According to Eq. (5-34),

$$\Delta\mathscr{E}_{\text{tet}}(trans) = -2\Delta\mathscr{E}_{\text{tet}}(cis) \tag{8-1}$$

The spectra of the dichloro compounds are given in Figs. 8-10 to 8-13, showing the component transitions with $^1A_{1g} \rightarrow {}^1T_{1g}$ cubic parentage determined by Gaussian analysis [17]. The slight deviation from the relationship in Eq. (8-1) for the ethylenediamine complexes is thought to be due to a configurational interaction term.

The method has been applied to many systems, including multidentates (for example, see Refs. 8, 32) and has proved to be most useful. One multidentate system is considered here. Two geometrical isomers are possible for the complex $[CoEDDAen]^+$, where EDDA is ethylenediaminediacetate, with the two acetate groups coordinated either cis or trans to each other, that is, with one acetate arm equatorial and the other apical, or with the two arms apical. The absorption spectrum shown in Fig. 8-14, which shows a relatively large splitting of the two components of the transition with $^1A_{1g} \rightarrow {}^1T_{1g}$ parentage, is consistent with that expected for the trans isomer [29].

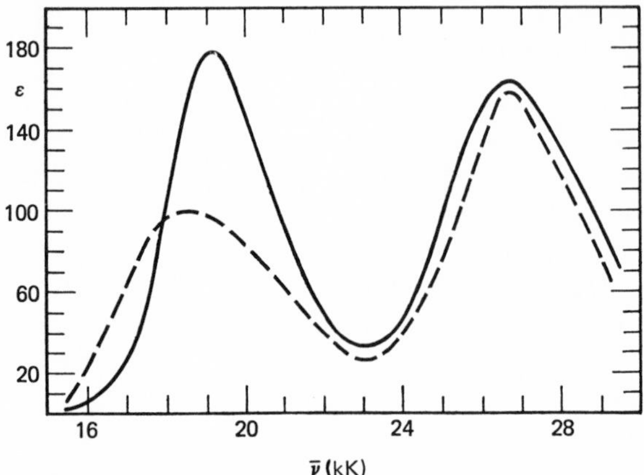

Fig. 8-9 Absorption spectra of *cis*- (——) and *trans*- (- - -) [Co(gly)$_3$] [7].

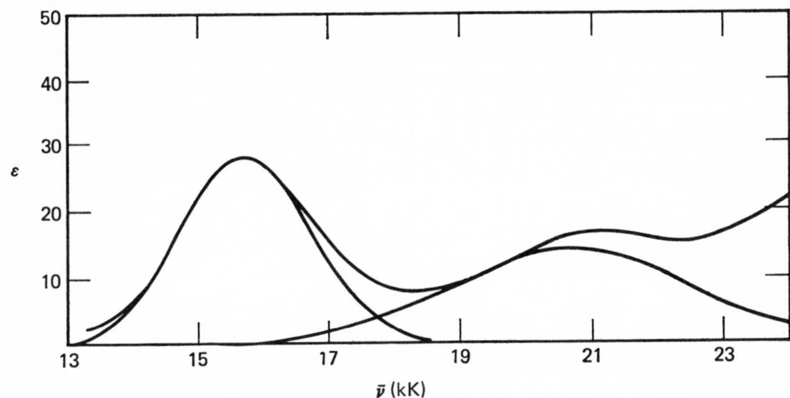

Fig. 8-10 Absorption spectrum of *trans*-$[Co(NH_3)_4Cl_2]HSO_4$ in dimethylsulfoxide showing $^1A_{1g} \rightarrow {}^1E_g$ (at lower energy) and $^1A_{1g} \rightarrow {}^1A_{2g}$ tetragonal components of $^1A_{1g} \rightarrow {}^1T_{1g}$ cubic absorption band [17].

Yasui and Shimura have attempted to distinguish the geometrical isomers of bis(α-aminocarboxylato)copper(II) complexes by their diffuse reflectance spectra [48]. Their assignments were based mainly on a proposal that the cis form has its *d–d* absorption band at a lower energy than that of the trans, and the cis shows a shoulder on the low-energy side of its band. They correctly predicted that $[Cu(\text{L-isoleu})_2] \cdot H_2O$ has the cis structure [47]. Gillard and

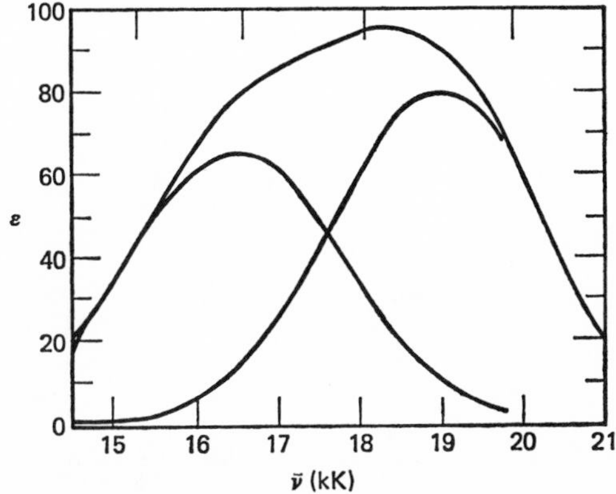

Fig. 8-11 Absorption spectrum of *cis*-$[Co(NH_3)_4Cl_2]Cl$ in dimethysulfoxide, showing $^1A_{1g} \rightarrow {}^1A_{2g}$ (at lower energy) and $^1A_{1g} \rightarrow {}^1E_g$ tetragonal components of $^1A_{1g} \rightarrow {}^1T_{1g}$ absorption band [17].

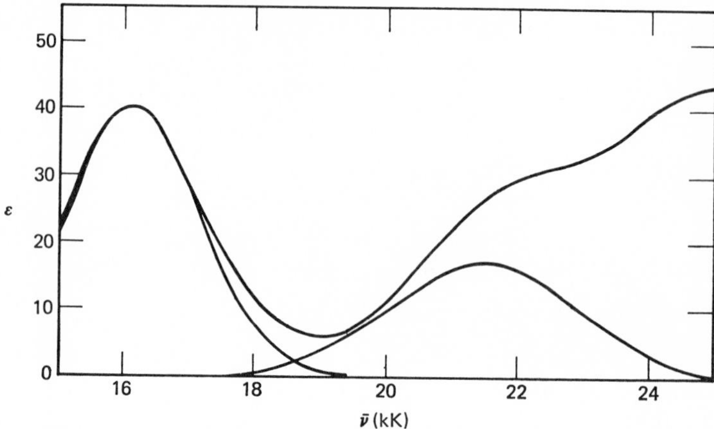

Fig. 8-12 Absorption spectrum of *trans*-[$Coen_2Cl_2$]Cl in dimethylsulfoxide, showing $^1A_{1g} \rightarrow {}^1E_g$ (at lower energy) and $^1A_{1g} \rightarrow {}^1A_{2g}$ tetragonal components of $^1A_{1g} \rightarrow {}^1T_{1g}$ cubic absorption band [17].

his coworkers, however, were unable to observe the features above in the diffuse reflectance spectra of *cis*- and *trans*-[Cu(L-ala)$_2$], whose structures had been determined by X-ray analysis [16]. One of the difficulties with this kind of system is that the absorption band is broad for both isomers. Further, the groups (if any) that occupy the fifth and sixth octahedral positions in the crystalline state are not necessarily identical for the two isomers, and this could significantly affect the splitting of the energy levels.

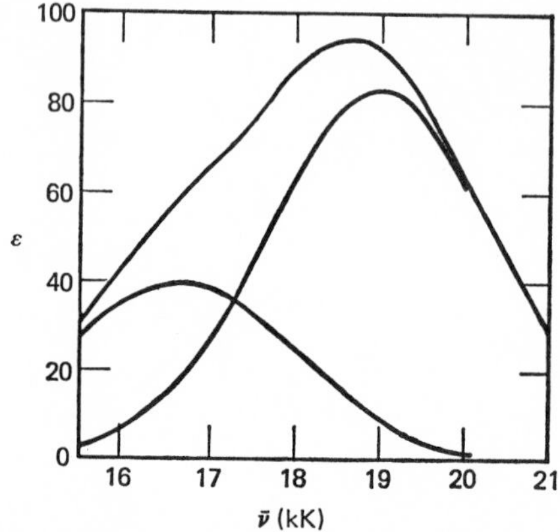

Fig. 8-13 Absorption spectrum of *cis*-[$Coen_2Cl_2$]Cl in dimethylsulfoxide, showing $^1A_{1g} \rightarrow {}^1A_{2g}$ (at lower energy) and $^1A_{1g} \rightarrow {}^1E_g$ tetragonal components of $^1A_{1g} \rightarrow {}^1T_{1g}$ cubic absorption band [17].

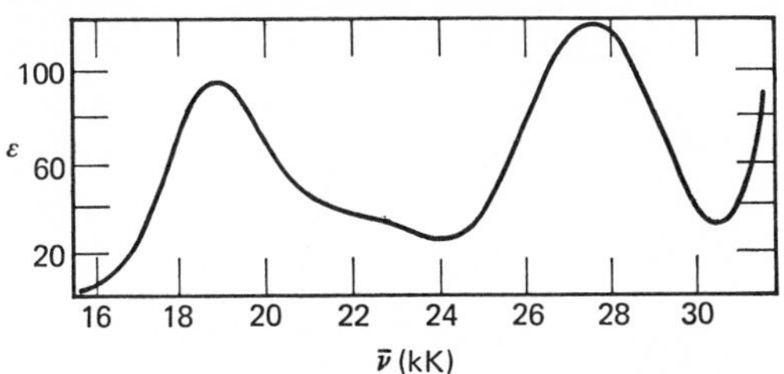

Fig. 8-14 Absorption spectrum of *trans*-[CoEDDAen]$^+$ [29].

The charge-transfer bands have also been used to distinguish between geometrical isomers [6,45]. It was found that the charge-transfer bands for some trans complexes were at lower energy than the corresponding absorption for the cis. This is particularly important for dinitrotetramminecobalt(III) and other $CoN_4(NO_2)_2{}^+$ complexes, because the method based on the *d–d* transitions is unable to distinguish between the two isomers. The spectra of the two isomers of $[Co(NH_3)_4(NO_2)_2]^+$ are shown in Fig. 8-15, and the difference in the position of the charge-transfer absorption bands at about 350 and 250 nm is easily seen. In contrast, the *d–d* absorption bands at about 440 nm show

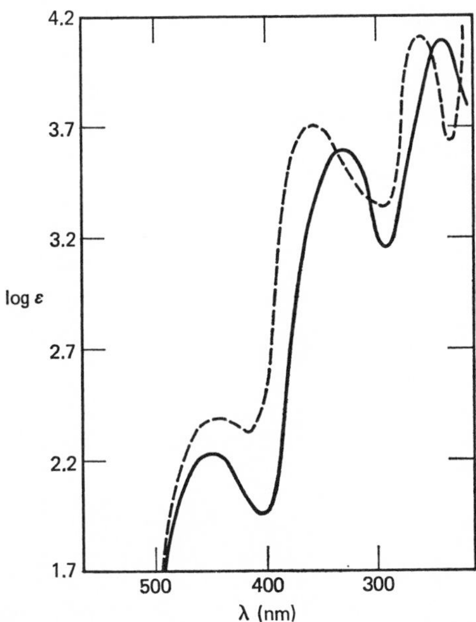

Fig. 8-15 Absorption spectra of *cis*- (——) and *trans*- (– – –) $[Co(NH_3)_4(NO_2)_2]Cl$ in an aqueous methanol solution [6].

little difference. Tsuchida and his coworkers have used a semiempirical molecular-orbital calculation to rationalize the energy difference for these ligand → metal charge-transfer bands [38]. A similar relationship to the above has also been claimed to hold for square-planar platinum(II) complexes [3].

OTHER METHODS

Dipole-moment measurements have been traditionally mentioned as a suitable technique for assigning geometrical isomers (for example, see Ref. 2). The method has been little used, however, for metal complexes. This probably derives mainly from experimental difficulties, such as low solubility in the nonpolar solvents necessary for the measurements.

The successful resolution of a chelate-containing compound has sometimes been used as an indication of the cis structure. In some instances—Men_2X_2, for example—this is unambiguous. Often, however, this distinction is not possible. Take, for example, the $MtrienX_2$ system. All three geometrical isomers (**XXVI**, **XXVII**, and **XXVIII**), even the trans, are capable of resolution [10].

Chemical reactions have been successfully used to distinguish geometrical isomers of planar complexes. Except for a few isolated examples, however, chemical assignments of configuration are unreliable for octahedral complexes. One example for each stereochemistry is considered here.

When the two isomers of $[Pt(NH_3)_2Cl_2]$ were reacted with thiourea (th), one was found to yield tetrathioureaplatinum(II) and the other diamminedithioureaplatinum(II) [27,30,39]. Because chlorine lies above ammonia in the trans effect series, and thiourea above chlorine, the results indicate, as shown in the following reaction scheme, where * designates the group labilized by a trans effect, that the former has the cis configuration and the latter the trans:

$$\begin{array}{ccc} H_3N^* & & Cl \\ & \searrow\ \swarrow & \\ & Pt & \\ & \nearrow\ \nwarrow & \\ H_3N & & Cl \end{array} \xrightarrow{th} \begin{array}{ccc} th & & Cl \\ & \searrow\ \swarrow & \\ & Pt & \\ & \nearrow\ \nwarrow & \\ H_3N & & Cl^* \end{array} \xrightarrow{th} \begin{array}{ccc} th & & Cl \\ & \searrow\ \swarrow & \\ & Pt & \\ & \nearrow\ \nwarrow & \\ H_3N^* & & th \end{array}$$

$$\downarrow th$$

$$\begin{array}{ccc} th & & th \\ & \searrow\ \swarrow & \\ & Pt & \\ & \nearrow\ \nwarrow & \\ th & & th \end{array} \xleftarrow{th} \begin{array}{ccc} th & & Cl^* \\ & \searrow\ \swarrow & \\ & Pt & \\ & \nearrow\ \nwarrow & \\ th & & th \end{array}$$

$$\begin{array}{ccc} H_3N & & Cl \\ & \searrow\ \swarrow & \\ & Pt & \\ & \nearrow\ \nwarrow & \\ Cl^* & & NH_3 \end{array} \xrightarrow{th} \begin{array}{ccc} H_3N & & Cl^* \\ & \searrow\ \swarrow & \\ & Pt & \\ & \nearrow\ \nwarrow & \\ th & & NH_3 \end{array} \xrightarrow{th} \begin{array}{ccc} H_3N & & th \\ & \searrow\ \swarrow & \\ & Pt & \\ & \nearrow\ \nwarrow & \\ th & & NH_3 \end{array}$$

The rather large trans-labilizing influence of the sulfito ligand has been used to distinguish the cis and trans isomers of $[Co(NH_3)_4(SO_3)_2]^-$ [4,25]. If the configuration is cis, the two ammonia ligands trans to the sulfito groups would be distinctly more reactive than the other two and should be readily replaced by ethylenediamine to give bis(sulfito)diammine(ethylenediamine)-cobalt(III). In contrast, if the two sulfites are trans, the four ammonia groups are equivalent, and all would be replaced by ethylenediamine with equal ease.

GENERAL READING

1. Nakamoto, K., and P. J. McCarthy (eds.), "Spectroscopy and Structure of Metal Chelate Compounds," Wiley, New York, 1968.
2. Wilkins, R. G., and M. J. G. Williams, in J. Lewis and R. G. Wilkins (eds.), "Modern Coordination Chemistry," pp. 187–202, Interscience, New York, 1960.

REFERENCES

3. Babaeva, A. V., *Chem. Abstracts*, **50**:9151e (1956).
4. Bailar, J. C., and D. F. Peppard, *J. Am. Chem. Soc.*, **62**:105 (1940).
5. Baldwin, M. E., *J. Chem. Soc.*, **1960**:4369.
6. Basolo, F., *J. Am. Chem. Soc.*, **72**:4393 (1950).
7. Basolo, F., C. J. Ballhausen, and J. Bjerrum, *Acta Chem. Scand.*, **9**:810 (1955).
8. Bosnich, B., C. K. Poon, and M. L. Tobe, *Inorg. Chem.*, **4**:1102 (1965).
9. Buckingham, D. A., and D. Jones, *Inorg. Chem.*, **4**:1387 (1965).
10. Buckingham, D. A., P. A. Marzilli, and A. M. Sargeson, *Inorg. Chem.*, **6**:1032 (1967).
11. Chamberlain, M. M., and J. C. Bailar, *J. Am. Chem. Soc.*, **81**:6412 (1959).
12. Clark, R. J. H., *Spectrochim. Acta*, **21**:955 (1965).
13. Collman, J. P., and P. W. Schneider, *Inorg. Chem.*, **5**:1380 (1966).
14. Condrate, R. A., and K. Nakamoto, *J. Chem. Phys.*, **42**:2590 (1965).
15. Faust, J. P., and Quagliano, J. V., *J. Am. Chem. Soc.*, **76**:5346 (1954).
16. Gillard, R. D., H. M. Irving, R. M. Perkins, N. C. Payne, and L. D. Pettit, *J. Chem. Soc., A.*, **1966**:1159.
17. Hawkins, C. J., J. Niethe, and C. L. Wong, *Aust. J. Chem.*, Submitted for publication.
18. House, D. A., and C. S. Garner, *J. Am. Chem. Soc.*, **88**:2156 (1966).
19. House, D. A., and C. S. Garner, *Inorg. Chem.*, **5**:2097 (1966).
20. Hughes, M. N., and W. R. McWhinnie, *J. Inorg. Nucl. Chem.*, **28**:1659 (1966).
21. Hughes, M. N., and W. R. McWhinnie, *J. Chem. Soc., A.*, **1967**:592.
22. Johnson, S. A., and F. Basolo, *Inorg. Chem.*, **1**:925 (1962).
23. Kida, S., *Bull. Chem. Soc. Japan*, **39**:2415 (1966).
24. Kieft, J. A., and K. Nakamoto, *J. Inorg. Nucl. Chem.*, **29**:2561 (1967).
25. Klement, R., *Z. Anorg. Allgem. Chem.*, **150**:117 (1926).
26. Kling, O., and H. L. Schlaefer, *Z. Anorg. Allgem. Chem.*, **313**:187 (1961).
27. Kurnakov, N. S., *J. Prakt. Chem.*, (2) **50**:483, 498 (1894).
28. Lane, T. J., J. A. Durkin, and R. J. Hooper, *Spectrochim. Acta*, **20**:1013 (1964).
29. Legg, J. I., D. W. Cooke, and B. E. Douglas, *Inorg. Chem.*, **6**:700 (1967).

30. Lutton, J. M., and R. W. Parry, *J. Am. Chem. Soc.*, **76**:4271 (1954).
31. McLean, J. A., A. F. Schreiner, and A. F. Laethem, *J. Inorg. Nucl. Chem.*, **26**:1245 (1964).
32. Melson, G. A., and R. G. Wilkins, *J. Chem. Soc.*, **1963**:2662.
33. Merritt, P. E., and S. E. Wiberley, *J. Phys. Chem.*, **59**:55 (1955).
34. Morris, M. L., and D. H. Busch, *J. Am. Chem. Soc.*, **82**:1521 (1960).
35. Nakagawa, I., *Bunko Kagaku*, **1965**:87.
36. Nakagawa, I., personal communication, 1969.
37. Nakamoto, K., in Ref. 1, pp. 216–285.
38. Nakamoto, K., J. Fujita, M. Kobayashi, and R. Tsuchida, *J. Chem. Phys.*, **27**:439 (1957).
39. Quagliano, J. V., and L. Schubert, *Chem. Revs.*, **50**:201 (1952).
40. Rigg, J. M., and E. Sherwin, *J. Inorg. Nucl. Chem.*, **27**:653 (1965).
41. Rosenberg, A., *Acta Chem. Scand.*, **10**:840 (1956).
42. Saito, Y., in Ref. 1, pp. 1–72.
43. Saraceno, A. J., I. Nakagawa, S. Mizushima, C. Curran, and J. V. Quagliano, *J. Am. Chem. Soc.*, **80**:5018 (1958).
44. Selbin, J., and J. C. Bailar, *J. Am. Chem. Soc.*, **82**:1524 (1960).
45. Shimura, Y., *J. Am. Chem. Soc.*, **73**:5079 (1951).
46. Tomita, K., *Bull. Chem. Soc. Japan*, **34**:280 (1961).
47. Weeks, C. M., A. Cooper, and D. A. Norton, *Acta Cryst.*, **B25**:443 (1969).
48. Yasui, T., and Y. Shimura, *Bull. Chem. Soc. Japan*, **39**:604 (1966).

Index